Magnetically Confined Fusion Plasma Physics, Volume 2

Multifluid theory

Magnetically Confined Fusion Plasma Physics, Volume 2

Multifluid theory

Linjin Zheng
The University of Texas at Austin, Texas, USA

IOP Publishing, Bristol, UK

ISBN 978-0-7503-3575-1 (ebook)
ISBN 978-0-7503-3573-7 (print)
ISBN 978-0-7503-3576-8 (myPrint)
ISBN 978-0-7503-3574-4 (mobi)

DOI 10.1088/978-0-7503-3575-1

Version: 20200901

IOP ebooks

British Library Cataloguing-in-Publication Data: A catalogue record for this book is available from the British Library.

Published by IOP Publishing, wholly owned by The Institute of Physics, London

IOP Publishing, Temple Circus, Temple Way, Bristol, BS1 6HG, UK

US Office: IOP Publishing, Inc., 190 North Independence Mall West, Suite 601, Philadelphia, PA 19106, USA

This volume is dedicated with love to my wife and children.

Contents

Part III Fluid theory in the collisionless or weak collision limit

Preface

This book continues the author's previous book *Magnetically Confined Fusion Plasma Physics: Ideal MHD theory* to describe the subsequent topic: multifluid theory. First, the conventional two-fluid theory by S I Braginskii is described with the underlying physical thoughts explained and the derivation processes detailed. Next, the newly developed perpendicular magnetofluid theory in the collisionless limit is introduced. The applications of the theory are described systematically so that the connections to the kinetic theory are elucidated. The fluid-kinetic hybrid theories are also briefly reviewed. The book is intended to fill the gap between the textbooks and the up-to-date advanced research developments in this field.

A famous quote from Albert Einstein is 'The most incomprehensible thing about the world is that it is comprehensible'. Surprisingly, nature is not only governed by laws, but the laws also exhibit intrinsic simplicity and beauty—His wisdom. Physics is not only to discover the laws but also to quest for simplicity and beauty. The fusion plasma system is a many-particle system. One might conceive that there was a supercomputer that could track the movements of each of the individual particles of number 10^n. However, how to specify the initial and boundary conditions for the 10^n individual particles makes this idea unfeasible. Actually, magnetic confinement of fusion plasmas never intends to actively control each of the individual particles. Instead, the concept strives for the collective restraint of hot plasmas. Facing this challenge, fluid theory defines its position. It is simple but nonetheless captures the essential physics.

By convention, the Braginskii two-fluid theory is often employed in this field. It assumes the collisionality to be dominant. In reality, however, the magnetically confined plasmas in thermonuclear fusion tend to be collisionless. Some particles are trapped by the inhomogeneous equilibrium field or by the waves, while some others remain in circulation. One cannot expect charged particles to move together as a fluid cell along the magnetic field lines. Consequently, the fluid parameters relating to the parallel motion, such as parallel fluid velocity, become inadequate. One may expect the spatial localization in the perpendicular direction, while the finite Larmor radius effects are taken into account in the next order. In the parallel direction, however, the particle mobility features need to be fully retained. This leads the current book to present extensively the perpendicular magnetofluid theory, as well as the fluid-kinetic hybrid theories.

In view of the importance of multifluid theory as a bridge between the ideal magnetohydrodynamics and kinetic theory, I hope this book will help to further advance the scientific exploration of peaceful use of fusion energy.

Dr Linjin Zheng
Institute for Fusion Studies
The University of Texas at Austin
Texas, USA

May 27, 2020

Acknowledgements

This book was written when the author worked at Institute for Fusion Studies, The University of Texas at Austin, with support by the U.S. Department of Energy, Office of Fusion Energy Science. Many research results in this book have benefited from the collaborations with colleagues in this institute, including Drs M T Kotschenreuther, R D Hazeltine, F L Waelbroeck, H L Berk, B Breizman, R Fitzpatrick, W Horton *et al* and J W Van Dam, who is now in the U.S. Department of Energy. The perpendicular magnetofluid theory particularly benefited from helpful discussion and suggestions from Professor R D Hazeltine. The author is also grateful for collaborations with many other colleagues in this field.

Author biography

Linjin Zheng

Dr Linjin Zheng is a theoretical physicist for controlled thermonuclear fusion plasmas. He received his MS degree from The University of Science and Technology of China and PhD from Institute of Physics – Beijing, Chinese Academy of Sciences. He is currently a research scientist at Institute for Fusion Studies, The University of Texas at Austin. He published more than one hundred scientific papers for examples in *Physics Review Letters, Physics Letters, Nuclear Fusion, Physics of Plasmas,* and major conferences. He has edited one and authored three books. His research covers ideal/resistive magnetohydrodynamics, two-fluid, and kinetic theories for both equilibrium and stability. His major contributions with his colleagues include the reformulation of gyrokinetic theory, the development of the perpendicular magnetofluid theory, the theoretical interpretation for the so-called edge localized modes, the invention of the free boundary ballooning representation, the discoveries of second toroidal Alfvén eigenmodes and the current interchange tearing modes, etc. With his colleagues, he also developed the tokamak equilibrium code ATEQ and stability codes AEGIS and AEGIS-K.

Part I

Introduction

Chapter 1

Principle of the fluid description of magnetically confined plasmas

In this chapter, the principle of the fluid description of magnetically confined plasmas is delineated. First, the background for the fluid description of plasmas is briefly described. The necessity for the perpendicular fluid formulation or the fluid-kinetic hybrid description is then explained. Next, the theoretical frameworks and underlying thoughts of the fluid description are outlined.

1.1 Framework of conventional fluid theory

In physics and engineering, fluid dynamics, or more generally fluid mechanics, refers usually to the physics branch that describes the motion of fluids—liquids and gases. In particular, it includes hydrodynamics (the study of liquids in motion) and aerodynamics (the study of air and other gases in motion).

Different from a solid, in which the relative positions of molecules remain essentially fixed under non-destructive conditions, a fluid is instead a substance in which the constituent molecules or particles are considerably free to move relative to each other. Nevertheless, the particles in a fluid are not completely free. They are expected to move together in cells and within each cell, a local thermal equilibrium is established. Fluids are continuous, rather than discrete as particles, and therefore obey the so-called continuum assumption.

Since a fluid cell is assumed to be in the local thermal equilibrium, a fluid system can be determined by the macroscopic fluid thermodynamic variables, such as the density n, velocity $\mathbf{V}$, and temperature T. The equations for fluid thermodynamic variables are obtained by constructing the moments of the kinetic equations. Average over the velocity space is performed, leaving the fluid variables to vary in space and evolve in time. Nevertheless, the change rate is assumed to be much smaller than the relaxation rate to re-establish the local thermal equilibrium in a

doi:10.1088/978-0-7503-3575-1ch1

fluid cell. It is this time scale separation that makes it possible to describe the system with the fluid variables.

The fluid moment equations are not closed. The continuity equation for density involves the higher order moment: velocity. The momentum equation for velocity involves the even higher order moment: pressure. In the fluid theory, the closure of the moment equations is facilitated by introducing the so-called transport coefficients, which relate the higher order moments to the lower order moments and their gradients.

In physics, the variations of fluid thermodynamic variables are affected by the interaction between the fluid cells. The particle diffusibility gives the particle flux from one cell to the other; the pressure and pressure tensor specify the force exerted from the neighboring cells; the heat conductivity represents the energy flux from one cell to the other, and so on. Besides, the interaction of different particle species can be 'internal', like the friction force and heat generation. These are referred to as the transport processes. They result from the departure from the 'global' equilibrium although the local one is established relatively. The departure from the 'global' equilibrium is characterized by the gradients of fluid thermodynamic variables, such as ∇n, ∇T, etc. They are referred to as the thermodynamic forces, $A_i(\mathbf{x}, t)$ with $i = 1, \ldots, n$. Assuming the forces to be sufficiently weak, the fluxes can be supposed to vary linearly with the corresponding forces:

$$\Gamma_i = \sum_j L_{ij}\mathbf{A}_j,$$

Here, the quantities L_{ij} are referred to as transport coefficients. A typical example is the heat flux: $\mathbf{q} = \kappa\nabla T$. In fluid theory, the closure of the fluid moment equations is realized by computing the transport coefficients L_{ij} with the kinetic theory, which relates L_{ij} to the lower order moments. This completes the fluid description of the system.

1.2 Fluid description of magnetically confined plasmas

Plasma is regarded as the fourth state beyond solid, liquid, and gas. The fluid description of magnetically confined plasmas evolves from the fluid theory of liquids and gases. An important theory for describing the magnetically confined plasmas is the so-called magnetohydrodynamics [1]. The terminology 'magnetohydrodynamics' derives exactly from 'hydrodynamics' in the fluid mechanics. In principle, the magnetically confined plasmas differ quite a lot from conventional fluids, like water or neutral gases. Nonetheless, the 'magneto'—i.e. the magnetic field—makes the plasmas fluid-like, in particular in the direction perpendicular to the magnetic field.

The mathematical theory of nonuniform gases is given in detail by Chapman and Cowling [2]. It describes the method for obtaining the transport equations from the kinetic equations, i.e. the fluid moment equations. The theory is also extended to describe the ionized gases in this book. The stress tensor, heat flux, and electric conductivity are all derived in it. Nevertheless, before Braginskii's work in references

[3] and [4], the transport theories for plasmas were mainly limited to the cases with the local distribution very close to Maxwellian, for example in references [5]–[8].

In extending the theory by Chapman and Cowling, Braginskii derived a system of transport equations for a plasma consisting of electrons and one kind of positive ions immersed in a strong magnetic field [3, 4]. The equation system contains the continuity equations, equations of motion, and the equation of heat transport for electron and ion fluids. The electron and ion temperatures can be different. Reference [3] claims that the case of arbitrary ratio of the particle collision frequency to gyrofrequency is included, as the gyroviscous tensor for the collisionless case is obtained. It is actually not, as will be explained later. The Braginskii fluid theory basically belongs to the classical transport theory and applies only to collision-dominant plasmas.

The claim that Braginskii's two-fluid equations apply to the low collisional frequency case is invalid, simply because the parallel fluid velocity is present in the formulation. Note that the magnetically confined plasmas in thermonuclear fusion tend to be collisionless. Some particles are trapped by the inhomogeneous equilibrium field or by the waves, while some others remain in circulation. One cannot expect charged particles to move together as a fluid cell along the magnetic field lines. Consequently, the fluid variables relating to the parallel motion, such as parallel fluid velocity, become inadequate. It is due to the presence of parallel fluid velocity that makes the Braginskii gyroviscous tensor inapplicable for the collisionless case.

In realizing that the fluid description of magnetically confined fusion plasmas is only good in the direction perpendicular to the magnetic field [9, 10], numerous efforts have been made in this field to improve the fluid model. The simplest case is the so-called perpendicular MHD model [1]. The Landau-fluid model [11, 12] and gyrofluid model [13]–[17], etc are developed later on. Readers are referred to other books for more extended reviews, for example in references [18]–[21].

References [22] and [23] proposed to construct only the perpendicular moments of the kinetic equation or to divide charged particles with different parallel velocities into multifluid to preserve the parallel kinetic description. In this formalism, the particle spatial localization is solo based on the strong magnetic field in the perpendicular direction, while the finite Larmor radius (FLR) effects are taken into account in the next order. In the parallel direction, the particle mobility features are fully retained. The theory is based on the systematic and rigorous analyses and naturally contains Landau damping, particle trapping, etc without requiring ad hoc modelling. It is referred to as the perpendicular magnetofluid theory.

As will be extensively presented in this book, the perpendicular magnetofluid description of fusion plasmas is required as well for weakly collisional effects. The like-particle collisions can change the density distributions for individual fluid components with different parallel velocities. The collisions between the trapped and circulating particles can affect, for example, the stability of trapped particle modes [24].

Another effort is the so-called fluid-kinetic hybrid models. Two types of efforts have been made: the pressure moment approach, for example in reference [25], and the current moment approach [26, 27]. They are intended to close the moment equation chain by directly coupling the suitable moments to the solutions of kinetic equations. The kinetic equations become a part of the complete set of hybrid equations.

In the end, we point out that the kinetic description may often be regarded as an advantageous and more complete description of magnetized plasma, but in reality, it may not always be the case. Determining the distribution function requires not only the governing differential equations but also the initial and boundary conditions. One may be able to simulate a large number of particles with supercomputers. However, one is unable to specify the initial and boundary conditions for each individual particle. The fluid thermodynamic variables are often introduced to specify these conditions, for example, loading the Maxwellian distribution initially involves the density n, velocity $\mathbf{V}$, and temperature T. This indicates that the fluid method has its value, not mentioning the simplicity.

References

[1] Freidberg J P 1987 *Ideal Magnetohydrodynamics* (New York: Plenum Press)

[2] Chapman S and Cowling T G 1953 *Mathematical Theory of Nonuniform Gases* (New York: Cambridge University Press)

[3] Braginskii S I 1957 *ZhETF (J. Exp. Theor. Phys. USSR)* **23** 459
Braginskii S I 1958 Transport phenomena in a completely Ionized two-temperature plasma *Soviet Phys. JETP* **6** 358–69

[4] Braginskii S I 1966 Transport processes *Reviews of Plasma Physics* vol 1 ed M Leontovich (New York: Consultants Bureau) 205–311

[5] Cowling T G 1945 The electrical conductivity of an ionized gas in a magnetic field, with applications to the solar atmosphere and the ionosphere *Proc. R. Soc.* A 183 453–79

[6] Spitzer L Jr and Härm R 1953 Transport phenomena in a completely ionized gas *Phys. Rev.* **89** 977–81

[7] Longmire C L and Rosenbluth M N 1956 Diffusion of charged particles across a magnetic field *Phys. Rev.* **103** 507–10

[8] Kaufman A N 1960 Plasma viscosity in a magnetic field *Phys. Fluids* **3** 610–6

[9] Hazeltine R D 1998 Transport theory in the collisionless limit *Phys. Plasmas* **5** 3282–6

[10] Hinton F L and Hazeltine R D 1976 Theory of plasma transport in toroidal confinement systems *Rev. Mod. Phys.* **48** 239–308

[11] Hammet G W and Perkins F W 1990 Fluid moment models for Landau damping with application to the ion-temperature-gradient instability *Phys. Rev. Lett.* **64** 3019–22

[12] Snyder P B and Hammett G W 2001 A Landau fluid model for electromagnetic plasma microturbulence *Phys. Plasmas* **8** 3199–216

[13] Dorland W and Hammett G W 1993 Gyrofluid turbulence models with kinetic effects *Phys. Fluids* **5** 812–35

[14] Waltz R E, Kerbel G D and Milovich J 1994 Toroidal gyro-Landau fluid model turbulence simulations in a nonlinear ballooning mode representation with radial modes *Phys. Plasmas* **1** 2229–44

[15] Ramos J J 2008 Finite-Larmor-radius kinetic theory of a magnetized plasma in the macroscopic flow reference frame *Phys. Plasmas* **15** 082106

[16] Waelbroeck F L, Hazeltine H D and Morrison P J 2009 A Hamiltonian electromagnetic gyrofluid model *Phys. Plasmas* **16** 032109

[17] Keramidas Charidakos I, Waelbroeck F L and Morrison P J 2015 A Hamiltonian five-field gyrofluid model *Phys. Plasmas* **22** 112113

[18] Hazeltine R D and Meiss J D 1992 *Plasma Confinement* (New York: Addison-Wesley)

[19] Hazeltine R D and Waelbroeck F L 2018 *The Framework of Plasma Physics* (Boca Raton, FL: CRC Press)

[20] Weiland J 2000 *Collective Modes in Inhomogeneous Plasma* (Bristol: Institute of Physics Publishing)

[21] Helander P and Sigmar D 2002 *Collisional Transport in Magnetized Plasmas* (New York: Cambridge University Press)

[22] Zheng 2020 Perpendicular magnetofluid theory for magnetically confined plasmas in the collisionless limit *Phys. Plasmas* **27** 052503

[23] Zheng L 2019 *Magnetically Confined Fusion Plasma Physics: Ideal MHD theory IOP Concise Physics* (San Rafael, CA: Morgan & Claypool)

[24] Tang W M 1978 Microinstability theory in tokamaks *Nucl. Fusion* **18** 1089–160

[25] Kulsrud R M 1983 MHD description of plasma *Basic Plasma Physics I (Handbook of Plasma Physics)* ed A A Galeev and R N Sudan (Amsterdam: North-Holland) 1–31 pp

[26] Zheng L J, Kotschenreuther M T and Van Dam J W 2007 Revisiting linear gyrokinetics to recover ideal magnetohydrodynamics and missing finite Larmor radius effects *Phys. Plasmas* **14** 072505

[27] Zheng L 2015 *Advanced Tokamak Stability TheoryIOP Concise Physics* (San Rafael: Morgan & Claypool)

Part II

Fluid theory in the collision-dominant regime

Chapter 2

Braginskii two-fluid theory

The Braginskiis two-fluid theory [1, 2] is widely used in this field to investigate the stability and transport behaviors of magnetized plasmas both in the analytical analyses and numerical simulations. It is considered to be the most direct extension of the ideal MHD single fluid theory. Nonetheless, because of the highly concise feature of Braginskii's papers, the detailed derivations of Braginskii's equations are mainly absent. Also, although it is known that the particle localization assumption in the direction parallel to the magnetic field in the Braginskii transport theory is not good for fusion plasmas as the collision frequency is not high enough to bind the charged particles together, the theoretical framework for constructing fluid equations is valuable. We therefore anyway describe the theory in detail in this chapter. The non-localization issue will be addressed in the next chapter.

We first describe the derivation of fluid moment equations of the kinetic equation. The moment equations involve three lowest moments: density n, velocity $\mathbf{V}$, and temperature T. They are coupled to the higher order moments. Instead of resorting to the higher order moment equations, these higher order moments are evaluated perturbedly by solving the kinetic equation of next order to relate them to the lower order moments and their gradients. This process is first explained by the estimation and interpretation of the transport coefficients. Then, the kinetic equations for ions and electrons of first order are solved, with the collisional operators reduced. The solutions are used to compute the transport coefficients: such as the friction force, pressure tensor, and heat flux, etc. Both the collisional and collisionless contributions to the transport coefficients are obtained to conclude the closure of moment equations. The conditions of applicability are discussed at the end.

2.1 Transport equations

In this section, the equations of fluid motion are derived in accordance with the Braginskii two-fluid theory [1, 2]. The Boltzmann equation is first introduced. The moment equations are then constructed. The continuity, momentum, and energy

doi:10.1088/978-0-7503-3575-1ch2

transport equations are obtained afterwards. The closures of fluid equations and transport coefficients are discussed at the end. Only the so-called simple plasma, which contains only one type of ions together with electrons, is considered as required by the Braginskii theory [1, 2].

2.1.1 Boltzmann equation

In this subsection, we briefly introduce the Boltzmann equation [3]. A classical many-particle system is considered, in which particle motions are governed by Newton's second law of motion.

Note that Newton's second law is a second order differential equation in time t, the particle motion in the next time step is determined by the current values of position $\mathbf{x}$ and velocity $\mathbf{v}$. Therefore, to specify the state of a many-particle system with the total number of particles N, one needs to determine the probability density function $f_N(\mathbf{x}_1, \mathbf{v}_1, \cdots, \mathbf{x}_N, \mathbf{v}_N)$ in the $6N$ dimensional phase space (three spatial and three velocity coordinates per particle). As is known, one may employ either the BBGKY or Klimontovich formalisms to describe the many-particle statistical problem [4, 5].

Instead of considering the $6N$ dimensional phase space, the Boltzmann equation formalism invokes the particle distribution functions in the six-dimensional phase space: $f_a(t, \mathbf{x}, \mathbf{v})$, where the subscript a is introduced to label the particle species with $a = i, e$ denoting ion and electron species, respectively. The Boltzmann distribution functions specify the particle density in the phase space at a given time t. Therefore, $f(t, \mathbf{x}, \mathbf{v})d\mathbf{x}d\mathbf{v}$ is the number of particles in the six-dimensional volume element $d\mathbf{x}d\mathbf{v}$. This is a reduction from the full probability density function f_N. The particle correlation effects are then treated as the collisional operator in the Boltzmann equation.

The Boltzmann equation is given as follows [3]

$$\frac{\partial f_a}{\partial t} + \mathbf{v}_a \cdot \frac{\partial f_a}{\partial \mathbf{x}} + \mathbf{a}_a \cdot \frac{\partial f_a}{\partial \mathbf{v}} = C_a, \tag{2.1}$$

where $\mathbf{a}$ is the acceleration, which can be related to the force by Newton's second law of motion. In the presence of electromagnetic field $\mathbf{E}$ and $\mathbf{B}$

$$\mathbf{a}_a = \frac{e_a}{m_a}(\mathbf{E} + \mathbf{v}_a \times \mathbf{B}),$$

where e_a denotes the charge and m_a the mass. In equation (2.1) C_a on the right-hand side is the so-called collisional operator. Because the force related to the acceleration $\mathbf{a}$ term in equation (2.1) is a macroscopic one, averaged (or 'smoothed') over a volume containing many particles in space and over time long compared with the appropriate time of particle flight, the collisional operator needs to be introduced to account for the rapidly fluctuating microforces, arising when particles come very close to each other. This can also be understood as to count in the remaining effects from reducing the 6N dimensional description f_N to the six-dimensional one f [4, 5]. Because particles of species a can collide with each other or with other species, the collisional operator can be expressed as

$$C_a = \sum_b C_{ab}(f_a, f_b) \tag{2.2}$$

with $b = a$ included as well. Here, $C_{ab}(f_a, f_b)$ represents the change of the distribution function per unit time from the collisions of species a with the species b.

There are some basic conservation properties relating to the collisional process, which can be obtained without detailed knowledge of the collisional operator. It is assumed that there are no species production and annihilation, for example, ionization and charged particle recombination. In this case one has

$$\int C_{ab}\, d\mathbf{v}_a = 0, \tag{2.3}$$

$$\int m_a \mathbf{v}_a C_{aa}\, d\mathbf{v}_a = 0, \tag{2.4}$$

$$\int \frac{m_a v_a^2}{2} C_{aa}\, d\mathbf{v}_a = 0. \tag{2.5}$$

Here, equation (2.3) just tells us that the particle density of species 'a' does not change in the collisions, and equations (2.4) and (2.5) indicate the momentum and energy are conserved in the like-particle collisional processes. One can also show that total momentum and energy are conserved in the collisional processes between two species:

$$\int m_a v_a C_{ab}\, d\mathbf{v}_a + \int m_b v_b C_{ba}\, d\mathbf{v}_b = 0, \tag{2.6}$$

$$\int \frac{m_a v_a^2}{2} C_{ab}\, d\mathbf{v}_a + \int \frac{m_b v_b^2}{2} C_{ba}\, d\mathbf{v}_b = 0. \tag{2.7}$$

The Landau collisional operator [6] is employed in the Braginskii two-fluid theory:

$$\begin{aligned} C_{ab}(f_a, f_b) = &-\frac{2\pi\lambda e_a^2 e_b^2}{m_a}\frac{\partial}{\partial v_{a\beta}} \\ &\times \int \left\{ \frac{f_a(\mathbf{v}_a)}{m_b}\frac{\partial f_b(\mathbf{v}_b)}{\partial v_{b\gamma}} - \frac{f_b(\mathbf{v}_b)}{m_a}\frac{\partial f_a(\mathbf{v}_a)}{\partial v_{a\gamma}} \right\} U_{\beta\gamma}\, d\mathbf{v}_b, \end{aligned} \tag{2.8}$$

where the Einstein summation convention is implied for same indices,

$$U_{\beta\gamma} = \frac{1}{u^3}\left(u^2\delta_{\beta\gamma} - u_\beta u_\gamma\right), \quad \text{with } u_\beta = v_{a\beta} - v_{b\beta} \tag{2.9}$$

and λ is the so-called Coulomb logarithm, which equals the logarithm of the ratio of maximum to minimum impact parameters $\lambda = \log(P_{\max}/P_{\min})$. The minimum impact parameter is taken to be the distance that causes a deflection of 90° degrees, which is proved to be $P_{\min} \sim e^2/3T$ in the classical electrodynamics. The maximum impactor parameter is assumed to be the Debye length $P_{\max} \sim \delta_D$, where $\delta_D = (T_e/e^2n)^{1/2}$. Here, T denotes temperature and n is density.

Here, it should be noted that $U_{\alpha\gamma}$ depends only on the relative velocities between two species u_β. This leads to the collisional operator being invariant to the frame change. Note also that the ratio of the first to the second terms in equation (2.84) is of order $\sqrt{m_a/m_b}$. This is useful since electron and ion masses differ vastly. One can tap this ordering in evaluating the collision effects.

A collision-dominant plasma is considered in the Braginskii fluid theory. The mean free path is assumed to be much smaller than the macroscopic scale length and the collision time is shorter than the time scale to be considered. In this case, local Maxwellian distribution is established in the lowest order:

$$f_a^0 = \frac{n_a}{(2\pi T_a/m_a)^{3/2}} e^{-\frac{m_a}{2T_a}(\mathbf{v}_a - \mathbf{V}_a)^2}, \tag{2.10}$$

where $\mathbf{V}$ is the average velocity and the temperature is expressed in the energy unit so that the Boltzmann constant is left out. We also define the random velocity $\mathbf{v}'_a = \mathbf{v}_a - \mathbf{V}_a$.

It is well-known that the Maxwellian distribution leads the collision operator to vanish, regardless of the type of collision operators. In contrast, according to the H-theorem of Boltzmann, no matter what initial conditions are, the collision effects will lead the distribution function to approach a Maxwellian in due course. The latter is referred to as the relaxation process, which generally occurs in the collisional time scale.

The Braginskii two-fluid theory deals with transport phenomena in plasmas, involving the movement of various quantities, such as mass, momentum, or energy, through a plasma medium. The equilibrium non-uniformity can cause the thermal transports and the collision effects result in the collisional transports.

2.1.2 Moments of the kinetic equations

In this subsection, we outline the derivation of moment equations [1, 2]. The moment equation approach will be compared to the perpendicular magnetofluid theory in the next chapter.

For any phase space quantity $\eta_a(\mathbf{x}_a, \mathbf{v}_a, t)$, the corresponding moment of the distribution function is defined as follows

$$\langle \eta_a \rangle = \frac{1}{n_a} \int \eta_a f_a \, d\mathbf{v}_a,$$

where $n_\alpha = \langle 1 \rangle$ is simply the fluid density. Multiplying equation (2.1) with η_α and integrating it over the velocity space, one can derive the general moment equation. Noting that

$$\partial_{v\alpha} a_\alpha = \frac{e}{m}\partial_{v\alpha}\left(e_{\alpha\beta\gamma} v_\beta B_\gamma\right) = \frac{e}{m} e_{\alpha\beta\gamma}\delta_{\alpha\beta} B_\gamma = \frac{e}{m} e_{\alpha\alpha\gamma} B_\gamma = 0, \tag{2.11}$$

where $\partial_v = \partial/\partial v_\alpha$ and $e_{\alpha\beta\gamma}$ is the Levi-Civita symbol or the alternating tensor as explained in appendix A, one has

$$\int \eta a_\alpha \partial_{v\alpha} f \, d\mathbf{v} = -\int f a_\alpha \partial_{v\alpha} \eta \, d\mathbf{v}.$$

Therefore, one obtains the general moment equation:

$$\begin{aligned} &\frac{\partial}{\partial t}(n_a\langle \eta_a \rangle) - n_a\left\langle \frac{\partial \eta_a}{\partial t} \right\rangle + \frac{\partial}{\partial \mathbf{x}_a} \cdot n_a \langle \mathbf{v}_a \eta_a \rangle - n_a\left\langle \mathbf{v}_a \cdot \frac{\partial \eta_a}{\partial \mathbf{x}_a} \right\rangle \\ &- n_a\left\langle \mathbf{a}_a \cdot \frac{\partial \eta_a}{\partial \mathbf{v}_a} \right\rangle = \sum_b \int \eta_a C_{ab}(f_a, f_b) d\mathbf{v}_a. \end{aligned} \tag{2.12}$$

Using equation (2.12) one can derive the fluid equations, such as continuity, momentum, and energy transport equations, in the following subsubsections.

Continuity equation

Letting $\eta_a = 1$ with $a = e, i$ in equation (2.12), one obtains the continuity equations for electrons and ions as follows

$$\frac{\partial n_e}{\partial t} + \frac{\partial}{\partial x_\alpha}(n_e V_{e\alpha}) = 0, \tag{2.13}$$

$$\frac{\partial n_i}{\partial t} + \frac{\partial}{\partial x_\alpha}(n_i V_{i\alpha}) = 0, \tag{2.14}$$

where the fluid velocity is defined as follows

$$V_{a\alpha} = \frac{1}{n_{a\alpha}} \int v_{a\alpha} f_a d\mathbf{v}_a.$$

Here, equation (2.3) has been used to prove that the collision term contribution vanishes.

Momentum equation

To derive the momentum equation, we let $\eta_a = m_a \mathbf{v}_a$ with $a = e, i$ in equation (2.12). Noting that

$$\partial_t(nV_\alpha) = n\partial_t V_\alpha + V_\alpha \partial_t n = n\partial_t V_\alpha - V_\alpha \partial_\beta(nV_\beta),$$

$$\begin{aligned}\partial_\beta\left(n\langle v_\beta v_\alpha\rangle\right) &= \partial_\beta\left[n\left\langle (v'_\beta + V_\beta)(v'_\alpha + V_\alpha)\right\rangle\right] \\ &= \partial_\beta\left(n\left\langle v'_\beta v'_\alpha + v'_\beta V_\alpha + v'_\alpha V_\beta + V_\beta V_\alpha\right\rangle\right) \\ &= \partial_\beta\left(n\left\langle v'_\beta v'_\alpha\right\rangle\right) + nV_\beta \partial_\beta V_\alpha + V_\alpha \partial_\beta(nV_\beta),\end{aligned}$$

$$nm\left\langle v'_\beta v'_\alpha\right\rangle = nm\left\langle \frac{1}{3}v'^2\delta_{\alpha\beta} + \left(v'_\alpha v'_\beta - \frac{1}{3}v'^2\delta_{\alpha\beta}\right)\right\rangle,$$

where for brevity we denote $\partial_\alpha = \partial/\partial x_\alpha$ and $\partial_t = \partial/\partial t$, one obtains the momentum transport equations for electrons and ions as follows

$$n_e m_e \frac{d_e V_{e\alpha}}{dt} = -\frac{\partial p_e}{\partial x_\alpha} - \frac{\partial \pi_{e\alpha\beta}}{\partial x_\beta} - n_e e(E_\alpha + e_{\alpha\beta\gamma}V_{e\beta}B_\gamma) + R_{e\alpha}, \tag{2.15}$$

$$n_i m_i \frac{d_i V_{i\alpha}}{dt} = -\frac{\partial p_i}{\partial x_\alpha} - \frac{\partial \pi_{i\alpha\beta}}{\partial x_\beta} + n_i e(E_\alpha + e_{\alpha\beta\gamma}V_{i\beta}B_\gamma) + R_{i\alpha}, \tag{2.16}$$

where the pressure $p_a = n_a T_a$ and

$$\begin{aligned} T_a &= \frac{1}{3}m_a\langle v'^2\rangle, \\ \pi_{a\alpha\beta} &= \left\langle v'_{a\alpha}v'_{a\beta} - \frac{1}{3}v'^2_a\delta_{\alpha\beta}\right\rangle, \end{aligned} \tag{2.17}$$

$$\begin{aligned} R_{a\alpha} &= \sum_b R_{ab\alpha} = \sum_b \int m_a v'_{a\alpha} C_{ab}\, d\mathbf{v}_a, \\ \frac{d_a}{dt} &= \frac{\partial}{\partial t} + V_{a\beta}\frac{\partial}{\partial x_\beta}. \end{aligned} \tag{2.18}$$

Here, equation (2.4) has been used to reduce the momentum transfer vector $R_{a\alpha}$, so that only $v'_{a\alpha}$, instead of $v_{a\alpha}$, remains in the integrand. Using equation (2.6), one can also prove that

$$R_{ei\alpha} = -R_{ie\alpha}. \tag{2.19}$$

We also note that the viscosity tensor $\pi_{\alpha\beta}$ is symmetric, i.e. $\pi_{\alpha\beta} = \pi_{\beta\alpha}$.

Here, the momentum transfer vector $\mathbf{R}$ reflects the momentum transfer from one species to the other due to the collisions and the stress tenor or viscosity tensor $\pi_{\alpha\beta}$ is the contribution of the non-isotropic part of distribution to the pressure.

Energy transport equation

To derive the energy transport equation, we let $\eta_a = \frac{1}{2}m_a v_a^2$ with $a = e, i$ in equation (2.12). Noting that

$$
\begin{aligned}
\frac{1}{2}mn\langle v^2\rangle &= \frac{1}{2}mn\left\langle \left(v'_\alpha + V_\alpha\right)^2 \right\rangle \\
&= \frac{1}{2}mn\langle v'^2 + V^2\rangle \\
&= \frac{3}{2}nT + \frac{1}{2}mnV^2, \\
\frac{1}{2}m(n\langle v^2 v_\alpha\rangle) &= \frac{1}{2}m\partial_\alpha\left[n\left\langle\left(v'^2 + 2v'_\beta V_\beta + V^2\right)\left(v'_\alpha + V_\alpha\right)\right\rangle\right] \\
&= \frac{1}{2}m\left[n\left\langle v'^2 v'_\alpha + 2v'_\alpha v'_\beta V_\beta + v'^2 V_\alpha + V^2 V_\alpha\right\rangle\right] \\
&= \frac{1}{2}mn\left\langle v'^2 v'_\alpha\right\rangle + mn\left(\frac{1}{2}V^2 + \frac{5}{2}p\right)V_\alpha + \pi_{\alpha\beta}V_\alpha, \\
\int \frac{1}{2}m_a v_a^2 C_{ab}(f_a, f_b)d\mathbf{v}_a &= \int \frac{1}{2}m_a\left(v'^2_a + 2v'_a V_a\right)C_{ab}(f_a, f_b)d\mathbf{v}_a \\
&= \int \frac{m_a v'^2_a}{2} C_{ab}\, d\mathbf{v}_a + R_{ab\beta}V_{a\beta},
\end{aligned}
\tag{2.20}
$$

where equation (2.5) has been used, one obtains the energy equations for electrons and ions as follows

$$
\begin{aligned}
&\frac{\partial}{\partial t}\left(\frac{m_e n_e}{2}V_e^2 + \frac{3}{2}n_e T_e\right) + \frac{\partial}{\partial x_\beta}\left[\left(\frac{m_e n_e}{2}V_e^2 + \frac{3}{2}n_e T_e\right)V_{e\beta}\right. \\
&\left. + \left(p_e\delta_{\alpha\beta} + \pi_{e\alpha\beta}\right)V_{e\alpha} + q_{e\beta}\right] = e_e n_e E_\beta V_{e\beta} + R_{e\beta}V_{e\beta} + Q_e,
\end{aligned}
\tag{2.21}
$$

$$
\begin{aligned}
&\frac{\partial}{\partial t}\left(\frac{m_i n_i}{2}V_i^2 + \frac{3}{2}n_i T_i\right) + \frac{\partial}{\partial x_\beta}\left[\left(\frac{m_i n_i}{2}V_i^2 + \frac{3}{2}n_i T_i\right)V_{i\beta}\right. \\
&\left. + \left(p_i\delta_{\alpha\beta} + \pi_{i\alpha\beta}\right)V_{i\alpha} + q_{i\beta}\right] = e_i n_i E_\beta V_{i\beta} + R_{i\beta}V_{i\beta} + Q_i,
\end{aligned}
\tag{2.22}
$$

where

$$
q_{a\alpha} = n_a m_a\left\langle \frac{1}{2}v'^2_a v_{a\alpha}\right\rangle, \tag{2.23}
$$

$$
Q_a = \sum_b Q_{ab} = \sum_b \int \frac{m_a v'^2_a}{2} C_{ab}\, d\mathbf{v}_a. \tag{2.24}
$$

Here, the vector **q** is the heat flux carried by a given species. It represents the energy transport associated with the random motion at a rest frame. The quantity Q is related to the heating resulting from the collisions with other species.

The kinetic energy due to the macroscopic velocity V_a in the energy transport equations (2.21) and (2.22) can be eliminated by using the momentum equations

(2.15) and (2.16), dot-multiplied by **V**, and the continuity equations (2.13) and (2.14). Noting especially that

$$\frac{3}{2}\frac{\partial n_a T_a}{\partial t} + \frac{\partial}{\partial x_\beta}\left(\frac{3}{2} n_a T_a V_{a\beta}\right) = \frac{3}{2} n_a \frac{d_a T_a}{dt},$$

the energy transport equations (2.21) and (2.22) can be transformed to

$$\frac{3}{2} n_e \frac{d_e T_e}{dt} + p_e \frac{\partial V_{e\alpha}}{\partial x_\alpha} = -\frac{\partial q_{e\alpha}}{\partial x_\alpha} - \pi_{i\alpha\beta} \frac{\partial V_{e\alpha}}{\partial x_\beta} + Q_e, \tag{2.25}$$

$$\frac{3}{2} n_i \frac{d_i T_i}{dt} + p_i \frac{\partial V_{i\alpha}}{\partial x_\alpha} = -\frac{\partial q_{i\alpha}}{\partial x_\alpha} - \pi_{i\alpha\beta} \frac{\partial V_{i\alpha}}{\partial x_\beta} + Q_i. \tag{2.26}$$

Define the quantity

$$s_a \equiv \ln\left(T_a^{3/2}/n_a\right) = \ln\left(p_a^{3/2}/n_a^{5/2}\right),$$

which actually represents the entropy per particle within an unimportant constant. Again using the continuity equations (2.13) and (2.14), the entropy equations can be obtained from the energy transport equations (2.21) and (2.22)

$$\begin{aligned} n_e T_e \frac{d_e s_e}{dt} &= T_e\left[\frac{\partial n_e s_e}{\partial t} + \partial_\beta(n_e s_e V_{e\beta})\right] \\ &= -\frac{\partial q_{e\alpha}}{\partial x_\alpha} - \pi_{i\alpha\beta} \frac{\partial V_{e\alpha}}{\partial x_\beta} + Q_e, \end{aligned} \tag{2.27}$$

$$\begin{aligned} n_i T_i \frac{d_i s_i}{dt} &= T_i\left[\frac{\partial n_i s_i}{\partial t} + \partial_\beta(n_i s_i V_{i\beta})\right] \\ &= -\frac{\partial q_{i\alpha}}{\partial x_\alpha} - \pi_{i\alpha\beta} \frac{\partial V_{i\alpha}}{\partial x_\beta} + Q_i. \end{aligned} \tag{2.28}$$

Noting the energy conservation in equation (2.7) and using equation (2.20), one has

$$\left(Q_{ei} + R_{ei\beta} V_{e\beta}\right) + \left(Q_{ie} + R_{ie\beta} V_{i\beta}\right) = 0.$$

Using equation (2.19), this equation is further reduced to

$$Q_{ei} + Q_{ie} = -R_{ei\beta} V_{e\beta} - R_{ie\beta} V_{i\beta} = -R_{ei\beta}\left(V_{e\beta} - V_{i\beta}\right). \tag{2.29}$$

Maxwell equations
To constitute the complete set of equations the Maxwell equations are also required.

$$\nabla \cdot \mathbf{E} = \frac{1}{\epsilon_0} \sum_a n_a e_a,$$
$$\nabla \times \mathbf{E} = -\frac{\partial \mathbf{B}}{\partial t},$$
$$\nabla \cdot \mathbf{B} = 0,$$
$$\nabla \times \mathbf{B} = \mu_0 \left(\mathbf{J} + \epsilon_0 \frac{\partial \mathbf{E}}{\partial t} \right),$$

where ϵ_0 and μ_0 are, respectively, the permittivity and the permeability of free space. Here, the Gauss law is often replaced by the quasineutrality condition.

2.1.3 On the moment closure and transport coefficients

In this subsection, we outline the procedure for the fluid moment closure and list the transport coefficients. Estimation and interpretation of transport coefficients will be given in section 2.2. The evaluation of transport coefficients through kinetic theory will be detailed in section 2.3.

The continuity equations (2.13) and (2.14), the momentum equations (2.15) and (2.16), and the energy transport equations (2.25) and (2.26) can be used to determine the three lowest moments: densities n_a, velocities $\mathbf{V}_a$, and temperatures T_a for electron and ion fluids. However, this set of equations is incomplete. One needs to know the higher order moments: the stress tensor $\pi_{\alpha\beta}$ and the heat flux $\mathbf{q}$, and the collisional effects: the momentum transfer $\mathbf{R}$ and the heat generation Q. One cannot close the fluid moment equations by simply developing higher order moment equations. It will generate even higher order transport coefficients. Therefore, one needs to suitably evaluate the moments: $\pi_{\alpha\beta}$, $\mathbf{R}$, $\mathbf{q}$, and Q for moment equation closure. They are related to lower order moments and their gradients through the so-called transport coefficients.

To evaluate the transport coefficients for moment closure, one needs to solve the kinetic equation under a suitable ordering scheme. In the Braginskii two-fluid theory the collision frequency and particle gyrofrequency ω_c are assumed to be much higher than the frequency of interest, and the particle mean free path and the Larmor radius are much shorter than the scale length of the problems to be investigated. In principle, a true Maxwellian equilibrium state applies only in the homogeneous cases, i.e. there are no gradients for the equilibrium quantities. The equilibrium in the absolute homogeneous case can last forever. However, in the cases with slow variations in time and in space one can still seek for a local transient Maxwellian equilibrium distribution as the lowest order solution:

$$f_a^0(\mathbf{x}, \mathbf{v}_a, t) = \frac{n_a(\mathbf{x}, t)}{(2\pi T_a(\mathbf{x}, t)/m_a)^{3/2}} e^{-\frac{m_a}{2T_a(\mathbf{x}, t)}(\mathbf{v}_a - \mathbf{V}_a(\mathbf{x}, t))^2}, \qquad (2.30)$$

in which the time and spatial dependences of the distribution function are solo through the three lowest order moments: n_a, $\mathbf{V}_a$, and T_a. The full distribution function is then decomposed as

$$f_a(\mathbf{x}, \mathbf{v}_a, t) = f_a^0(\mathbf{x}, \mathbf{v}_a, t) + f_a^1, \quad \text{with} f_a^1 \ll f_a^0. \tag{2.31}$$

With this ordering separation one can solve the kinetic equation for next order distribution function f_a^1. With f_a^1 one can evaluate the transport coefficients to close the moment equations. In this formalism, the three lowest order moments n_a, $\mathbf{V}_a$, and T_a are governed by the corresponding three moment equations, With n_a, $\mathbf{V}_a$, and T_a determined, one can obtain the distribution function f_a^0 in the lowest order. Using f_a^0 one can solve perturbedly the next order distribution function f_a^1. In return, f_a^1 can be used to evaluate the transport coefficients to determine the higher order moments and the collisional effects in the moment equations. Before detailing how to implement this procedure in section 2.3, for the purpose of references here, let us first summarize the results for transport coefficients, as given in references [1, 2].

The electron and ion collision times are given as follows:

$$\tau_e = \frac{3\sqrt{m_e}T_e^{3/2}}{4\sqrt{2\pi}\lambda e^4 Z^2 n_i}, \tag{2.32}$$

$$\tau_i = \frac{3\sqrt{m_i}T_i^{3/2}}{4\sqrt{\pi}\lambda e^4 Z^4 n_i}. \tag{2.33}$$

The momentum transfer is

$$\mathbf{R}_{ei} = -\mathbf{R}_{ie} \equiv \mathbf{R} = \mathbf{R}_u + \mathbf{R}_T. \tag{2.34}$$

The friction force

$$\mathbf{R}_u = -\frac{m_e n_e}{\tau_e}\left(0.5\mathbf{u}_\parallel + \mathbf{u}_\perp\right) = en\left(\frac{\mathbf{j}_\parallel}{\sigma_\parallel} + \frac{\mathbf{j}_\perp}{\sigma_\perp}\right), \tag{2.35}$$

where $\mathbf{u} = \mathbf{V}_e - \mathbf{V}_i$ and the electrical conductivities are

$$\sigma_\perp = \frac{e^2 n_e \tau_e}{m_e}, \qquad \sigma_\parallel = 1.96\sigma_\perp. \tag{2.36}$$

The thermal force

$$\mathbf{R}_T = -0.71 n_e \nabla_\parallel T_e - \frac{3}{2}\frac{n_e}{\omega_{ce}\tau_e}\mathbf{b} \times \nabla T_e. \tag{2.37}$$

The stress tensor in the absence of magnetic field is given by

$$\pi_{\alpha\beta} = -\eta_0 W_{\alpha\beta}, \tag{2.38}$$

where the rate-of-strain tensor is

$$W_{\alpha\beta} = \frac{\partial V_\alpha}{\partial x_\beta} + \frac{\partial V_\beta}{\partial x_\alpha} - \frac{2}{3}\delta_{\alpha\beta}\nabla \cdot \mathbf{V}. \tag{2.39}$$

In the presence of strong magnetic field $\omega_c\tau \gg 1$ the stress tensor is given in tensor notation as follows

$$\pi_{\alpha\beta} = -\eta_0 W_{0\alpha\beta} - \eta_1 W_{1\alpha\beta} - \eta_2 W_{2\alpha\beta} + \eta_3 W_{3\alpha\beta} + \eta_4 W_{4\alpha\beta} \tag{2.40}$$

where

$$W_{0\alpha\beta} = \frac{3}{2}\left(h_\alpha h_\beta - \frac{1}{2}\delta_{\alpha\beta}\right)\left(h_\mu h_\nu - \frac{1}{3}\delta_{\mu\nu}\right)W_{\mu\nu}, \tag{2.41}$$

$$W_{1\alpha\beta} = \left(\delta^{\perp}_{\alpha\mu}\delta^{\perp}_{\beta\nu} + \frac{1}{2}\delta^{\perp}_{\alpha\beta}h_\mu h_\nu\right)W_{\mu\nu}, \tag{2.42}$$

$$W_{2\alpha\beta} = \left(\delta^{\perp}_{\alpha\mu}h_\beta h_\nu + \delta^{\perp}_{\beta\nu}h_\alpha h_\mu\right)W_{\mu\nu}, \tag{2.43}$$

$$W_{3\alpha\beta} = \frac{1}{2}\left(\delta^{\perp}_{\alpha\mu}e_{\beta\gamma\nu} + \delta^{\perp}_{\beta\nu}e_{\alpha\gamma\mu}\right)h_\gamma W_{\mu\nu}, \tag{2.44}$$

$$W_{4\alpha\beta} = \left(h_\alpha h_m u e_{\beta\gamma\nu} + h_\beta h_\nu e_{\alpha\gamma\mu}\right)h_\gamma W_{\mu\nu}, \tag{2.45}$$

where $\delta^{\perp}_{\alpha\beta} = \delta_{\alpha\beta} - h_\alpha h_\beta$ and $e_{\alpha\beta\gamma}$ is the alternating tensor as given in appendix A. In the orthogonal coordinate system (x, y, z) with magnetic field aligned along the z-axis, the stress tensor can be expressed in components as follows

$$\pi_{zz} = -\eta_0 W_{zz}, \tag{2.46}$$

$$\pi_{xx} = -\eta_0\frac{1}{2}\left(W_{xx} + W_{yy}\right) - \eta_1\frac{1}{2}\left(W_{xx} - W_{yy}\right) - \eta_3 W_{xy}, \tag{2.47}$$

$$\pi_{yy} = -\eta_0\frac{1}{2}\left(W_{xx} + W_{yy}\right) - \eta_1\frac{1}{2}\left(W_{xx} - W_{yy}\right) + \eta_3 W_{xy}, \tag{2.48}$$

$$\pi_{xy} = \pi_{yx} = -\eta_1\frac{1}{2}W_{xx} + \eta_3\frac{1}{2}\left(W_{xx} - W_{yy}\right), \tag{2.49}$$

$$\pi_{zx} = \pi_{xz} = -\eta_2 W_{xz} - \eta_4 W_{yz}, \tag{2.50}$$

$$\pi_{yz} = \pi_{zy} = -\eta_2 W_{xz} + \eta_4 W_{yz}. \tag{2.51}$$

Here, the stress tensor expressions apply for both ion and electron species only with different velocities and viscosity coefficients, as one can expect. The ion viscosity coefficients are given as follows

$$\eta_0^i = 0.96 n_i T_i \tau_i, \tag{2.52}$$

$$\eta_1^i = \frac{3}{10}\frac{n_i T_i}{\omega_{ci}^2\tau_i}, \qquad \eta_2^i = 4\eta_1^i, \tag{2.53}$$

$$\eta_3^i = \frac{1}{2}\frac{n_i T_i}{\omega_{ci}}, \qquad \eta_4^i = 2\eta_3^i. \tag{2.54}$$

The electron viscosity coefficients are

$$\eta_0^e = 0.736 n_e T_e \tau_e, \tag{2.55}$$

$$\eta_1^e = 0.51\frac{n_e T_e}{\omega_{ce}^2 \tau_e}, \qquad \eta_2^e = 4\eta_1^e, \tag{2.56}$$

$$\eta_3^e = -\frac{1}{2}\frac{n_e T_e}{\omega_{ce}}, \qquad \eta_4^e = 2\eta_3^e. \tag{2.57}$$

The electron heat flux is

$$\mathbf{q}^e = \mathbf{q}_u^e + \mathbf{q}_T^e, \tag{2.58}$$

where the collisional heat flux is

$$\mathbf{q}_u^e = 0.71 n_e T_e \mathbf{u}_\parallel + \frac{3}{2}\frac{n_e T_e}{\omega_e \tau_e}\mathbf{b} \times \mathbf{u} \tag{2.59}$$

and the thermal conductivity

$$\mathbf{q}_T^e = -\kappa_\parallel^e \nabla_\parallel T_e - \kappa_\perp^e \nabla_\perp T_e - \frac{5}{2}\frac{n_e T_e}{eB}\mathbf{b} \times \nabla T_e, \tag{2.60}$$

with

$$\kappa_\parallel^e = 3.16\frac{n_e T_e \tau_e}{m_e}, \tag{2.61}$$

$$\kappa_\perp^e = 4.66\frac{n_e T_e}{m_e \omega_{ce}^2 \tau_e}. \tag{2.62}$$

Because of the large difference between ion and electron masses, the collision heat flux from electrons to ions is negligible. The ion heat flux is therefore contributed mainly by the thermal gradient heat flux

$$\mathbf{q}^i = \mathbf{q}_T^i = -\kappa_\parallel^i \nabla_\parallel T_i - \kappa_\perp^i \nabla_\perp T_i + \frac{5}{2}\frac{n_i T_i}{ZeB}\mathbf{b} \times \nabla T_i, \tag{2.63}$$

where

$$\kappa_\parallel^e = 3.9\frac{n_i T_i \tau_i}{m_i}, \tag{2.64}$$

$$\kappa_\perp^i = 2\frac{n_i T_i}{m_i \omega_{ci}^2 \tau_i}. \tag{2.65}$$

The heat acquired by the ions due to the collision with electrons is

$$Q_i = Q_\Delta = \frac{3m_e}{m_i}\frac{n_e}{\tau_e}(T_e - T_i). \tag{2.66}$$

According to equation (2.29), the heat acquired by the electrons from collisions with ions is then

$$Q_e = -\mathbf{R}\cdot\mathbf{u} - Q_\Delta = \frac{j_\parallel^2}{\sigma_\parallel} + \frac{j_\perp^2}{\sigma_\perp} + \frac{1}{en_e}\mathbf{j}\cdot\mathbf{R}_T - \frac{3m_e}{m_i}\frac{n_e}{\tau_e}(T_e - T_i). \tag{2.67}$$

2.2 Estimation and interpretation of transport coefficients

In this section, we estimate the transport coefficients in section 2.1 and give their physics interpretations as given in reference [2]. They are helpful for understanding the kinetic theory to be described in section 2.3.

The Coulomb collision is a binary elastic collision between two charged particles. It rarely results in a large deflection in a plasma. The accumulative effect of many small angle collisions usually is larger than that of a few large angle collisions. A qualitative description requires only the characteristic time intervals between two effective collisions. This time is conventionally taken to the time required for the accumulated deflections of a particle's trajectories to be 90°. As evaluated in many plasma transport works [7–9], τ_e and τ_i in equations (2.32) and (2.33) are actually the electron and ion collision times, respectively. They actually characterize the respective times for momentum transfers for electron–electron and ion–ion collisions. Because an ion's mass is much bigger than an electron's, the characteristic time for electrons gaining their momentum from colliding with ions τ_{ei}^m is about the same as the electron time τ_e.

We also introduce the energy equilibrium times: τ_{ee}, τ_{ii}, and τ_{ei}^ε for electron–electron, ion–ion, and electron–ion collisions, respectively. Because for like-particle collisions the velocities are exchanged in a single collision, one can conclude that the times for energy transfers are the same as those for momentum transfers for like-particle collisions, i.e.

$$\tau_{ee} \sim \tau_e, \qquad \tau_{ii} \sim \tau_i. \tag{2.68}$$

When the temperatures of electrons and ions are of the same order, the energy equilibrium times take the following orderings:

$$\tau_{ee} : \tau_{ii} : \tau_{ei}^\varepsilon = 1 : (m_i/m_e)^{1/2} : (m_i/m_e). \tag{2.69}$$

This can be interpreted as follows. Because collision time is proportional to the thermal speed and the ratio of electron and ion thermal speeds is of order $(m_i/m_e)^{1/2}$, one has $\tau_e/\tau_i \sim (m_e/m_i)^{1/2}$ as shown in equations (2.32) and (2.33). This justifies the first ordering sequence in equation (2.69), by noting equation (2.68). For a collision between two particles with vastly different masses, the energy change in a single collision is proportional to their mass ratio. Note also that τ_{ei}^m is of the same order as τ_e, one obtains the second order sequence in equation (2.69).

The ordering sequences in equation (2.69) lead the electrons and ions to be able to establish their own equilibria, before considering the transport between electrons and ions using their individual fluid thermodynamic variables.

2.2.1 Random walks

Before detailing the estimation of transport coefficients, let us first review the random walk approximation in the usual kinetic theory of gases [2, 10]. As will be seen later on, several transport phenomena can be interpreted under this picture.

Let us consider one-dimensional density transport. As shown in figure 2.1, there is a density gradient along the x axis, such that

$$n(x) = n(x_0) + \left.\frac{dn}{dx}\right|_{x=x_0} (x - x_0), \tag{2.70}$$

where x_0 denotes the point of observation. Because there is a density gradient, the particle fluxes from the right to left (i_{right}) and from the left to right (i_{left}) at the point x_0 are different. The particles within the mean free path from the observation point x_0 can cross the point x_0 during the random walks. The net density flux is, therefore,

$$i = -\int_{-\Delta x}^{\Delta x} \left.\frac{dn}{dx}\right|_{x=x_0} (x - x_0)\frac{dx}{\tau} = -D\frac{dn}{dx}, \tag{2.71}$$

where

$$D = \frac{(\Delta x)^2}{2\tau} = \frac{l^2}{2\tau} = \frac{v_T^2 \tau}{2} = \frac{T\tau}{m}. \tag{2.72}$$

Here, Δx has been replaced by the mean free path $l = v_T\tau$, v_T is the thermal velocity, and τ is the collision time in the collisional transport without magnetic field or along the magnetic field line. Equation (2.71) can be extended to the general scenario, as soon as Δx and τ are attributed to the proper values. For example, when Δx is

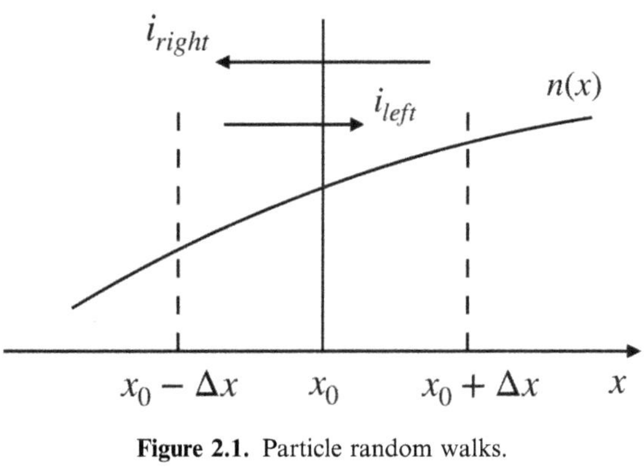

Figure 2.1. Particle random walks.

regarded as the Larmor radius and $1/\tau$ the gyrofrequency, equation (2.71) describes the density transport perpendicular to the magnetic field.

2.2.2 Momentum transfer

Let us now consider the momentum transport between electrons and ions $\mathbf{R}$. As is known, the momentum transfer from ions to electrons and from electrons to ions are just opposite to each other. We therefore discuss only the momentum transfer from ions to electrons. It contains two parts: the collisional transport and thermal transport. In each of two cases, there are two scenarios: parallel or perpendicular to the magnetic field. One can imagine that the transport coefficients in the parallel direction are the same as those in the absence of the magnetic field.

Friction force

Let us first discuss the collisional momentum transfer in the direction parallel to the magnetic field, i.e. the parallel friction force in equation (2.35): $\mathbf{R}_{u\parallel}$. The friction force is due to the relative motion of electrons and ions, and is obviously related to the electrical conductivity of the plasma. Because the ion mass is much larger than the electron one, in the electron–ion collision the electrons only undergo a random change of velocity direction, while the velocity magnitude remains basically unchanged. Consequently, the electron loses its momentum $m_e\mathbf{u} = m_e(\mathbf{V}_e - \mathbf{V}_i)$ in a single collision, i.e. in the electron collision time τ_e. Therefore, for a fluid element with density n_e the parallel friction force is roughly

$$\mathbf{R}_{u\parallel} \sim (m_e n_e/\tau_e)\mathbf{u}_\parallel. \tag{2.73}$$

This agrees with equation (2.36).

The above analyses have not taken into consideration a subtle factor about the dependence of Coulomb collision cross section on the particle speed. From equation (2.32) one can see that $\tau_e \sim T_e^{3/2} \sim v_T^3$. This indicates that the fast electrons experience less friction and therefore contribute more to the current. The distribution function, in this case, diverts from the shifted Maxwellian one. The next order correction is needed to account for the additional contribution from fast electrons to the current. Because the fast electrons have less Coulomb collision cross section, one can imagine the friction force should be reduced. For $Z = 1$ case the reduction factor is 0.51. It becomes higher as Z increases. This reduction factor has been absorbed in the definition of τ_e, so that $\sigma_\parallel$ in equation (2.36) remains a simple form. However, this simplification will affect the definitions of other related transport parameters, as will be seen later on.

Next, let us discuss the collisional moment transfer in the direction perpendicular to the magnetic field, i.e. the perpendicular friction force in equation (2.35): $\mathbf{R}_{u\perp}$. For the same argument as the parallel case, in lowest order one can expect that the perpendicular friction force is roughly $(m_e n_e/\tau_e)\mathbf{u}_\perp$. However, because of the strong magnetic field, i.e. $\omega_{ce}\tau_e \gg 1$, the next order correction is of order $(\omega_{ce}\tau_e)^{-1}$ and is therefore negligible. Because the reduction factor 0.51 ($\approx 1/2$) is absorbed in the definition for $\mathbf{R}_{u\parallel}$, one has $\sigma_\parallel \approx 2\sigma_\perp$ for the $Z = 1$ case, as shown in equation (2.36).

Thermal force

In the friction force case, we have considered the force due to the fluid macroscopic motion. One needs to consider the non-averaged effects from individual particles as well. This happens when there is temperature gradient. For simplicity we consider it in a rest frame $\mathbf{V}_e = 0$.

Let us first discuss the electron thermal momentum transfer in the direction parallel to the magnetic field, i.e. the parallel thermal force in equation (2.37): $\mathbf{R}_{T\|}$. Figure 2.2 is introduced to explain it. The temperature gradient is in the x direction. The mean free path is denoted by $l = v_{Te}\tau_e$ with v_{Te} being the electron thermal velocity. Because the electron momentum change takes place in a single effective collision with ions. One only needs to consider the particles within the interval l on the left and on the right of the point to be examined: x_0. The difference of thermal speeds on the right and on the left is $\Delta v_{Te} \sim l(dv_{Te}/dx) \sim lv_{Te}(d \ln T_e/dx)$. The thermal force in the parallel direction is, therefore, the net force from the right to the left:

$$\mathbf{R}_{T\|} \sim \frac{m_e n_e \Delta v_{Te}}{\tau_e} \sim lv_{Te}\frac{d \ln T_e}{dx}\frac{m_e n_e}{\tau_e} \sim n_e\frac{dT_e}{dx}. \tag{2.74}$$

This agrees with equation (2.37). Although the collision time does not appear in the final result, it is the collisional effects that cause the thermal force along the magnetic field line. Comparing with equation (2.73), one can see that equation (2.74) is obtained by replacing $\mathbf{u}$ with Δv_{Te}.

Next, let us discuss the thermal momentum transfer in the direction perpendicular to the magnetic field, i.e. the perpendicular thermal force in equation (2.37): $\mathbf{R}_{T\perp}$. Figure 2.3 is introduced to explain it. The temperature gradient is still in the x direction. The magnetic field is assumed to be in the z direction. The difference of thermal speeds on the left and on the right of the observation location x_0 is $\Delta v_{Te} \sim r_e(dv_{Te}/dx) \sim r_e v_{Te}(d \ln T_e/dx)$. Different from the parallel case, here $r_e \sim v_{Te}/\omega_{ce}$ is taken to be the electron gyroradius. Therefore, The thermal force in the perpendicular direction becomes

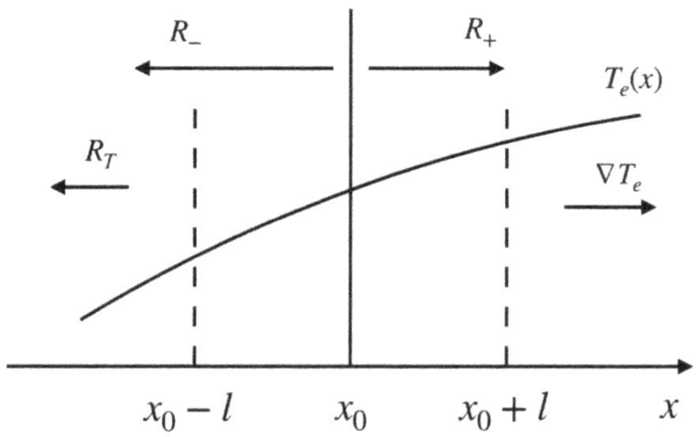

Figure 2.2. Thermal moment transfer in the parallel direction.

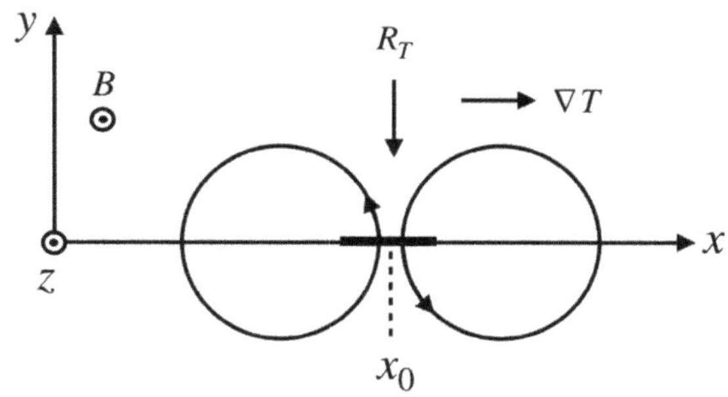

Figure 2.3. Thermal moment transfer in the perpendicular direction.

$$\mathbf{R}_{T\perp} \sim \frac{m_e n_e \Delta v_{Te}}{\tau_e} \sim r_e v_{Te} \frac{d \ln T_e}{dx} \frac{m_e n_e}{\tau_e} \sim \frac{n_e T_e}{\omega_{ce}\tau_e} \frac{dT_e}{dx}. \tag{2.75}$$

This agrees with equation (2.37). The direction for thermal force can be verified as well using figure 2.3.

2.2.3 Viscosity

In this subsection, we consider the viscosity in equation (2.40). We only focus on the collisional viscosities in the parallel and perpendicular direction: η_0 and $\eta_{1,2}$. The other two parameters $\eta_{3,4}$ are the so-called gyroviscosity, which is related to the finite Larmor radius effects. They can be obtained by properly replacing the mean free path with the gyroradius and the collision frequency with the gyrofrequency. The discussion proceeds without distinguishing the ion and electron species since they are based on the same physics. Their difference will be indicated at the end.

In the parallel direction, one can use the random walk picture in section 2.2.1 to obtain the parallel viscosity. In the particle flux case, the density flux: $(\Delta x/\tau)n$ is sought, while in the viscosity case, the momentum flux related to the fluid velocity $(\Delta x/\tau)nm\mathbf{V}$ is looked for. Note that the pressure tensor has multiple components. Let us consider, for example, the component π_{yx} with the velocity in y direction and its gradient in x direction. The magnetic field line is assumed in the z direction. In this case $\Delta x \sim l \sim v_T\tau$. Comparing with equations (2.71) and (2.72), one obtains

$$\pi_{yx} = -\eta_0 \frac{\partial V_y}{\partial x},$$

where

$$\eta_0 \sim \frac{nm(\Delta x)^2}{\tau} \sim nT\tau = nmD.$$

Similar analyses apply to other components. This agrees with equations (2.52) and (2.55).

In the perpendicular direction, one can still use the random walk picture discussed in section 2.2.1 to estimate the perpendicular viscosity. It is assumed that $\omega_c\tau \gg 1$. Note that the pressure tensor has multiple components. Let us consider, for example, the component π_{zx} with the velocity in z direction and its gradient in x direction. The magnetic field line is assumed in the z direction as well. In this case $\Delta x \sim r \sim v_T/\omega_c$. Comparing with equations (2.71) and (2.72), one obtains

$$\pi_{yx} = -\eta_\perp \frac{\partial V_y}{\partial x},$$

where

$$\eta_\perp \sim \eta_{1,2} \sim \frac{nm(\Delta x)^2}{\tau} \sim \frac{nT}{\tau\omega_c^2} = \frac{\eta_0}{(\omega_c\tau)^2}.$$

Similar analyses apply to other components. This agrees with equations (2.53) and (2.56).

It is noted that, in the case with $\omega_{ci}\tau_i \gg 1$ and $T_i \sim T_e$, the ion collisional viscosity is much larger than the electron one:

$$\eta_0^i \sim \left(\frac{m_i}{m_e}\right)^{1/2} \eta_0^e. \tag{2.76}$$

The ion gyroviscosity in equation (2.54) also is much larger than the electron gyroviscosity in equation (2.57), since they are inversely proportional to the gyrofrequency ω_c.

2.2.4 Heat flux

Let us now consider the heat flux. We consider only the heat transfer from ions to electrons $\mathbf{q}^e$. The heat transfer from electrons to ions is similar. It also contains two parts: collisional transport and thermal transport. In each of the two cases, there are two scenarios: parallel or perpendicular to the magnetic field. One can imagine that the transport coefficients in the parallel direction are the same as those in the absence of the magnetic field.

Collisional heat flux

Let us first discuss the electron heat flux in the direction parallel to the magnetic field, i.e. the parallel heat flux in equation (2.59): $\mathbf{q}_{u\|}^e$. From the discussion about $\mathbf{R}_{u\|}$ in section 2.2.2 one can find that the averaged friction force on an electron is $m_e u/\tau_e$. Noting that the mean free path is l, the work on the electron is $lm_e u_\|/\tau_e$. From the energy conservation, one knows that this work done by the friction force is equivalent to the heat gain by the electron. Note further that the heat flux is carried over by the electron particle flux $n_e v_{Te}$. Consequently, one obtains the estimation for the collisional electron heat flux in the parallel direction:

$$q_{u\|}^e \sim l\frac{m_e u_\|}{\tau_e} n_e v_{Te} \sim \tau_e v_{Te} \frac{m_e u_\|}{\tau_e} n_e v_{Te} \sim n_e T_e u_\|.$$

This agrees with equation (2.59). Although the collision time does not appear in the final result, it is the collisional effects that cause the heat flux along the magnetic field

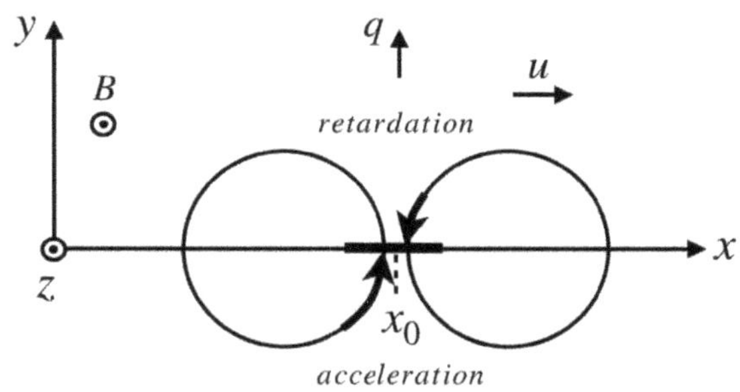

Figure 2.4. Collisional heat flux in the perpendicular direction.

line. This is similar to the parallel friction force. Also, based on the same reason for the reduction of parallel friction force, there is a reduction factor 0.71 here as well.

Next, let us discuss the electron heat flux in the direction perpendicular to the magnetic field, i.e. the perpendicular heat flux in equation (2.59): $\mathbf{q}_{u\perp}^{e}$. Figure 2.4 is introduced to explain it. Electrons gyrate around the magnetic field lines. Since the friction force is along the '-x' direction, electrons on the upper and lower halves of gyromotion orbits experience different friction force. As discussed in section 2.2.2, the net perpendicular friction force on an electron is $m_e u_\perp / \tau_e$. Noting that the scale length is v_{Te}/ω_{ce} for the interaction of perpendicular friction force on the electrons, the work on the electrons is $(v_{Te}/\omega_{ce})(m_e u/\tau_e)$. From the energy conservation, one knows that this work done by the friction force is equivalent to the heat gain by the electrons. Note further that the heat flux is carried over by the electron particle flux $n_e v_{Te}$. Consequently, one has the estimation for the collisional electron heat flux in the perpendicular direction:

$$q_{u\perp}^{e} \sim l\frac{m_e u_\perp}{\tau_e} n_e v_{Te} \sim \frac{v_{Te}}{\omega_{ce}} \frac{m_e u_\perp}{\tau_e} n_e v_{Te} \sim \frac{n_e T_e}{\omega_{ce}\tau_e} u_\perp.$$

This agrees with equation (2.59). The heat flux direction can be verified as well using figure 2.4.

Thermal conductivity

The thermal heat flux is referred to as thermal conductivity. From equations (2.60) and (2.63), one can see that thermal conductivity appears in three directions: parallel, perpendicular, and transverse (i.e. perpendicular both to the magnetic field and temperature gradient) directions. Let us discuss them one by one.

In the parallel direction, one can use the random walk picture in section 2.2.1 to obtain the parallel thermal conductivity. In the particle flux case, the density flux, $(\Delta x/\tau)n$, is sought, while in the thermal conductivity case the heat flux $(\Delta x/\tau)nT$ is looked for. Using equations (2.71) and (2.72) and assuming that the magnetic field line is in the x direction, one obtains

$$q_\parallel = -\kappa_\parallel \frac{\partial T}{\partial x},$$

where

$$\kappa_\parallel \sim \frac{n(\Delta x)^2}{\tau} \sim \frac{nT\tau}{m} = nD.$$

In the perpendicular direction, one can still use the random walk picture discussed in section 2.2.1 to estimate the thermal conductivity. It is assumed that $\omega_c\tau \gg 1$ and the magnetic field line is in the z direction. The distance Δx should be replaced by the Larmor radius $r = v_T/\omega_c$. Therefore, one has

$$q_\perp = -\kappa_\perp \frac{\partial T}{\partial x}, \tag{2.77}$$

where

$$\kappa_\perp \sim \frac{n(\Delta x)^2}{\tau} \sim \frac{nv_T^2}{\tau\omega_c^2} \sim \frac{nT}{m\tau\omega_c^2} \sim (\omega_c\tau)^{-2}\kappa_\parallel. \tag{2.78}$$

In the transverse direction as shown in figure 2.5, because there is temperature gradient, charged particles gyrate faster on the higher temperature side than on the lower side. The difference is $r\ \partial T/\partial x$. Note further that the thermal energy is carried over by the particle flux nv_T. Therefore, the transverse heat flux becomes

$$q_y \sim nv_T r\frac{\partial T}{\partial x} \sim \frac{nT}{eB}\frac{\partial T}{\partial x}.$$

Note that the fluxes for ions and electrons are in opposite directions. Also, they carry heat in the direction perpendicular to the temperature gradient, i.e. along the isotherm. Therefore, they do not cool the plasma or increase its entropy.

Thermal heat fluxes in various directions estimated here are in agreement with equations (2.60) and (2.63).

2.2.5 Heat generation

Let us discuss the heat generation in this subsection. The heat generation is closely related to the heat flux: one is the interior effect and the other is the exterior effect. The heat generation also contains the collisional and thermal parts

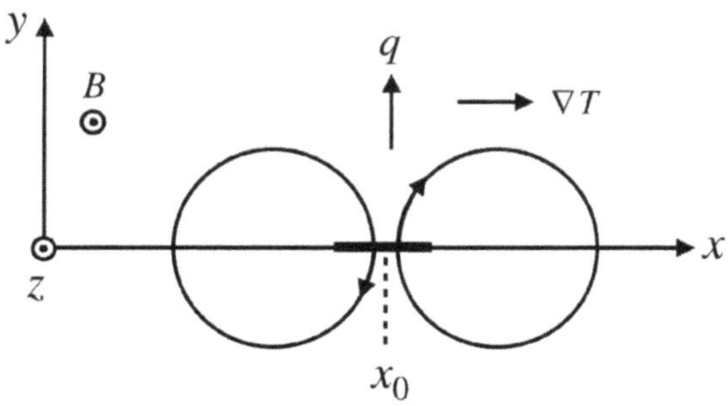

Figure 2.5. Thermal conductivity in the transverse direction.

Collisional heat generation
We first consider the electron heat generation due to the collisions with ions. Because the ion mass is much larger than that of the electron, the relative ion energy change is much smaller than for the electron. In this case, we can assume that the ions are at rest on average: $\mathbf{V}_i = 0$. In each effective collision the electrons are turned away by 90° and, therefore, are randomized. The kinetic energy related to the ordered velocity: $\mathbf{u} = \mathbf{V}_e - \mathbf{V}_i$ is converted into the thermal energy, i.e. heat. The heat generated in the electrons can be simply evaluated through the work done by the friction force exerted on the electrons by the ions, i.e.

$$Q_e = -\mathbf{R} \cdot \mathbf{u}.$$

Thermal heat generation
One needs also to consider the thermal heat generation. When ions and electrons have the same temperatures, i.e. $T_i = T_e$, they are in thermal equilibrium and there is no heat transfer between them. However, if the electron temperature is higher than that of the ion: $T_e > T_i$, heat can transfer from electrons to ions. As discussed at the beginning of this section, when the light electron collides with a heavy ion, the energy transfer is of order $\mathcal{O}(m_e/m_i)$. Reference [6] shows that the energy exchanged per unit time between electrons and ions is

$$Q_\Delta = \frac{n_e}{\tau_e}\frac{2m_e}{m_i}\frac{3}{2}(T_e - T_i). \tag{2.79}$$

These agree with equation (2.67).

2.3 Kinetic theory evaluation of transport coefficients for moment closure

In this section, the transport coefficients will be evaluated by the kinetic theory, which is based on references [1] and [2]. From section 2.1.3 one can see that there are two types of transport coefficients: one depends directly on the collisional operators and the other depends on the distribution functions, such as $\mathbf{R}_u$ and $\mathbf{R}_T$. To obtain the transport coefficients related to the collisional operators, we will reduce the Landau collision operator both for electron–ion and ion–electron collisions. To derive the transport coefficients related to the distribution functions, we will solve the kinetic equations both for electron and ion species to the first order. These results are then used to compute the transport coefficients. This is basically the structure of this section.

The procedure in references [1] and [2] by Braginskii basically extends the Chapman and Cowling approach in reference [10]. The collision frequencies are assumed to be dominant. Therefore, the lowest order distribution function becomes a Maxwellian, f^0, with the three moments, n, $\mathbf{V}$, and T, as parameters, as given in equation (2.30). These thermal parameters are assumed to be slowly varying functions of space and time. Then, the distribution function can be expanded as follows:

$$f = f^0 + f^1 + f^2 + \cdots. \tag{2.80}$$

In solving the kinetic equation perturbedly, the collisional and gyromotion terms are also assumed to be dominant. We limit ourselves only to the first order solutions. For collisional operators, therefore, only the linear parts, i.e. $C(f^0, f^1)$ and $C(f^1, f^0)$ need to be considered, noting that the lowest order contribution vanishes for a Maxwellian. The kinetic equations can be solved perturbedly order by order. With these results, both the collisional and thermal transport parameters can be evaluated. Since the first order solutions are derived from the local Maxwellian as the zeroth order solution, the transport coefficients depend on the gradients of moments, n, $\mathbf{V}$, and T, as well as the moments themselves.

2.3.1 Kinetic equations in the center of mass

In this subsection, the kinetic equation (2.1) in section 2.1.1 is expressed in the center of mass, $\mathbf{V}_a$, for electron and ion species with $a = i, e$. It is convenient to include the fluid velocity effects in the kinetic theory and, nevertheless, does not induce any difficulty for collisional operator treatment; it will be shown later that the collisional operator is invariant to the boost transform [8, 9].

The kinetic equation (2.1) can be rewritten in the component form as follows:

$$\frac{\partial f_a}{\partial t} + \frac{\partial}{\partial x_\beta}(v_\beta f_a) + \frac{\partial}{\partial v_\beta}\left(\frac{F_{a\beta}}{m_a} f_a\right) = C_a, \tag{2.81}$$

where $\nabla_v \cdot \mathbf{F}_a = e_a \partial_{v\alpha}\left(E_\alpha + e_{\alpha\beta\gamma} v_{a\beta} B_\gamma\right) = e_a e_{\alpha\beta\gamma} \delta_{\alpha\beta} B_\gamma = 0$ for Lorentz force has been used. In the center-of-mass frame, the particle velocity $\mathbf{v}$ is transformed to the random velocity: $\mathbf{v}_a' = \mathbf{v}_a - \mathbf{V}_a$. Without causing ambiguity, the prime denoting the random velocity is dropped for brevity later in this section. With f_a expressed in the random velocity one has

$$\frac{\partial f_a}{\partial t} = \frac{\partial f_a}{\partial t} - \frac{\partial V_{a\beta}}{\partial t}\frac{\partial f_a}{\partial v_{a\beta}},$$

$$\begin{aligned}
\frac{\partial}{\partial x_\beta}(v_\beta f_a) &= v_\beta \frac{\partial f_a}{\partial x_\beta}, \\
&= v_{a\beta}\frac{\partial f_a}{\partial x_\beta} + V_{a\beta}\frac{\partial f_a}{\partial x_\beta} - v_{a\alpha}\frac{\partial V_{a\beta}}{\partial x_\alpha}\frac{\partial f_a}{\partial v_{a\beta}} - V_{a\alpha}\frac{\partial V_{a\beta}}{\partial x_\alpha}\frac{\partial f_a}{\partial v_{a\beta}},
\end{aligned}$$

$$\begin{aligned}
\frac{\partial}{\partial v_\beta}\left(\frac{F_{a\beta}}{m_a} f_a\right) &= \frac{e_a}{m_a}\frac{\partial}{\partial v_{a\beta}}\left(E_\beta f_a + \frac{e_{\beta\gamma\alpha} v_\gamma B_\beta}{c} f_a\right) \\
&= \frac{e_a}{m_a}\left\{E_\beta \frac{\partial f_a}{\partial v_{a\beta}} + \frac{e_{\beta\gamma\alpha} V_{a\gamma} B_\alpha}{c}\frac{\partial f_a}{\partial v_{a\beta}} + \frac{e_{\beta\gamma\alpha} v_{a\gamma} B_\alpha}{c}\frac{\partial f_a}{\partial v_{a\beta}}\right\}.
\end{aligned}$$

Using these results equation (2.81) is transformed to

$$\begin{aligned} &\frac{d_a f_a}{dt} + v_{a\beta}\frac{\partial f_a}{\partial x_\beta} - \left(\frac{e_a}{m_a}E^*_{a\beta} - \frac{d_a V_{a\beta}}{dt}\right)\frac{\partial f_a}{\partial v_{a\beta}} \\ &= \frac{\partial V_{a\gamma}}{\partial x_\beta}v_{a\beta}\frac{\partial f_a}{\partial v_{a\gamma}} + \frac{e_a}{m_a}e_{\beta\gamma\alpha}v_{a\gamma}B_\alpha\frac{\partial f_a}{\partial v_{a\beta}} - \sum_b C_{ab}(f_a, f_b). \end{aligned} \tag{2.82}$$

where

$$\frac{d_a}{dt} \equiv \frac{\partial}{\partial t} + \mathbf{V}_a \cdot \nabla, \quad \mathbf{E}^*_a \equiv \mathbf{E} + \mathbf{V}_a \times \mathbf{B}, \tag{2.83}$$

and the Landau collisional operator is given as follows

$$\begin{aligned} C_{ab}(f_a, f_b) = &-\frac{2\pi\lambda e_a^2 e_b^2}{m_a}\frac{\partial}{\partial v_{a\beta}} \\ &\times \int\left\{\frac{f_a(\mathbf{v}_a)}{m_b}\frac{\partial f_b(\mathbf{v}_b)}{\partial v_{b\gamma}} - \frac{f_b(\mathbf{v}_b)}{m_a}\frac{\partial f_a(\mathbf{v}_a)}{\partial v_{a\gamma}}\right\}U_{\beta\gamma}\, d\mathbf{v}_b \end{aligned} \tag{2.84}$$

with

$$U_{\beta\gamma} = \frac{1}{u^3}\left(u^2\delta_{\beta\gamma} - u_\beta u_\gamma\right), \quad u_\beta = v_{a\beta} - v_{b\beta}. \tag{2.85}$$

The boost invariance of the collisional operator can be simply proved as follows [8, 9]. Let $\bar{\mathbf{v}} = \mathbf{v} + \mathbf{V}$ for a velocity boost $\mathbf{V}$. The scalar nature of the distribution function yields

$$\bar{f}_a(\bar{\mathbf{v}}_a) = f_a(\mathbf{v}_a + \mathbf{V}).$$

Note that

$$\frac{\partial}{\partial v_\alpha} = \frac{\partial}{\partial \bar{v}_\alpha}, \quad U_{\beta\gamma}(\mathbf{v}_a, \mathbf{v}_b) = U_{\beta\gamma}(\bar{\mathbf{v}}_a, \bar{\mathbf{v}}_b) = \bar{U}_{\beta\gamma},$$

where it has been noted that $U_{\alpha\beta}$ depends only on the relative velocity $\mathbf{v}_a$—$\mathbf{v}_b$.

$$\begin{aligned} &C_{ab}(f_a, f_b)\big|_{\mathbf{v}\to\mathbf{v}+\mathbf{V}} \\ &= -\frac{2\pi\lambda e_a^2 e_b^2}{m_a}\frac{\partial}{\partial \bar{v}_{a\beta}}\int\left\{\frac{f_a(\mathbf{v}_a + \mathbf{V})}{m_b}\frac{\partial f_b(\mathbf{v}_b)}{\partial v_{b\gamma}} - \frac{f_b(\mathbf{v}_b)}{m_a}\frac{\partial f_a(\mathbf{v}_a + \mathbf{V})}{\partial \bar{v}_{a\gamma}}\right\}U_{\beta\gamma}\big|_{\mathbf{v}_a\to\mathbf{v}_a+\mathbf{V}}\, d\mathbf{v}_b \\ &= -\frac{2\pi\lambda e_a^2 e_b^2}{m_a}\frac{\partial}{\partial \bar{v}_{a\beta}}\int\left\{\frac{f_a(\mathbf{v}_a + \mathbf{V})}{m_b}\frac{\partial f_b(\mathbf{v}_b + \mathbf{V})}{\partial \bar{v}_{b\gamma}} - \frac{f_b(\mathbf{v}_b + \mathbf{V})}{m_a}\frac{\partial f_a(\mathbf{v}_a + \mathbf{V})}{\partial \bar{v}_{a\gamma}}\right\}\bar{U}_{\beta\gamma}\, d\bar{\mathbf{v}}_b \\ &= -\frac{2\pi\lambda e_a^2 e_b^2}{m_a}\frac{\partial}{\partial \bar{v}_{a\beta}}\int\left\{\frac{f_a(\bar{\mathbf{v}}_a)}{m_b}\frac{\partial f_b(\bar{\mathbf{v}}_b)}{\partial \bar{v}_{b\gamma}} - \frac{f_b(\bar{\mathbf{v}}_b)}{m_a}\frac{\partial f_a(\bar{\mathbf{v}}_a)}{\partial \bar{v}_{a\gamma}}\right\}\bar{U}_{\beta\gamma}\, d\bar{\mathbf{v}}_b \\ &= C_{ab}(\bar{f}_a, \bar{f}_b)(\bar{\mathbf{v}}). \end{aligned}$$

This proves the invariance.

Noting that $v_a/v_b \sim \sqrt{m_b/m_a}$ for the $T_a \sim T_b$ case, one has the ratio of the first to the second terms in the collisional operator in equation (2.84) being of order $\mathcal{O}\left(\sqrt{m_a/m_b}\right)$. This ordering can be used to simplify the collision operator.

From the collision time ordering in equation (2.69) for simple plasma, one can see that electrons and ions can reach their own equilibria before elections and ions establish the equilibrium. This leads the fluid thermodynamic variables: n, V, and T etc, to be introduced for ion and electron fluids individually, while investigating the transport processes.

2.3.2 Electron–ion cross collisional operator

We first study the electron–ion cross collisional operator for the process with electrons being scattered by ions. From equation (2.84), the electron–ion cross collisional operator can be expressed as

$$C_{ei}(f_e, f_i) = -\frac{2\pi\lambda e_e^2 e_i^2}{m_e}\frac{\partial}{\partial v_{e\beta}} \times \int\left\{\frac{f_e(\mathbf{v}_e)}{m_i}\frac{\partial f_i(\mathbf{v}_i)}{\partial v_{i\gamma}} - \frac{f_i(\mathbf{v}_i)}{m_e}\frac{\partial f_e(\mathbf{v}_e)}{\partial v_{e\gamma}}\right\}U_{\beta\gamma}\, d\mathbf{v}_i, \tag{2.86}$$

where

$$U_{\beta\gamma} = \frac{1}{u^3}\left(u^2\delta_{\beta\gamma} - u_\beta u_\gamma\right), \quad u_\beta = v_{e\beta} - v_{i\beta}. \tag{2.87}$$

There is freedom to choose the frame to evaluate the collisional operator. It is convenient to reduce the electron–ion collisional operator $C_{ei}(f_i, f_e)$ in the mean ion velocity ($\mathbf{V}_i$) frame.

Note that the mass ratio of electron to ion is small. The first term in equation (2.86) is of order m_e/m_i smaller than the second term. Noting that the ion speed is much smaller than electron speed by order $\mathcal{O}\sqrt{m_e/m_i}$, we also expand the parameter $U_{\beta\gamma}$ in the collisional operator in equation (2.87) in terms of ion velocity till order m_e/m_i:

$$U_{\alpha\beta} = V_{e\alpha\beta} - \frac{\partial V_{e\alpha\beta}}{\partial v_{e\gamma}}v_{i\gamma} + \frac{1}{2}\frac{\partial^2 V_{e\alpha\beta}}{\partial v_{e\gamma}\partial v_{e\delta}}v_{i\gamma}v_{i\delta} + \cdots, \tag{2.88}$$

where

$$V_{e\alpha\beta} = U_{\alpha\beta}|_{v_i=0} = \frac{1}{v_e^3}\left(v_e^2\delta_{\beta\gamma} - v_{e\beta}v_{e\gamma}\right).$$

Here, it has been noted that $u_\alpha = v_{e\alpha} - v_{i\alpha}$, which leads to $\partial v_{i\gamma} = -\partial v_{e\gamma}$. We will consider the contributions of each term in the expansion equation (2.88) to equation (2.86) one by one.

Let us first consider the contribution of the first term on the right-hand side of equation (2.88). Noting that $V_{e\beta\gamma}$ is independent of $\mathbf{v}_i$, the first part of the collisional operator in equation (2.86) can be reduced to

$$C_{ei}^{I1} = -\frac{2\pi\lambda e_i^2 e_e^2}{m_e}\frac{\partial}{\partial v_{e\beta}}\int\left\{\frac{f_e(\mathbf{v}_e)}{m_i}\frac{\partial f_i(\mathbf{v}_i)}{\partial v_{i\gamma}}\right\}V_{e\beta\gamma}\, d\mathbf{v}_i = 0. \tag{2.89}$$

Here, integration by part has been performed and the boundary condition $f_i \to 0$ as $\mathbf{v}_i \to \infty$ has been used. The second part of the collisional operator in equation (2.86) can be reduced to

$$\begin{aligned} C_{ei}^{I2} &= \frac{2\lambda\pi e_i^2 e_e^2}{m_e}\frac{\partial}{\partial v_{e\beta}}\int\left\{\frac{f_i(\mathbf{v}_i)}{m_e}\frac{\partial f_e(\mathbf{v}_e)}{\partial v_{e\gamma}}\right\}V_{e\beta\gamma}\, d\mathbf{v}_i \\ &= \frac{2\pi\lambda e_i^2 e_e^2 n_i}{m_e^2}\frac{\partial}{\partial v_{e\beta}}\left(V_{e\beta\gamma}\frac{\partial f_e(\mathbf{v}_e)}{\partial v_{e\gamma}}\right) \\ &= \frac{3\sqrt{\pi}}{8}\frac{1}{\tau_e}\left(\frac{2T_e}{m_e}\right)^{3/2}\frac{\partial}{\partial v_{e\beta}}\left(V_{e\beta\gamma}\frac{\partial f_e(\mathbf{v}_e)}{\partial v_{e\gamma}}\right), \end{aligned} \tag{2.90}$$

where τ_e is given in equation (2.32).

Next, let us consider the contribution of the second term on the right-hand side of equation (2.88). Note that

$$\begin{aligned} \frac{\partial V_{e\beta\gamma}}{\partial v_{e\gamma}} &= \frac{\partial}{\partial v_{e\gamma}}\left\{\frac{1}{v_e}\delta_{\beta\gamma} - \frac{1}{v_e^3}v_{e\beta}v_{e\gamma}\right\} \\ &= \frac{\partial}{\partial v_{e\beta}}\frac{1}{v_e} - v_{e\beta}v_{e\gamma}\frac{\partial}{\partial v_{e\gamma}}\frac{1}{v_e^3} - \frac{3}{v_e^3}v_{e\beta} - \frac{1}{v_e^3}v_{e\gamma}\delta_{\beta\gamma} \\ &= -\frac{2}{v_e^3}v_{e\beta}, \end{aligned} \tag{2.91}$$

where it has been noted that $\partial v/\partial v_\beta = v_\beta/v$. Using this result, the first part of the collisional operator in equation (2.86) can be reduced to

$$\begin{aligned} C_{ei}^{II1} &= -\frac{2\pi e_i^2 e_e^2}{m_e}\frac{\partial}{\partial v_{e\beta}}\int\left\{\frac{f_e(\mathbf{v}_e)}{m_i}\frac{\partial f_i(\mathbf{v}_i)}{\partial v_{i\gamma}}\right\}(-1)\frac{\partial V_{e\beta\gamma}}{\partial v_{e\lambda}}v_{i\lambda}\, d\mathbf{v}_i \\ &= -\frac{3\sqrt{\pi}}{8}\frac{1}{\tau_e}\left(\frac{2T_e}{m_e}\right)^{3/2}\frac{m_e}{m_i}\frac{\partial}{\partial v_{e\beta}}\left(\frac{\partial V_{e\beta\gamma}}{\partial v_{e\lambda}}f_e(\mathbf{v}_e)\delta_{\lambda\gamma}\right) \\ &= \frac{3\sqrt{\pi}}{8}\frac{1}{\tau_e}\left(\frac{2T_e}{m_e}\right)^{3/2}\frac{m_e}{m_i}\frac{\partial}{\partial v_{e\beta}}\left(\frac{2v_{e\beta}}{v_e^3}f_e(\mathbf{v}_e)\right). \end{aligned} \tag{2.92}$$

The second part of the collisional operator in equation (2.86) can be reduced as follows

$$C_{ei}^{II2} = \frac{2\pi e_i^2 e_e^2}{m_e}\frac{\partial}{\partial v_{e\beta}}\int\left\{\frac{f_i(\mathbf{v}_i)}{m_e}\frac{\partial f_e(\mathbf{v}_e)}{\partial v_{e\gamma}}\right\}\frac{\partial V_{e\beta\gamma}}{\partial v_{e\lambda}}v_{i\lambda}\, d\mathbf{v}_i = 0.$$

Here, the integration vanishes since the integrand is odd.

Furthermore, let us consider the contribution of the third term on the right-hand side of equation (2.88). Noting that $V_{e\beta\gamma}$ is independent of $\mathbf{v}_i$, the first part of the collisional operator in equation (2.86) can be reduced to

$$C_{ei}^{III1} = -\frac{2\pi e_i^2 e_e^2}{m_e}\frac{\partial}{\partial v_{e\beta}}\int \frac{f_e(\mathbf{v}_e)}{m_i}\frac{\partial f_i(\mathbf{v}_i)}{\partial v_{i\gamma}}\frac{1}{2}\frac{\partial^2 V_{e\beta\gamma}}{\partial v_{e\lambda}\partial v_{e\zeta}} v_{i\lambda}v_{i\zeta}\, d\mathbf{v}_i = 0.$$

Here, the integration vanishes since the integrand is odd. For the contribution of the second part of the collisional operator in equation (2.86), it is noted that

$$\begin{aligned}
\frac{\partial V_{e\beta\gamma}}{\partial v_{e\lambda}} &= \frac{\partial}{\partial v_{e\lambda}}\left\{\frac{1}{v_e}\delta_{\beta\gamma} - \frac{1}{v_e^3}v_{e\beta}v_{e\gamma}\right\}\\
&= -\frac{v_{e\lambda}}{v_e^3}\delta_{\beta\gamma} + \frac{3}{v_e^5}v_{e\lambda}v_{e\beta}v_{e\gamma} - \frac{1}{v_e^3}\delta_{\lambda\beta}v_{e\gamma} - \frac{1}{v_e^3}v_{e\beta}\,\delta_{\lambda\gamma},\\
\frac{\partial^2 V_{e\beta\gamma}}{\partial v_{e\zeta}\,\partial v_{e\lambda}} &= \frac{\partial}{\partial v_{e\zeta}}\left\{-\frac{v_{e\lambda}}{v_e^3}\delta_{\beta\gamma} + \frac{3}{v_e^5}v_{e\lambda}v_{e\beta}v_{e\gamma} - \frac{1}{v_e^3}\delta_{\lambda\beta}v_{e\gamma} - \frac{1}{v_e^3}v_{e\beta}\,\delta_{\lambda\gamma}\right\}\\
&= \frac{3v_{e\zeta}v_{e\lambda}}{v_e^5}\delta_{\beta\gamma} - \underbrace{\frac{1}{v_e^3}\delta_{\zeta\lambda}\delta_{\beta\gamma}} - \frac{15}{v_e^7}v_{e\zeta}v_{e\lambda}v_{e\beta}v_{e\gamma} + \underbrace{\frac{3}{v_e^5}\delta_{\zeta\lambda}v_{e\beta}v_{e\gamma}}\\
&\quad + \frac{3}{v_e^5}v_{e\lambda}\delta_{\zeta\beta}v_{e\gamma} + \frac{3}{v_e^5}v_{e\lambda}v_{e\beta}\delta_{\zeta\gamma} + \frac{1}{v_e^5}\delta_{\lambda\beta}v_{e\zeta}v_{e\gamma}\\
&\quad - \frac{1}{v_e^3}\delta_{\lambda\beta}\,\delta_{\zeta\gamma} + \frac{1}{v_e^5}v_{e\zeta}v_{e\beta}\,\delta_{\lambda\gamma} - \frac{1}{v_e^3}\delta_{\zeta\beta}\,\delta_{\lambda\gamma}.
\end{aligned} \tag{2.93}$$

Using these results and noting that only two of the terms with underbraces give rise to non-vanishing contributions, the second part of the collisional operator in equation (2.86) can be reduced to

$$\begin{aligned}
C_{ei}^{III2} &= -\frac{2\pi e_i^2 e_e^2}{m_e}\frac{\partial}{\partial v_{e\beta}}\int \frac{f_i(\mathbf{v}_i)}{m_e}\frac{\partial f_e(\mathbf{v}_e)}{\partial v_{e\gamma}}\frac{1}{2}\frac{\partial^2 V_{e\beta\gamma}}{\partial v_{e\lambda}\partial v_{e\zeta}} v_{i\lambda}v_{i\zeta}\, d\mathbf{v}_i\\
&= \frac{3\sqrt{\pi}}{8}\frac{1}{\tau_e}\left(\frac{2T_e}{m_e}\right)^{3/2}\frac{\partial}{\partial v_{e\beta}}\left(\frac{T_i}{m_i}\frac{3v_{e\alpha}v_{e\beta} - v_e^2\delta_{\alpha\beta}}{v_e^5}\frac{\partial f_e}{\partial v_{e\beta}}\right).
\end{aligned} \tag{2.94}$$

Finally, combining the contributions from C_{ei}^{I2}, C_{ei}^{II1}, and C_{ei}^{III2} respectively in equations (2.90), (2.92), and (2.94), one obtains

$$C_{ei} = C_{ei}' + C_{ei}'', \tag{2.95}$$

where

$$C_{ei}' = \frac{3\sqrt{\pi}}{8}\frac{1}{\tau_e}\left(\frac{2T_e}{m_e}\right)^{3/2}\frac{\partial}{\partial v_{e\beta}}V_{e\beta\gamma}\frac{\partial f_e}{\partial v_{e\gamma}}, \tag{2.96}$$

$$
\begin{aligned}
C''_{ei} = {} & \frac{3\sqrt{\pi}}{8}\frac{1}{\tau_e}\left(\frac{2T_e}{m_e}\right)^{3/2} \\
& \times \frac{m_e}{m_i}\frac{\partial}{\partial v_{e\beta}}\left(\frac{2v_{e\beta}}{v_e^3}f_e + \frac{T_i}{m_e}\frac{3v_{e\alpha}v_{e\beta} - v_e^2\delta_{\alpha\beta}}{v_e^5}\frac{\partial f_e}{\partial v_{e\beta}}\right).
\end{aligned} \tag{2.97}
$$

Here, C''_{ei} is the $\mathcal{O}(m_e/m_i)$ correction as compared to C'_{ei}. It should be noted that the electron distribution function here is computed on the ion velocity ($\mathbf{V}_i$) frame. As we have expected, the cross collisional term $C_{ei}(f_e, f_i)$ is independent of the detailed ion distribution function, but only on the ion fluid variables: n_i, $\mathbf{V}_i$, and T_i.

Since the electron–ion collisional operator C_{ei} appears in the electron fluid equations as shown in section 2.1.2, one needs to change C_{ei} in equation (2.95) back to the electron mean velocity frame. This transformation can result in a high order correction. Noting that C''_{ei} is of order $\mathcal{O}(m_e/m_i)$ smaller, only the correction on C'_{ei} is needed. We decompose the electron–ion collisional operator in the ion mean velocity frame as

$$
C'_{ei}(f_e, f_i) = C'^{0}_{ei}(f_e, f_i) + C'^{1}_{ei}(f_e, f_i), \tag{2.98}
$$

where $C'^{0}_{ei}(f_e, f_i)$ and $C'^{1}_{ei}(f_e, f_i)$ are the lowest and first order parts, respectively.

The lowest order part $C'^{0}_{ei}(f_e, f_i)$ assumes the same expression as $C'_{ei}(f_e, f_i)$ in equation (2.96) except with the parameters being in the electron frame. To obtain $C'^{1}_{ei}(f_e, f_i)$, one needs to actually carry out the transformation for the electron distribution function in the ion mean velocity frame $f(\mathbf{v}_e - \mathbf{V}_i)$ back to the electron frame: $f(\mathbf{v}_e - \mathbf{V}_e)$. Note that $\mathbf{v}_e - \mathbf{V}_i = \mathbf{v}_e - \mathbf{V}_e - (\mathbf{V}_e - \mathbf{V}_i) = \mathbf{v}'_e + \mathbf{U}$, where $\mathbf{v}'_e = \mathbf{v}_e - \mathbf{V}_e$ and $\mathbf{U} = \mathbf{V}_e - \mathbf{V}_i$. As it is usually true, it is assumed that the difference of ion and electron velocities U is much smaller than the electron velocity v_e. In view of that the distribution function of lowest order is generally Maxwellian in the fluid description, one can transform the electron distribution function back to the mean electron velocity frame as follows:

$$
\begin{aligned}
f_e(\mathbf{v}_e - \mathbf{V}_i) &= \frac{n_e}{(2\pi T_e/m_e)^{3/2}} e^{-\frac{m_e}{2T_e}(\mathbf{v}'_e + \mathbf{U})^2} \\
&\approx f_e^0(\mathbf{v}'_e)\left(1 - \frac{m_e}{T_e}U_\gamma v'_{e\gamma}\right).
\end{aligned} \tag{2.99}
$$

One needs also to transform $V_{e\beta\gamma}$ as follows

$$
\begin{aligned}
V_{e\beta\gamma} &= (\mathbf{v}_e + \mathbf{U})^2\delta_{\beta\gamma} - (v_e + U)_\beta(v_e + U)_\gamma \\
&= V'_{e\beta\gamma} + \delta V_{e\beta\gamma},
\end{aligned} \tag{2.100}
$$

where $V'_{e\beta\gamma} = v'^2\delta_{\beta\gamma} - v'_{e\beta}v'_{e\gamma}$ and $\delta V_{e\beta\gamma} = 2v_{e\alpha}U_\alpha\delta_{\beta\gamma} - v_{e\beta}U_\gamma - v_{e\gamma}U_\beta$. Without causing ambiguity the prime on $\mathbf{v}'_e$ will be dropped again for brevity.

Using equations (2.99) and (2.100), one can transform C'_{ei} in equation (2.96) to the mean electron velocity frame as follows

$$\begin{aligned}
C'_{ei} &= \frac{3\sqrt{\pi}}{8}\frac{1}{\tau_e}\left(\frac{2T_e}{m_e}\right)^{3/2}\frac{\partial}{\partial v_{e\beta}}V_{e\beta\gamma}\frac{\partial f_e}{\partial v_{e\gamma}} \\
&= \frac{3\sqrt{\pi}}{8}\frac{1}{\tau_e}\left(\frac{2T_e}{m_e}\right)^{3/2} \\
&\times\frac{\partial}{\partial v_{e\beta}}\left[\left(V_{e\beta\gamma}+\delta V_{e\beta\gamma}\right)\frac{\partial f_e^0(\mathbf{v}_e)}{\partial v_{e\gamma}}\left(1-\frac{m_e}{T_e}U_\gamma v_{e\gamma}\right)+f_e^0(\mathbf{v}_e)\frac{m_e}{T_e}V_{e\beta\gamma}U_\gamma\right] \\
&= \frac{3\sqrt{\pi}}{8}\frac{1}{\tau_e}\left(\frac{2T_e}{m_e}\right)^{3/2} \\
&\times\frac{\partial}{\partial v_{e\beta}}\left[\frac{m_e}{T_e}f_e^0(\mathbf{v}_e)\left(V_{e\beta\gamma}+\delta V_{e\beta\gamma}\right)v_{e\gamma}\left(1-\frac{m_e}{T_e}U_\gamma v_{e\gamma}\right)+f_e^0(\mathbf{v}_e)\frac{m_e}{T_e}V_{e\beta\gamma}U_\gamma\right] \\
&= \frac{3\sqrt{\pi}}{8}\frac{1}{\tau_e}\left(\frac{2T_e}{m_e}\right)^{3/2}\frac{\partial}{\partial v_{e\beta}}\left[f_e^0(\mathbf{v}_e)\frac{m_e}{T_e}V_{e\beta\gamma}U_\gamma\right] \\
&= -\frac{3\sqrt{\pi}}{2}\frac{1}{\tau_e}\left(\frac{2T_e}{m_e}\right)^{1/2}f_e^0(\mathbf{v}_e)\frac{1}{v_e^3}v_{e\gamma}U_\gamma \equiv C'^{1}_{ei}.
\end{aligned} \tag{2.101}$$

Here, equation (2.91) has been used and it has been noted that

$$V_{e\beta\gamma}v_{e\gamma} = V_{e\beta\gamma}v_{e\beta} = 0,$$
$$\frac{\partial}{\partial v_{e\beta}}\left(f_e^0(\mathbf{v}_e)\delta V_{e\beta\gamma}v_{e\gamma}\right) = 0.$$

Note that in equation (2.101) the lowest order solution C'^{0}_{ei} disappears. This should be the case since the Maxwellian distribution function must annihilate the lowest order collisional operator.

2.3.3 Ion–electron cross collisional operator

We next reduce the ion–electron cross collisional operator for the process with ions being scattered by electrons. From equation (2.84), the ion–electron cross collisional operator can be expressed as

$$\begin{aligned}
C_{ie}(f_i, f_e) = &-\frac{2\pi\lambda e_i^2 e_e^2}{m_i}\frac{\partial}{\partial v_{i\beta}} \\
&\times\int\left\{\frac{f_i(\mathbf{v}_i)}{m_e}\frac{\partial f_e(\mathbf{v}_e)}{\partial v_{e\gamma}}-\frac{f_e(\mathbf{v}_e)}{m_i}\frac{\partial f_i(\mathbf{v}_i)}{\partial v_{e\gamma}}\right\}U_{\beta\gamma}\,d\mathbf{v}_e,
\end{aligned} \tag{2.102}$$

where

$$U_{\beta\gamma} = \frac{1}{u^3}\left(u^2\delta_{\beta\gamma} - u_\beta u_\gamma\right), \quad u_\beta = v_{i\beta} - v_{e\beta}.$$

There is freedom to choose the frame to evaluate the collisional operator. It is convenient to reduce the ion–electron collisional operator $C_{ie}(f_i, f_e)$ in the ion mean velocity ($\mathbf{V}_i$) frame. Nevertheless, to use the local Maxwellian for electrons for the integration over the electron velocity space in equation (2.102) both $U_{\beta\gamma}$ and f_e need to be transformed back to the electron mean velocity frame.

In the electron mean velocity frame, $U_{\beta\gamma}$ can be expanded in terms of ion velocity, noting that the ion speed is much smaller than electron speed,

$$U_{\alpha\beta} = V_{e\alpha\beta} - \frac{\partial V_{e\alpha\beta}}{\partial v_{e\gamma}} v_{i\gamma} + \frac{1}{2}\frac{\partial^2 V_{e\alpha\beta}}{\partial v_{e\gamma}\partial v_{e\delta}} v_{i\gamma} v_{i\delta} + \cdots, \tag{2.103}$$

where

$$V_{e\alpha\beta} = U_{\alpha\beta}|_{v_i=0} = \frac{1}{v_e{}^3}\left(v_e{}^2\delta_{\beta\gamma} - v_{e\beta}v_{e\gamma}\right). \tag{2.104}$$

Here, it has been noted that $u_\alpha = v_{i\alpha} - v_{e\alpha}$, which leads to $\partial v_{i\gamma} = -\partial v_{e\gamma}$.

The transformation of the electron distribution function from the ion mean velocity frame $f(\mathbf{v}_e - \mathbf{V}_i)$ back to the electron frame: $f(\mathbf{v}_e - \mathbf{V}_e)$ has been done in equation (2.99). The result is as follows

$$f_e(\mathbf{v}_e - \mathbf{V}_i) = f_e^0(\mathbf{v}'_e)\left(1 - \frac{m_e}{T_e} U_\gamma v'_{e\gamma}\right). \tag{2.105}$$

Note that the second term is of order m_e/m_i smaller than the first term in the collisional operator, the next order modification in equation (2.105) is required only for the first term in equation (2.102). Without causing ambiguity, the prime on $\mathbf{v}'_e$ will be dropped again for brevity.

First, let us consider the contribution of the first term on the right-hand side of equation (2.103). The first part of the collisional operator in equation (2.102) can be reduced to

$$\begin{aligned} C_{ie}^{I1} &= \frac{2\lambda\pi e_i^2 e_e^2}{m_e}\frac{\partial}{\partial v_{i\beta}}\int \frac{f_i(\mathbf{v}_i)}{m_e}\frac{\partial f_e(\mathbf{v}_e)}{\partial v_{e\gamma}} V_{e\beta\gamma}\, d\mathbf{v}_e \\ &= -\frac{2\lambda\pi e_i^2 e_e^2}{m_e}\frac{\partial}{\partial v_{i\beta}}\frac{f_i(\mathbf{v}_i)}{m_e}\int f_e(\mathbf{v}_e)\frac{\partial V_{e\beta\gamma}}{\partial v_{e\gamma}}\, d\mathbf{v}_e \\ &= -\frac{2\lambda\pi e_i^2 e_e^2}{m_e}\frac{\partial}{\partial v_{i\beta}}\frac{f_i(\mathbf{v}_i)}{m_e}\int f_e(\mathbf{v}_e)\left(-\frac{2}{v_e{}^3}v_{e\beta}\right) d\mathbf{v}_e, \end{aligned} \tag{2.106}$$

where equation (2.91) has been used (noting that $U_{\alpha\beta}$ is invariant to the reference frame). Inserting equation (2.105), one can further reduce equation (2.106). Note that the contribution of the lowest order term $f_e^0(\mathbf{v}_e')$ vanishes, since the integrand is odd. Therefore, equation (2.106) is reduced to

$$\begin{aligned}
C_{ie}^{I1} &= -\frac{2\lambda\pi e_i^2 e_e^2}{T_e}\frac{\partial}{\partial v_{i\beta}}\frac{f_i(\mathbf{v}_i)}{m_e}\int f_e^0(\mathbf{v}_e)U_\gamma v_{e\gamma}\frac{2}{v_e^3}v_{e\beta}\,d\mathbf{v}_e \\
&= -\frac{4\lambda\pi e_i^2 e_e^2}{3T_e}\frac{\partial}{\partial v_{i\beta}}\frac{f_i(\mathbf{v}_i)}{m_e}\int f_e^0(\mathbf{v}_e)U_\gamma\frac{\delta_{\gamma\beta}}{v_e}\,d\mathbf{v}_e \\
&= -\frac{4\lambda\pi e_i^2 e_e^2}{3T_e}U_\beta\frac{\partial}{\partial v_{i\beta}}\frac{f_i(\mathbf{v}_i)}{m_e}\int f_e^0(\mathbf{v}_e)\frac{1}{v_e}\,d\mathbf{v}_e \\
&\equiv \frac{1}{n_i m_i}\mathbf{R}_i\cdot\nabla_v f_i\,.
\end{aligned} \tag{2.107}$$

Here, the approximation $\overline{v_\beta v_\gamma} = (v^2/3)\delta_{\beta\gamma}$ has been used. This can be viewed as the direction average. Since there are three directions x, y, z, one has the factor 1/3. From the friction force on the electron fluid to be derived later in equation (2.113), one can see the definition about $\mathbf{R}_i$ here is consistent with that of $\mathbf{R}_e$, noting that $\mathbf{R}_i = -\mathbf{R}_e$ as shown in equation (2.19).

The second part of the collisional operator in equation (2.102) can be reduced to

$$\begin{aligned}
C_{ie}^{I2} &= -\frac{2\pi\lambda e_i^2 e_e^2}{m_e}\frac{\partial}{\partial v_{i\beta}}\int\frac{f_e(\mathbf{v}_e)}{m_i}\frac{\partial f_i(\mathbf{v}_i)}{\partial v_{i\gamma}}V_{e\beta\gamma}\,d\mathbf{v}_e \\
&= -\frac{2\pi\lambda e_i^2 e_e^2}{m_e}\frac{\partial}{\partial v_{i\beta}}\frac{\partial f_i(\mathbf{v}_i)}{\partial v_{i\gamma}}\int\frac{f_e^0(\mathbf{v}_e)}{m_i}\left(\frac{1}{v_e}\delta_{\beta\gamma} - \frac{1}{v^3}v_{e\beta}v_{e\gamma}\right)d\mathbf{v}_e \\
&= -\frac{2\pi\lambda e_i^2 e_e^2}{m_e}\frac{\partial}{\partial v_{i\beta}}\frac{\partial f_i(\mathbf{v}_i)}{\partial v_{i\beta}}\int\frac{f_e^0(\mathbf{v}_e)}{m_i}\frac{1}{v_e}\,d\mathbf{v}_e \\
&= \frac{m_e n_e}{m_i n_i}\frac{1}{\tau_e}\frac{\partial}{\partial v_{i\alpha}}\left(\frac{T_e}{m_i}\frac{\partial f_i}{\partial v_{i\alpha}}\right).
\end{aligned} \tag{2.108}$$

Next, let us consider the contribution of the second term on the right-hand side of equation (2.103). The first part of the collisional operator in equation (2.102) can be reduced as follows.

$$
\begin{aligned}
C_{ie}^{II1} &= \frac{2\lambda\pi e_i^2 e_e^2}{m_e}\frac{\partial}{\partial v_{i\beta}}\int \frac{f_i(\mathbf{v}_i)}{m_e}\frac{\partial f_e(\mathbf{v}_e)}{\partial v_{e\gamma}} v_{i\lambda}\frac{\partial V_{e\beta\gamma}}{\partial v_{e\lambda}} d\mathbf{v}_e = \frac{2\lambda\pi e_i^2 e_e^2}{m_e}\frac{\partial}{\partial v_{i\beta}} v_{i\lambda}\frac{f_i(\mathbf{v}_i)}{m_e} \\
&\times \int \frac{\partial f_e(\mathbf{v}_e)}{\partial v_{e\gamma}}\left(-\frac{v_{e\lambda}}{v_e^3}\delta_{\beta\gamma} + \frac{3}{v_e^5}v_\lambda v_{e\beta} v_{e\gamma} - \frac{1}{v_e^3}\delta_{\lambda\beta} v_{e\gamma} - \frac{1}{v_e^3} v_{e\beta}\,\delta_{\lambda\gamma}\right) d\mathbf{v}_e \\
&= \frac{2\lambda\pi e_i^2 e_e^2}{m_e}\frac{\partial}{\partial v_{i\beta}}\frac{f_i(\mathbf{v}_i)}{m_e}\int\left[v_{i\lambda}\frac{\partial f_e(\mathbf{v}_e)}{\partial v_{e\beta}}\left(-\frac{v_{e\lambda}}{v_e^3}\right)\right. \\
&\left. + v_{i\beta}\frac{\partial f_e(\mathbf{v}_e)}{\partial v_{e\gamma}}\left(-\frac{1}{v_e^3}v_{e\gamma}\right) + v_{i\lambda}\frac{\partial f_e(\mathbf{v}_e)}{\partial v_{e\lambda}}\left(-\frac{1}{v_e^3}v_{e\beta}\right)\right] d\mathbf{v}_e \\
&= \frac{2\lambda\pi e_i^2 e_e^2}{m_e}\frac{\partial}{\partial v_{i\beta}}\frac{f_i(\mathbf{v}_i)}{m_e}\int f_e(\mathbf{v}_e) \\
&\times \left\{ v_{i\beta}\left[\frac{1}{v_e^3} + \left(\frac{3}{v_e^3} - 3\frac{v_{e\beta}v_{e\beta}}{v_e^5}\right) + \frac{1}{v_e^3}\right] + v_{i\lambda}\frac{v_{e\beta}v_{e\lambda}}{v_e^5}\right\} d\mathbf{v}_e \\
&= \frac{2\lambda\pi e_i^2 e_e^2}{m_e}\frac{\partial}{\partial v_{i\beta}}\left(v_{i\beta}\frac{f_i(\mathbf{v}_i)}{m_e}\right)\int f_e^0(\mathbf{v}_e)\left(\frac{2}{v_e^3} + \frac{1}{3v_e^3}\right) d\mathbf{v}_e \\
&= \frac{m_e n_e}{m_i n_i}\frac{1}{\tau_e}\frac{\partial}{\partial v_{i\beta}}\left(v_{i\beta} f_i(\mathbf{v}_i)\right).
\end{aligned}
\tag{2.109}
$$

Here, equation (2.93) and the approximation $\overline{v_\beta v_\lambda} = (v^2/3)\delta_{\beta\lambda}$ have been used.

The second part of the collisional operator in equation (2.102) can be reduced to

$$
C_{ie}^{II2} = -\frac{2\pi\lambda e_i^2 e_e^2}{m_e}\frac{\partial}{\partial v_{i\beta}}\int \frac{f_e(\mathbf{v}_e)}{m_i}\frac{\partial f_i(\mathbf{v}_i)}{\partial v_{i\gamma}} v_{i\lambda}\frac{\partial V_{e\beta\gamma}}{\partial v_{e\lambda}} d\mathbf{v}_e = 0. \tag{2.110}
$$

Here, it has been considered that the integrand is odd.

Finally, combining the contributions of C_{ie}^{I1}, C_{ie}^{I2} and C_{ie}^{II1} respectively in equations (2.107), (2.108) and (2.109) yields

$$
C_{ie} = \frac{m_e n_e}{m_i n_i}\frac{1}{\tau_e}\frac{\partial}{\partial v_{i\beta}}\left(v_{i\beta} f_i(\mathbf{v}_i) + \frac{T_e}{m_i}\frac{\partial f_i(\mathbf{v}_i)}{\partial v_{i\beta}}\right) + \frac{1}{n_i m_i} R_{i\beta}\frac{\partial f_i(\mathbf{v}_i)}{\partial v_{i\beta}}. \tag{2.111}
$$

Note here that the ion distribution function is calculated in the ion mean velocity frame.

2.3.4 Transport coefficients based on the collisional operators

Having reduced the electron–ion and ion–electron cross collisional operators in subsections 2.3.2 and 2.3.3, one can compute in this subsection the transport parameters based on the collisional operators: $\mathbf{R}_u$ and Q as given in equations (2.18) and (2.24), respectively, as in references [1] and [2].

Friction force on the electron fluid

First, we use the results in subsections 2.3.2 to compute the friction force on the electron fluid by ions. According to equation (2.19), the friction force on the ion fluid by electrons is just opposite to the friction force on the electron fluid. From equation (2.18) we can write the friction force on the electron fluid as follows

$$R_{ea} = \int m_e v_{ea} C_{ei}\, d\mathbf{v}_e. \tag{2.112}$$

Here, it should be pointed out that R_{ea} is defined in the electron mean velocity frame.

To evaluate the integration in equation (2.112) one needs C_{ei} expressed in the mean electron velocity frame. This has been done in equations (2.95), (2.98), and (2.101). Neglecting order $\mathcal{O}(m_e/m_i)$ effects (C''_{ei} etc) and inserting equation (2.101) into equation (2.112), one obtains

$$\begin{aligned} R_{e\lambda} &= \int m_e v_{e\lambda} C_{ei}\, d\mathbf{v}_e \\ &= -\frac{3\sqrt{\pi}}{2}\frac{1}{\tau_e}\left(\frac{2m_e}{T_e}\right)^{1/2} \int m_e v_{e\lambda} \frac{2}{v_e^3} v_{e\beta} U_\beta f_e^0(\mathbf{v}_e) d\mathbf{v}_e \\ &= -\frac{3\sqrt{\pi}}{2}\frac{1}{\tau_e}\left(\frac{2m_e}{T_e}\right)^{1/2} U_\lambda \frac{2}{3} \int m_e \frac{1}{v_e} f_e^0(\mathbf{v}_e) d\mathbf{v}_e \\ &= -\frac{n_e m_e}{\tau_e} U_\lambda. \end{aligned} \tag{2.113}$$

Here, the approximation for the average $\overline{v_\lambda v_\beta} = (v^2/3)\delta_{\lambda\beta}$ has been used. Equation (2.113) corresponds to the results in equation (2.35). The constant 0.51 in equation (2.35) has been discussed in section 2.2.2.

Heat generation

Next, we use the result in subsections 2.3.3 to compute the heat generation due to the collisions. From equation (2.24) we can write the ion heat generation due to the collisions with electrons as follows

$$Q_i = \int \frac{m_i v_i^2}{2} C_{ie}\, d\mathbf{v}_i. \tag{2.114}$$

Note here that Q_i is defined in the ion mean velocity frame, which is the same as the collisional operator C_{ie} in equation (2.111).

Inserting equation (2.111) into equation (2.114), one obtains

$$\begin{aligned} Q_i &= \frac{m_e n_e}{m_i n_i}\frac{1}{\tau_e}\int d\mathbf{v}_i \frac{m_i v_i^2}{2}\frac{\partial}{\partial v_{i\beta}}\left(v_{i\beta} f_i(\mathbf{v}_i) + \frac{T_e}{m_i}\frac{\partial f_i(\mathbf{v}_i)}{\partial v_{i\beta}}\right) \\ &= -\frac{m_e n_e}{m_i n_i}\frac{1}{\tau_e} m_i \int d\mathbf{v}_i v_{i\beta}\left(v_{i\beta} f_i(\mathbf{v}_i) + \frac{T_e}{m_i}\frac{\partial f_i(\mathbf{v}_i)}{\partial v_{i\beta}}\right) \\ &= \frac{3 m_e n_e}{m_i \tau_e}(T_e - T_i). \end{aligned} \tag{2.115}$$

Here, it has been noted that the contribution of the last term in equation (2.111) vanishes since the integrand is odd. Equation (2.115) recovers the results in equation (2.66).

For electron heat generation due to collisions with ions, one can use the energy conservation in equation (2.29) to obtain

$$Q_e = -\mathbf{R}_e \cdot \mathbf{U} - Q_i,$$

as given in equation (2.67).

2.3.5 The first order kinetic equations

In this subsection, we first derive the first order kinetic equations. After that, we proceed to solve them and to evaluate the transport coefficients that depend on the distribution functions, such as $\mathbf{R}_T$, $\mathbf{q}$, and $\pi_{\alpha\beta}$. The kinetic equations both for electron and ion species will be solved first. The evaluation of the transport coefficients will be described afterwards.

The first order electron kinetic equation

We first derive the first order electron kinetic equation. The kinetic equation in the center of mass in equation (2.82) can be written for electron species as follows

$$\begin{aligned} &C_{ee}(f_e, f_e). + C_{ei}^{\prime 0}(f_e, f_i) - e_{\beta\gamma\alpha} v_{e\gamma}\omega_{e\alpha}\frac{\partial f_e}{\partial v_{e\beta}} \\ &= \frac{d_e f_e}{dt} + v_{e\beta}\frac{\partial f_e}{\partial x_\beta} + \left(\frac{e_e}{m_e}E_{e\beta}^{*} - \frac{d_e V_{e\beta}}{dt}\right)\frac{\partial f_e}{\partial v_{e\beta}} - \frac{\partial V_{e\gamma}}{\partial x_\beta} v_{e\beta}\frac{\partial f_e}{\partial v_{e\gamma}} \\ &- C_{ei}^{\prime 1}(f_e, f_i). - C_{ei}^{\prime\prime}(f_e, f_i). \end{aligned} \tag{2.116}$$

Here, the decompositions for electron–ion collisional operator in equations (2.95), (2.98), and (2.101) have been used and the dominant terms are grouped on the left-hand side. The collisional operators $C_{ee}(f_e, f_e).$ and $C_{ei}^{\prime 0}(f_e, f_i)$ are put on the left-hand side, since the collisional frequency is assumed to be high. The collisional operators $C_{ei}^{\prime 1}(f_e, f_i).$and $C_{ei}^{\prime\prime}(f_e, f_i)$ are left on the right-hand side since they are high order contributions from the cross collisional operator. Since the gyrofrequency is also assumed to be dominant, the term $e_{\beta\gamma\alpha} v_{e\gamma}\omega_{e\alpha}\frac{\partial f_e}{\partial v_{e\beta}}$ is also put on the left.

In lowest order, equation (2.116) becomes

$$C_{ee}(f_e, f_e). + C'_{ei}(f_e, f_i') - e_{\beta\gamma\alpha} v_{e\gamma} \omega_{ea} \frac{\partial f_e}{\partial v_{e\beta}} = 0.$$

Its solution is simply the Maxwellian distribution in the mean electron velocity frame

$$f_e^0 = \frac{n_e}{(2\pi T_e/m_e)^{3/2}} \exp\left\{-\frac{m_e}{2T_e} v_e^2\right\}. \tag{2.117}$$

To obtain the next order equation from equation (2.116), the electron distribution function is expanded as follows $f_e = f_e^0(1 + \Phi_e)$, where Φ_e is a small correction. In this case, the linearization of the left-hand side of equation (2.116) becomes

$$I_{ee}(\Phi_e) + I_{ei}(\Phi_e) - f_e^0 e_{\beta\gamma\alpha} v_{a\gamma} \omega_{ea} \frac{\partial \Phi_e}{\partial v_{e\beta}}, \tag{2.118}$$

where

$$I_{ee}(\Phi_e) = C_{ee}\left(f_e^0, f_e^0 \Phi_e\right) + C_{ee}\left(f^0 \Phi_e, f_e^0\right),$$
$$I_{ei}(\Phi_e) = C'_{ei}\left(f_e^0 \Phi_e, f_i\right).$$

Since the right-hand side of equation (2.116) is small, one can use the lowest order solution, i.e. the Maxwellian distribution in equation (2.117), for the distribution function there. With this simplification, one can reduce the right-hand side of equation (2.116) term by term. Note that the first term on the right of equation (2.116) becomes

$$\frac{df_e^0}{dt} = f_e^0\left(-\frac{3}{2} + \frac{m_e}{2T_e} v_e^2\right)\frac{d_a \ln T_e}{dt} + f_e^0 \frac{d_a \ln n_e}{dt}. \tag{2.119}$$

It can be simplified by using the continuity and energy equations, equations (2.13) and (2.25), of lowest order:

$$\frac{d_a \ln n_e}{dt} + \nabla \cdot \mathbf{V}_e = 0,$$
$$\frac{3}{2}\frac{d_e \ln T_e}{dt} + \nabla \cdot \mathbf{V}_e = 0.$$

Here, in the energy transport equation, the terms related the transport coefficients, such as the heat flux, heat generation etc, are left out. They will be computed by the next order solution of the kinetic equation afterwards. Therefore, equation (2.119) becomes

$$\frac{df_e^0}{dt} = -f_e^0 \frac{m_e v_e^2}{3T_e} \nabla \cdot \mathbf{V}_e. \tag{2.120}$$

The fourth term on the right-hand side of equation (2.116) can be reduced as follows

$$-\frac{\partial V_{e\gamma}}{\partial x_\beta} v_{e\beta} \frac{\partial f_e^0}{\partial v_{e\gamma}} = f_e^0 \frac{m_e}{T_e} \frac{\partial V_{e\gamma}}{\partial x_\beta} v_{e\beta} v_{e\gamma}. \tag{2.121}$$

Combining equations (2.120) and (2.121) yields

$$\begin{aligned}
&\frac{df_e^0}{dt} - \frac{\partial V_{e\gamma}}{\partial x_\beta} v_{e\beta} \frac{\partial f_e^0}{\partial v_{e\gamma}} \\
&= f_e^0 \frac{m_e}{T_e}\left(v_{e\alpha} v_{e\beta} - \frac{v_e^2}{3}\delta_{\alpha\beta}\right)\left(\frac{\partial V_{e\alpha}}{\partial x_\beta}\right) \\
&= f_e^0 \frac{m_e}{2T_e}\left(v_{e\alpha} v_{e\beta} - \frac{v_e^2}{3}\delta_{\alpha\beta}\right)\left(\frac{\partial V_{e\alpha}}{\partial x_\beta} + \frac{\partial V_{e\beta}}{\partial x_\alpha}\right) \\
&= f_e^0 \frac{m_e}{2T_e}\left(v_{e\alpha} v_{e\beta} - \frac{v_e^2}{3}\delta_{\alpha\beta}\right)\left(\frac{\partial V_{e\alpha}}{\partial x_\beta} + \frac{\partial V_{e\beta}}{\partial x_\alpha} - \frac{2}{3}\delta_{\alpha\beta}\nabla \cdot \mathbf{V}_e\right).
\end{aligned} \tag{2.122}$$

Here, we have noted that $\delta_{\alpha\alpha} = 3$.

Next, we consider the second and third terms on the right-hand side of equation (2.116). Similarly, the momentum equation (2.15) of lowest order can be used for simplification. Leaving out the viscosity term $\pi_{e\alpha\beta}$, the momentum equation (2.15) becomes

$$n_e e_e \mathbf{E}^* - n_e m_e d\mathbf{V}_e/dt = \nabla(n_e T_e) - \mathbf{R}_e.$$

With this equation for simplification the second and third terms on the right-hand side of equation (2.116) can be reduced as follows

$$\begin{aligned}
&v_{e\beta}\frac{\partial f_e}{\partial x_\beta} + \left(\frac{e_e}{m_e}E_{e\beta}^* - \frac{d_e V_{e\beta}}{dt}\right)\frac{\partial f_e}{\partial v_{e\beta}} \\
&= f_e^0\left(\frac{m}{2T}v_e^2 - \frac{3}{2}\right)\mathbf{v}_e \cdot \nabla \ln T_e + f_e^0 \mathbf{v}_e \cdot \nabla \ln n_e \\
&- f_e^0 \frac{1}{n_e T_e}\mathbf{v}_e \cdot [\nabla(n_e T_e) - \mathbf{R}_e] \\
&= f_e^0\left(\frac{m}{2T}v_e^2 - \frac{5}{2}\right)\mathbf{v}_e \cdot \nabla \ln T_e + f_e^0 \frac{1}{n_e T_e}\mathbf{v}_e \cdot \mathbf{R}_e,
\end{aligned} \tag{2.123}$$

where $\mathbf{R}_e = \mathbf{R}_e^0 + \mathbf{R}_e^1$ and

$$\mathbf{R}^1 = \int m_e \mathbf{v}_e I_{ei}(\Phi_e) d\mathbf{v}_e. \tag{2.124}$$

One needs also to express the collisional terms in equation (2.116). First, note that $C_{ei}''(f_e, f_i)$ is of order $(m_e/m_i)^{1/2}$ as compared with $C_{ei}'^{1}$ and therefore can be neglected. Next, using equation (2.101) one has

$$I_{ei}(\Phi) = C_{ei}'^{0}(f_e, f_i) = \frac{3\sqrt{\pi}}{8}\frac{1}{\tau_e}\left(\frac{2T_e}{m_e}\right)^{3/2}\frac{\partial}{\partial v_{e\beta}}V_{e\beta\gamma}\frac{\partial\left(f_e^0\Phi\right)}{\partial v_{e\gamma}}. \tag{2.125}$$

Furthermore, from equation (2.101) one has

$$C_{ei}'^{1}(f_e, f_i) = -f_e^0\frac{3\sqrt{\pi}}{\sqrt{2}}\frac{(T_e/m_e)^{3/2}}{v_e^3}\frac{m_e}{T_e\tau_e}\mathbf{U}\cdot\mathbf{v}_e. \tag{2.126}$$

Collecting the results in equations (2.118)–(2.126), the first order kinetic equation for electrons, equation (2.116), is reduced to

$$\begin{aligned} &I_{ee}(\Phi_e) + I_{ei}(\Phi_e) - f_e^0\mathbf{v}_e\times\boldsymbol{\omega}_{ce}\cdot\nabla_v\Phi_e \\ &= f_e^0\left\{\left(\frac{m_e v_e^2}{2T_e} - \frac{5}{2}\right)\mathbf{v}_e\cdot\nabla\ln T_e\right. \\ &+\left[\frac{3\sqrt{\pi}}{\sqrt{2}}\frac{(T_e/m_e)^{3/2}}{v_e^3} - 1\right]\frac{m_e}{T_e\tau_e}\mathbf{U}\cdot\mathbf{v}_e + \frac{1}{n_eT_e}\mathbf{R}^1\cdot\mathbf{v}_e \\ &\left.+\frac{m_e}{2T_e}\left(v_{e\alpha}v_{e\beta} - \frac{v_e^2}{3}\delta_{\alpha\beta}\right)W_{e\alpha\beta}\right\}, \end{aligned} \tag{2.127}$$

where

$$W_{e\alpha\beta} = \frac{\partial V_{e\alpha}}{\partial x_\beta} + \frac{\partial V_{e\beta}}{\partial x_\alpha} - \frac{2}{3}\delta_{\alpha\beta}\nabla\cdot\mathbf{V}_e.$$

Here, $W_{\alpha\beta}$ is referred to as the rate-of-strain tensor.

The first order ion kinetic equation

We now derive the first order kinetic equation for ions. The kinetic equation in equation (2.82) can be rearranged as follows

$$\begin{aligned} &C_{ii}(f_i, f_i) - e_{\beta\gamma\alpha}v_{i\gamma}\omega_{ci\alpha}\frac{\partial f_i}{\partial v_{i\beta}} \\ &= \frac{d_i f_i}{dt} + v_{i\beta}\frac{\partial f_i}{\partial x_\beta} + \left(\frac{e_i}{m_i}E_{i\beta}^* - \frac{d_i V_{i\beta}}{dt}\right)\frac{\partial f_i}{\partial v_{i\beta}} - \frac{\partial V_{i\gamma}}{\partial x_\beta}v_{i\beta}\frac{\partial f_i}{\partial v_{i\gamma}} \\ &+ C_{ie}(f_i, f_e). \end{aligned} \tag{2.128}$$

Noting that the ion–ion collision effects are much stronger than the ion–electron ones, we have moved $C_{ie}(f_i, f_e)$ to the right-hand side, where the smaller terms are collected.

In the lowest order, equation (2.128) becomes

$$C_{ii}(f_i, f_i) + e_{\beta\gamma\alpha}v_{i\gamma}\omega_{ci\alpha}\frac{\partial f_i}{\partial v_{i\beta}} = 0.$$

Its solution is simply the Maxwellian distribution

$$f_i^0 = \frac{n_i}{(2\pi T_i/m_i)^{3/2}} \exp\left\{-\frac{m_i}{2T_i} v_i^2\right\}. \tag{2.129}$$

To derive the first order kinetic equation for ions, the ion distribution function is expanded as follows $f_i = f_i^0(1 + \Phi_i)$, where Φ_i is a small correction. In this case, the linearization of the left-hand side of equation (2.128) becomes

$$I_{ii}(\Phi_i) - f_i^0 e_{\beta\gamma\alpha} v_{i\gamma} \omega_{ci\alpha} \frac{\partial \Phi_i}{\partial v_{i\beta}}, \tag{2.130}$$

where

$$I_{ii}(\Phi_i) = C_{ii}\left(f_i^0, f_i^0 \Phi_i\right) + C_{ii}\left(f^0 \Phi_i, f_i^0\right).$$

Since the right-hand side of equation (2.128) is small, one can use the lowest order solution, i.e. the Maxwellian distribution in equation (2.129), for the distribution function there. With this simplification, one can reduce the right-hand side of equation (2.128) term by term. Similar to the derivation in the case for electrons in equation (2.122), one can prove that

$$\frac{d_i f_i}{dt} - \frac{\partial V_{i\gamma}}{\partial x_\beta} v_{i\beta} \frac{\partial f_i}{\partial v_{i\gamma}} = \frac{m_i}{2T_i}\left(v_{e\alpha} v_{e\beta} - \frac{v_i^2}{3}\delta_{\alpha\beta}\right) W_{i\alpha\beta}, \tag{2.131}$$

where

$$W_{i\alpha\beta} = \frac{\partial V_{i\alpha}}{\partial x_\beta} + \frac{\partial V_{i\beta}}{\partial x_\alpha} - \frac{2}{3}\delta_{\alpha\beta} \nabla \cdot \mathbf{V}_i.$$

Here, $W_{i\alpha\beta}$ is the ion rate-of-strain tensor.

Similar to the electron case in equation (2.123), one can introduce the ion momentum equation of lowest order for simplification:

$$n_i e_i \mathbf{E}^* - n_i m_i \, d\mathbf{V}_i / dt = \nabla(n_i T_i).$$

Here, the friction term $\mathbf{R}_i$ has been dropped. This can be justified by the ordering comparison with the electron equation (2.127). Note that the right hand of equation (2.127) is of the same order as the ion equation we are going to derive and that equation (2.127) only keeps the correction of order $\sqrt{m_e/m_i}$. Since $\mathbf{R}_i$ is of order (m_e/m_i) as compared with $\mathbf{R}_e$, its contribution can be neglected. Therefore, one has

$$v_{i\beta} \frac{\partial f_i}{\partial x_\beta} + \left(\frac{e_i}{m_i} E_{i\beta}^* - \frac{d_i V_{i\beta}}{dt}\right) \frac{\partial f_i}{\partial v_{i\beta}} = f_i^0 \left(\frac{m_i v_i^2}{2T_i} - \frac{5}{2}\right) \mathbf{v}_i \cdot \nabla \ln T_i. \tag{2.132}$$

Noting that the ion–electron collisional operator $C_{ie}(f_i, f_e)$ contribution is of order (m_e/m_i) smaller in comparison with the $C_{ei}(f_e, f_i)$ role in the electron equation, its contribution can be neglected. Therefore, collecting the results in equations (2.130)–(2.132), equation (2.128) is reduced to

$$I_{ii}(\Phi_i) - f_i^0 \mathbf{v}_i \times \boldsymbol{\omega}_{ci} \cdot \nabla_v \Phi_i$$
$$= f_i^0 \left\{ \left(\frac{m_i v_i^2}{2T_i} - \frac{5}{2} \right) \mathbf{v}_i \cdot \nabla \ln T_i + \frac{m_i}{2T_i} \left(v_{e\alpha} v_{e\beta} - \frac{v_i^2}{3} \delta_{\alpha\beta} \right) W_{i\alpha\beta} \right\}. \tag{2.133}$$

2.3.6 Transport coefficients based on the distribution functions

Having derived the first order kinetic equations for electrons and ions in the previous subsection, we are going solve the equations and use the solutions to evaluate the thermal transport coefficients in this subsection as in references [1] and [2]. Since there is a similarity between the electron and ion treatments, we only detail the derivations of the thermal friction force and heat flux for electrons and the viscosity tensor for ions. In computing the transport coefficients, the integration table in appendix B.1 is used.

Note that the first order kinetic equations for electrons and ions, equations (2.127) and (2.133), are linear integral equations. The effects of the non-equilibrium driving forces on the right-hand side are addable. One can solve individually for the responses $\Phi_{e,i}$ from the driving forces, such as the temperature gradient ∇T, the velocity shift $\mathbf{U}$ and the velocity inhomogeneity $W_{\alpha\beta}$. Since we only describe the derivations for the thermal friction force and heat flux for electrons and the viscosity tensor for ions, we will then solve the electron equation (2.127) for the responses from the source terms related to ∇T_e and $\mathbf{U}_e$ and solve the ion equation (2.133) for the responses from the source term related to $W_{i\alpha\beta}$.

Furthermore, noting the tensor invariance, one can assume the solutions to equations (2.127) and (2.149) to have the following form

$$\Phi(\mathbf{v}) = \Phi_\alpha(v^2) v_\alpha + \Phi_{\alpha\beta}(v^2) \left(v_\alpha v_\beta - \frac{v^2}{3} \delta_{\alpha\beta} \right). \tag{2.134}$$

Here, $\Phi_\alpha(v^2)$ and $\Phi_{\alpha\beta}(v^2)$ are isotropic in the velocity space. The first term on the right-hand side of equation (2.134) corresponds to the vector perturbations: ∇T and $\mathbf{U}$, while the second term is related to the tensor perturbation: $\pi_{\alpha\beta}$. Therefore, the friction force due to $\mathbf{R}^1$ and the heat flux $\mathbf{q}$ are determined by the vector $\Phi_\alpha(v^2)$, while the viscosity $\pi_{\alpha\beta}$ is specified by the tensor $\Phi_{\alpha\beta}(v^2)$.

There are some constraints for the solutions to the first order kinetic equations. In the fluid description, the three moments: the density n, the velocity $\mathbf{V}$, and the temperature T, are assumed to be fully determined by the lowest order distribution functions, i.e. the Maxwellian distribution functions. The first order solutions are assumed to contribute only to the transport coefficients for the closure. Therefore, one has

$$\int f^1 \Phi \, d\mathbf{v} = 0, \quad \int \mathbf{v} f^1 \Phi \, d\mathbf{v} \doteq 0, \quad \text{and} \quad \int v^2 f^1 \Phi \, d\mathbf{v} = 0. \tag{2.135}$$

Solution of the first order electron equation

We first proceed to solve the first order electron equation (2.127). We only solve for the responses from the source terms related to ∇T_e. This is an example to solve the vector part in equation (2.134). The procedure to solve the other vector part from $\mathbf{U}_e$ is similar and will not be detailed. The treatment for $W_{e\alpha\beta}$ contribution is similar to that for ions to be described later on. Leaving out the terms containing $\mathbf{U}$ and $W_{e\alpha\beta}$, the first order kinetic equation for electrons, equation (2.127), becomes

$$I_{ee}(\Phi_e) + I_{ei}(\Phi_e) - f_e^0 \mathbf{v}_e \times \boldsymbol{\omega}_{ce} \cdot \nabla_v \Phi_e = f_e^0 \left[\left(\frac{m_e v_e^2}{2T_e} - \frac{5}{2} \right) \mathbf{v}_e \cdot \nabla \ln T_e + \frac{1}{n_e T_e} \mathbf{R}^1 \cdot \mathbf{v}_e \right]. \tag{2.136}$$

According to equation (2.134), the vector part of the solution to equation (2.136) assumes the following form

$$\Phi_e = A_e(v^2)\mathbf{v} \cdot \nabla_\parallel \ln T_e + A_e'(v^2)\mathbf{v} \cdot \nabla_\perp \ln T_e + A_e''(v^2)\mathbf{v} \cdot (\boldsymbol{\omega}_{ce} \times \nabla_\perp \ln T_e).$$

Here, $A_e(v^2)\nabla_\parallel \ln T_e$, etc, correspond to the vector $\Phi_\alpha(v^2)$ in equation (2.134). From equation (2.136) one can see that if letting $\omega_{ce} = 0$, the solutions in response to $\nabla_\parallel \ln T_e$ and $\nabla_\perp \ln T_e$ are identical, i.e. $A_e(v^2) = A_e'(v^2)$ if $\omega_{ce} = 0$. Therefore, we can simply solve for the response to $\nabla_\perp \ln T_e$ and to obtain the response to $\nabla_\parallel \ln T_e$. $A_e(v^2)$, by letting $\omega_{ce} = 0$. It is then assumed that

$$\Phi_e = A_e'(v^2)\mathbf{v} \cdot \nabla_\perp \ln T_e + A_e''(v^2)\mathbf{v} \cdot (\boldsymbol{\omega}_{ce} \times \nabla_\perp \ln T_e). \tag{2.137}$$

Noting that $\mathbf{R}^1$ in equation (2.124) also is a function of the next order distribution function, Φ_e, we introduce the following representation as well

$$\mathbf{R}^1 = n_e T_e [K_e'(v^2)\nabla_\perp \ln T_e + K_e'(v^2)(\boldsymbol{\omega}_{ce} \times \nabla_\perp \ln T_e)]. \tag{2.138}$$

Inserting equations (2.137) and (2.138) into equation (2.136) yields

$$\begin{aligned} & I_{ee}(A_e'\mathbf{v}) \cdot \nabla_\perp \ln T_e + I_{ee}\left(A_e''\mathbf{v}\right) \cdot (\boldsymbol{\omega}_{ce} \times \nabla_\perp \ln T_e) \\ & + I_{ei}(A_e'\mathbf{v}) \cdot \nabla_\perp \ln T_e + I_{ei}\left(A_e''\mathbf{v}\right) \cdot (\boldsymbol{\omega}_{ce} \times \nabla_\perp \ln T_e) \\ & - f_e^0 A_e'\mathbf{v} \cdot (\boldsymbol{\omega}_{ce} \times \nabla_\perp \ln T_e) + \omega_{ce}^2 f_e^0 A_e''\mathbf{v} \cdot \nabla_\perp \ln T_e \\ & = f_e \left(\frac{m_e v^2}{2T_e} - \frac{5}{2} \right) \mathbf{v} \cdot \nabla_\perp \ln T_e \\ & + f_e K_e'\mathbf{v} \cdot \nabla_\perp \ln T_e + f_e K_e''\mathbf{v} \cdot (\boldsymbol{\omega}_{ce} \times \nabla_\perp \ln T_e). \end{aligned}$$

Sorting out the terms proportional to $\nabla_\perp \ln T_e$ and $(\boldsymbol{\omega}_{ce} \times \nabla_\perp \ln T_e)$, one obtains two independent equations. The terms proportional to $\nabla_\perp \ln T_e$ yield:

$$\begin{aligned} & I_{ee}(A_e'\mathbf{v}) + I_{ei}(A_e'\mathbf{v}) + \omega_{ce}^2 f_e^0 A_e''\mathbf{v} \\ & = f_e \left(\frac{m_e v^2}{2T_e} - \frac{5}{2} \right) \mathbf{v} + f_e K_e'\mathbf{v}. \end{aligned} \tag{2.139}$$

The terms proportional to $(\boldsymbol{\omega}_e \times \nabla_\perp \ln T_e)$ give rise to:

$$I_{ee}(A_e''\mathbf{v}) + I_{ei}(A_e''\mathbf{v}) - f_e^0 A_e'\mathbf{v} = f_e K_e''\mathbf{v}. \tag{2.140}$$

We define that

$$\hat{A}_e \equiv A_e' + i\omega_{ce}A_e'' \quad \text{and} \quad \hat{K}_e \equiv K_e' + i\omega_{ce}K_e''. \tag{2.141}$$

Combining equations (2.139) and (2.140) multiplied by $i\omega_{ce}$, one has

$$\begin{aligned} &I_{ee}(\hat{A}_e\mathbf{v}) + I_{ei}(\hat{A}_e\mathbf{v}) - i\omega_{ce}f_e^0 \hat{A}_e\mathbf{v} \\ &= f_e\left(\frac{m_e v^2}{2T_e} - \frac{5}{2}\right)\mathbf{v} + f_e\hat{K}_e\mathbf{v}. \end{aligned} \tag{2.142}$$

This is an integral equation. As usual, the decomposition in terms of orthogonal functions can be used to solve it. In view of the feature of the right-hand side of equation (2.142), the Laguerre polynomials $L_p^{(m)}(x)$ [11] can be used. The basic properties of Laguerre's polynomials are as follows. The generating function is

$$(1-\xi)^{-m-1}\exp\left(-\frac{x\xi}{1-\xi}\right) = \sum_{p=0}^{+\infty} \xi^p L_p^{(m)}(x). \tag{2.143}$$

The polynomials are orthogonal with the weight $x^m e^{-x}$ over the interval 0 to $+\infty$, i.e.

$$\int_0^{+\infty} x^m e^{-x} L_p^{(m)}(x) L_q^{(m)}(x) dx = \frac{(p+m)!}{p!}\delta_{pq}. \tag{2.144}$$

We specially note that the first two Laguerre polynomials are $L_0^{(m)}(x) = 1$ and $L_1^{(m)}(x) = 1 - x$, which can be used to express the right-hand side of equation (2.142):

$$\frac{m_e v^2}{2T_e} - \frac{5}{2} + \hat{K}_e = -L_1^{3/2}(x) - \hat{K}_e L_0^{3/2}, \quad \text{with } x = \frac{m_e v^2}{2T_e}. \tag{2.145}$$

This explains why the Laguerre polynomials are introduced to solve equation (2.142).

We can expand $\hat{A}_e(v^2)$ as follows

$$\hat{A}_e(v^2) = \tau_e \sum_{k=1}^{+\infty} a_k L_k^{(3/2)}(x). \tag{2.146}$$

Here, the $k = 0$ term on the right-hand side has been eliminated to assure the constraint $\int \mathbf{v} f_e^1 \Phi \, d\mathbf{v} = 0$ in equation (2.135). This is explained as follows. Noting that $\Phi \propto \hat{A}_e(v^2)\mathbf{s} \cdot \mathbf{v}$ and $d\mathbf{v} = 4\pi v^2 \, dv$ for the isotropic case, one has

$$\begin{aligned}\int \mathbf{v} f_e^0 \Phi \, d\mathbf{v} &= \mathbf{s} \cdot \int \hat{A}_e(v^2)\mathbf{v}\mathbf{v} f_e^0 \, d\mathbf{v} \\ &= \mathbf{s} \cdot \mathcal{I}\frac{1}{3} \int \hat{A}_e(v^2) v^2 f_e^0 \, d\mathbf{v} \\ &\propto \mathbf{s}\frac{1}{3} \int \hat{A}_e(x) x^{3/2} e^{-x} \, dx.\end{aligned} \tag{2.147}$$

Therefore, according to orthogonality relations, equation (2.144), the $k = 0$ term on the right-hand side of equation (2.146) should be eliminated.

Inserting the expansion equation (2.146) into equation (2.142), multiplying by the factor

$$-\frac{4}{15}\frac{1}{n_e}\frac{m_e}{2T_e}\mathbf{v}L_k^{3/2}\left(\frac{m_e v^e}{2T_e}\right)d\mathbf{v},$$

and integrating over the velocity space, one obtains an algebraic equation for the expansion coefficients

$$\sum_{l=1}^{+\infty}\left(\alpha_{kl} + \alpha_{kl}'\right)a_l + i\omega_{ce}\tau_e\frac{(k+3/2)!}{k!(3/2)!}a_k = \delta_{1k}, \quad k = 1, 2, \ldots, \tag{2.148}$$

where

$$\begin{aligned}\alpha_{kl} &= -\frac{4\tau_e}{15n_e}\frac{m_e}{2T_e}\int L_k^{3/2}(x)v_\beta I_{ee}\left(L_l^{(3/2)}(x)v_\beta\right)d\mathbf{v}, \\ \alpha_{kl}' &= -\frac{4\tau_e}{15n_e}\frac{m_e}{2T_e}\int L_k^{3/2}(x)v_\beta I_{ei}\left(L_l^{(3/2)}(x)v_\beta\right)d\mathbf{v}.\end{aligned}$$

The numerical scheme to compute α_{kl} and α_{kl}' has been detailed in reference [2] and will not be detailed here.

Equation (2.148) can be used to determine a_k in equation (2.146), so that $\hat{A}_e(v^2)$ is determined. According to equation (2.141), one can determine $A_e'(v^2)$ and $A_e''(v^2)$ by separating the real and imaginary parts of $\hat{A}_e(v^2)$. As discussed earlier, $A_e(v^2)$ can be obtained by letting $\omega_{ce} = 0$ in the solution $\hat{A}_e(v^2)$. This completes the solution for the electron equation (2.136).

Thermal force and heat flux

With the solution to the first order kinetic equation for electrons obtained, we can now discuss the evaluation of the transport coefficients based on the distribution functions. Here, we focus on the heat flux, thermal conductivity, and friction force. The viscosity will be discussed after solving the ion kinetic equation.

With equation (2.148) solved, $\Phi_e(v^2)$ in equation (2.137) can be determined. One can then determine the thermal force and heat flux due to the temperature gradient using equations (2.18) and (2.23):

$$\mathbf{R}_{eT} = -\frac{5}{2}n_e\sum_{k=1}^{\infty}\alpha_{0k}\left(a_k'\nabla_\perp T_e + a_k''\boldsymbol{\omega}_{ce}\times\nabla T_e\right),$$

$$\mathbf{q}_{eT} = -\frac{5}{2}\frac{n_e T_e \tau_e}{m_e}\left(a_1'\nabla_\perp T_e + a_1''\boldsymbol{\omega}_{ce}\times\nabla T_e\right).$$

Here, we have denoted $a_k = a_k' + i\omega_{ce}a_k''$ by analogy with equation (2.141).

Solution of the first order ion equation

Next, we proceed to solve the first order ion equation (2.133). We only solve for the responses from the source term tied to $W_{e\alpha\beta}$ for viscosity. This shows how to solve the tensor part in equation (2.134). The treatment for ∇T_i contributions is similar to those for electrons as discussed earlier. Leaving out the terms containing ∇T_i, the first order kinetic equation for ions, equation (2.133), becomes

$$I_{ii}(\Phi_i) - f_i^0\mathbf{v}_i\times\boldsymbol{\omega}_{cii}\cdot\nabla_v\Phi_i = f_i^0\frac{m_i}{2T_i}v_{e\alpha\beta}W_{i\alpha\beta}, \tag{2.149}$$

where

$$v_{\alpha\beta} = v_\alpha v_\beta - \frac{1}{3}v^2\,\delta_{\alpha\beta}.$$

To solve equation (2.149) we decompose $W_{i\alpha\beta}$ as follows

$$W_{i\alpha\beta} = W_{(0)} + W_{(1)} + W_{(2)}, \tag{2.150}$$

where

$$W_{(0)} = \begin{pmatrix} \frac{1}{2}(W_{xx}+W_{yy}) & 0 & 0 \\ 0 & \frac{1}{2}(W_{xx}+W_{yy}) & 0 \\ 0 & 0 & W_{zz} \end{pmatrix}, \tag{2.151}$$

$$W_{(1)} = \begin{pmatrix} \frac{1}{2}(W_{xx}-W_{yy}) & W_{xy} & 0 \\ W_{yx} & -\frac{1}{2}(W_{xx}-W_{yy}) & 0 \\ 0 & 0 & 0 \end{pmatrix}, \tag{2.152}$$

$$W_{(2)} = \begin{pmatrix} 0 & 0 & W_{xz} \\ 0 & 0 & W_{yz} \\ W_{zx} & W_{zy} & 0 \end{pmatrix}. \tag{2.153}$$

Noting that $W_{ixx} + W_{iyy} + W_{izz} = 0$ and $W_{i\alpha\beta} = W_{i\beta\alpha}$, one can see that these three matrices are symmetric and their traces vanish.

We also introduce the following accompanied matrices, respectively, to $W_{(1)}$ and $W_{(2)}$:

$$W_{(3)} = \begin{pmatrix} -W_{xy} & \frac{1}{2}(W_{xx} - W_{yy}) & 0 \\ \frac{1}{2}(W_{xx} - W_{yy}) & W_{xy} & 0 \\ 0 & 0 & 0 \end{pmatrix}, \tag{2.154}$$

$$W_{(4)} = \begin{pmatrix} 0 & 0 & -W_{yz} \\ 0 & 0 & W_{xz} \\ -W_{zy} & W_{zx} & 0 \end{pmatrix}. \tag{2.155}$$

These tensors have been chosen in such a way that their traces vanish and

$$\left(\mathbf{v} \times \mathbf{b} \cdot \nabla_v v_{\alpha\beta}\right) W_{(0)\alpha\beta} = 0, \tag{2.156}$$

$$\left(\mathbf{v} \times \mathbf{b} \cdot \nabla_v v_{\alpha\beta}\right) W_{(1)\alpha\beta} = 2 v_{\alpha\beta} W_{(3)\alpha\beta}, \tag{2.157}$$

$$\left(\mathbf{v} \times \mathbf{b} \cdot \nabla_v v_{\alpha\beta}\right) W_{(2)\alpha\beta} = v_{\alpha\beta} W_{(4)\alpha\beta}, \tag{2.158}$$

and vice versa

$$\left(\mathbf{v} \times \mathbf{b} \cdot \nabla_v v_{\alpha\beta}\right) W_{(3)\alpha\beta} = -2 v_{\alpha\beta} W_{(1)\alpha\beta}, \tag{2.159}$$

$$\left(\mathbf{v} \times \mathbf{b} \cdot \nabla_v v_{\alpha\beta}\right) W_{(4)\alpha\beta} = -v_{\alpha\beta} W_{(2)\alpha\beta}. \tag{2.160}$$

These can be proved for example as follows. Noting that

$$\mathbf{v} \times \mathbf{b} \cdot \nabla_v v_{\alpha\beta} = e_{3\alpha\gamma} v_\beta v_\gamma + e_{3\beta\gamma} v_\alpha v_\gamma,$$

one has

$$\begin{aligned}
&\left(\mathbf{v} \times \mathbf{b} \cdot \nabla_v v_{\alpha\beta}\right) W_{(1)\alpha\beta} \\
&= (e_{312} v_1 v_2 + e_{312} v_1 v_2) W_{(1)11} + (e_{321} v_2 v_1 + e_{321} v_2 v_1) W_{(1)22} \\
&+ (e_{312} v_2 v_2 + e_{321} v_1 v_1) W_{(1)12} + (e_{321} v_1 v_1 + e_{312} v_2 v_2) W_{(1)21} \\
&= (v_1 v_2 + v_1 v_2) W_{(1)11} - (v_2 v_1 + v_2 v_1) W_{(1)22} \\
&+ (v_2 v_2 - v_i v_1) W_{(1)12} + (-v_1 v_1 + v_2 v_2) W_{(1)21} \\
&= 2 v_{\alpha\beta} W_{(3)\alpha\beta}.
\end{aligned}$$

Here, the tensor notations in appendix A have been used. This proves equation (2.157). The proofs for other equations in equations (2.156)–(2.160) are similar.

With $W_{i\alpha\beta}$ decomposed and their accompanied matrices introduced on the right-hand side of equation (2.149), we can decompose Φ_i accordingly as well

$$\Phi_i = \Phi_{(0)} + \Phi_{(1)} + \Phi_{(2)}, \tag{2.161}$$

where

$$\begin{aligned}
\Phi_{(0)} &= - B_{(0)}(v^2) W_{(0)\alpha\beta} v_{\alpha\beta}, \\
\Phi_{(1)} &= - \left(B_{(1)}(v^2) W_{(1)\alpha\beta} + B_{(3)}(v^2) W_{(3)\alpha\beta}\right) v_{\alpha\beta}, \\
\Phi_{(2)} &= - \left(B_{(2)}(v^2) W_{(2)\alpha\beta} + B_{(4)}(v^2) W_{(4)\alpha\beta}\right) v_{\alpha\beta}.
\end{aligned}$$

Inserting equation (2.161) into equation (2.149), one can split equation (2.149) into three parts and solve them separately. For $\Phi_{(0)}$ part, one has

$$\begin{aligned}
&I_{ii}(-B_{(0)}(v^2) W_{(0)\alpha\beta} v_{\alpha\beta}) + f_0 \mathbf{v} \times \boldsymbol{\omega}_{ci} \cdot \nabla_v(-B_{(0)}(v^2) W_{(0)\alpha\beta} v_{\alpha\beta}) \\
&= -f_0 v_{\alpha\beta} W_{(0)\alpha\beta}.
\end{aligned}$$

Using equation (2.156), this equation is reduced to

$$I_{ii}(B_{(0)}(v^2) v_{\alpha\beta}) = f_0 v_{\alpha\beta}. \tag{2.162}$$

For $\Phi_{(1)}$ part, one has

$$\begin{aligned}
&I_{ii}\left(-\left(B_{(1)}(v^2) W_{(1)\alpha\beta} + B_{(3)}(v^2) W_{(3)\alpha\beta}\right) v_{\alpha\beta}\right) \\
&+ f_0 \mathbf{v} \times \boldsymbol{\omega}_{ci} \cdot \nabla_v\left[-\left(B_{(1)}(v^2) W_{(1)\alpha\beta} + B_{(3)}(v^2) W_{(3)\alpha\beta}\right) v_{\alpha\beta}\right] \\
&= -f_0 v_{\alpha\beta} W_{(1)\alpha\beta}.
\end{aligned}$$

Using equations (2.157) and (2.159), this equation is reduced to

$$\begin{aligned}
&I_{ii}\left(\left(B_{(1)}(v^2) W_{(1)\alpha\beta} + B_{(3)}(v^2) W_{(3)\alpha\beta}\right) v_{\alpha\beta}\right) \\
&+ \omega_{ci} f_0 \left(2B_{(1)}(v^2) W_{(3)\alpha\beta} - 2B_{(3)}(v^2) W_{(1)\alpha\beta}\right) v_{\alpha\beta} = f_0 v_{\alpha\beta} W_{(1)\alpha\beta}.
\end{aligned}$$

From this equation, one can obtain the following two equations, respectively, for $W_{(1)\alpha\beta}$ and $W_{(3)\alpha\beta}$ parts:

$$\begin{aligned}
I_{ii}\left(B_{(1)}(v^2) v_{\alpha\beta}\right) + 2i^2 \omega_{ci} f_0 B_{(3)}(v^2) v_{\alpha\beta} &= f_0 v_{\alpha\beta}, \\
I_{ii}\left(B_{(3)}(v^2) v_{\alpha\beta}\right) + 2\omega_{ci} f_0 B_{(1)}(v^2) v_{\alpha\beta} &= 0.
\end{aligned}$$

Letting $B_1 = B_{(1)} + iB_{(3)}$, these two equations can be combined to yield

$$I_{ii}\left(B_1(v^2) v_{\alpha\beta}\right) + 2i\omega_{ci} f_0 B_1 v_{\alpha\beta} = f_0 v_{\alpha\beta}. \tag{2.163}$$

For $\Phi_{(2)}$ part, one has

$$\begin{aligned}
&I_{ii}\left(-\left(B_{(2)}(v^2) W_{(2)\alpha\beta} + B_{(4)}(v^2) W_{(4)\alpha\beta}\right) v_{\alpha\beta}\right) \\
&+ f_0 \mathbf{v} \times \boldsymbol{\omega}_{ci} \cdot \nabla_v\left[-\left(B_{(2)}(v^2) W_{(1)\alpha\beta} + B_{(4)}(v^2) W_{(4)\alpha\beta}\right) v_{\alpha\beta}\right] \\
&= -f_0 v_{\alpha\beta} W_{(2)\alpha\beta}.
\end{aligned}$$

Using equations (2.158) and (2.160), this equation is reduced to

$$I_{ii}\Big(\big(B_{(2)}(v^2)W_{(2)\alpha\beta} + B_{(4)}(v^2)W_{(4)\alpha\beta}\big)v_{\alpha\beta}\Big)$$
$$+ \omega_{ci}f_0\big(B_{(4)}(v^2)W_{(4)\alpha\beta} - B_{(4)}(v^2)W_{(2)\alpha\beta}\big)v_{\alpha\beta} = f_0 v_{\alpha\beta} W_{(2)\alpha\beta}.$$

From this equation, one can obtain the following two equations, respectively, for $W_{(2)\alpha\beta}$ and $W_{(4)\alpha\beta}$ parts:

$$I_{ii}\big(B_{(2)}(v^2)v_{\alpha\beta}\big) + i^2\omega_{ci}f_0 B_{(4)}(v^2)v_{\alpha\beta} = f_0 v_{\alpha\beta},$$
$$I_{ii}\big(B_{(4)}(v^2)v_{\alpha\beta}\big) + \omega_{ci}f_0 B_{(2)}(v^2)v_{\alpha\beta} = 0.$$

Letting $B_2 = B_{(2)} + iB_{(4)}$, these two equations can be combined to yield

$$I_{ii}\big(B_2(v^2)v_{\alpha\beta}\big) + i\omega_{ci}f_0 B_2 v_{\alpha\beta} = f_0 v_{\alpha\beta}. \tag{2.164}$$

We can solve equations (2.162)–(2.164) for $B_{(0)}$, $B_{(1)}$, and $B_{(2)}$. Equations (2.162)–(2.164) have similarity. We can just solve equation (2.164) for $B_{(2)}$. The solution for equation (2.162), $B_{(0)}$, can be obtained from the solution for equation (2.164) by letting $\omega_{ci} \to 0$. The solution for equation (2.163), $B_{(1)}$, can be obtained from the solution for equation (2.164) by letting $\omega_{ci} \to 2\omega_{ci}$.

Equation (2.164) is an integral equation. Its solution can be obtained by the decomposition with the Laguerre polynomials of order 5/2:

$$B_2(x) = \tau_i \sum_{p=0}^{\infty} b_p L_p^{5/2}(x). \tag{2.165}$$

To reduce equation (2.164), we multiply equation (2.164) with $L_k^{5/2}(x)v_{\alpha\beta}$ with $x = m_i v^2/2T_i$ and integrate it over the velocity space. The orthogonal relation in equation (2.144) is used for simplification.

Noting that $d\mathbf{v} = 4\pi v^2\, dv$ and $v_{\alpha\beta}v_{\alpha\beta} = 2v^4/3$, one has

$$f_i^0 v_{\alpha\beta}v_{\alpha\beta}\, d\mathbf{v} = \frac{4n_i}{3\sqrt{\pi}}\left(\frac{2T_i}{m_i}\right)^2 \exp\{-x\}x^{5/2}\, dx.$$

This explains why the expansion with the Laguerre polynomials of order 5/2 in equation (2.165) is used. Noting that $L_0^{5/2} = 1$, the right-hand side of equation (2.164) is transformed to

$$\begin{aligned}\int d\mathbf{v} L_k^{5/2} v_{\alpha\beta} f_i^0 v_{\alpha\beta} &= \frac{4n_i}{3\sqrt{\pi}}\left(\frac{2T_i}{m_i}\right)^2 \int L_k^{5/2} \exp\{-x\}x^{5/2} L_0^{5/2}\, dx \\ &= \frac{4n_i}{3\sqrt{\pi}}\left(\frac{2T_i}{m_i}\right)^2 \frac{(k+5/2)!}{k!}\delta_{0k}.\end{aligned} \tag{2.166}$$

The second term on the left-hand side of (2.164) becomes

$$\begin{aligned}
&\int d\mathbf{v} L_k^{5/2} v_{\alpha\beta}\left(i\omega_{ci} f_i^0 B_2 v_{\alpha\beta}\right) \\
&= i\omega_{ci}\tau_i \frac{4n_i}{3\sqrt{\pi}}\left(\frac{2T_i}{m_i}\right)^2 \sum_p b_p \int L_k^{5/2} \exp\{-x\} x^{5/2} L_p^{5/2} dx \\
&= i\omega_{ci}\tau_i \frac{4n_i}{3\sqrt{\pi}}\left(\frac{2T_i}{m_i}\right)^2 \frac{(k+5/2)!}{k!} b_k.
\end{aligned}$$

Combining these results, one obtains

$$\sum_{l=0}^{+\infty} \beta_{kl} b_l + \frac{(k+5/2)!}{k!(5/2)!} i\omega_{ci}\tau_i b_k = \delta_{0k}, \tag{2.167}$$

where

$$\beta_{kl} = \tau_i \frac{3\sqrt{\pi}}{4n_i(5/2)!}\left(\frac{m_i}{2T_i}\right)^2 \int L_k^{5/2} v_{\alpha\beta} I_{ii}\left(L_l^{5/2} v_{\alpha\beta}\right) d\mathbf{v}.$$

The numerical scheme to compute β_{kl} has been detailed in reference [2] and will not be detailed here.

Equation (2.167) can be used to determine b_k in equation (2.165), so that $B_2(v^2)$ is determined. According to the definition of $B_2(v^2)$, one can determine $B_{(2)}$ and $B_{(4)}$ by separating the real and imaginary parts of B_2. As discussed earlier, the solution B_1 can be obtained from the solution B_2 by letting $\omega_{ci} \to 2\omega_{ci}$. Subsequently, we can obtain $B_{(1)}$ and $B_{(3)}$. The solution $B_{(0)}$ can be obtained from the solution B_2 by letting $\omega_{ci} \to 0$. Therefore, one obtains the complete solution Φ_i in equation (2.161).

Viscosity tensor

With the solution to the first order kinetic equation for ions obtained, we can now discuss the evaluation of the transport coefficients based on the distribution functions. Here, we focus on the viscosity for ions. The evaluations of heat flux, thermal conductivity, and friction force for ions are similar to those for electrons, as discussed earlier.

With equation (2.167) solved, the first order solution for ions, $\Phi_i(v^2)$, in equation (2.161) can be determined. One can then determine the viscosity using equation (2.17). Denoting $b_0 = b' + i\omega_{ci}\tau_i b''$ in equation (2.165), one can express the viscosity tensor as follows [1, 2]

$$\pi_{izz} = -0.96 n_i T_i \tau_i W_{zz},$$

$$\pi_{ixx} = n_i T_i \tau_i \left[0.48 W_{zz} - \frac{1}{2} b'(2\omega_{ci}) \left(W_{xx} - W_{yy} \right) - 2\omega_{ci} \tau_i b''(2\omega_{ci}) W_{xy} \right],$$

$$\pi_{iyy} = n_i T_i \tau_i \left[0.48 W_{zz} - \frac{1}{2} b'(2\omega_{ci}) \left(W_{xx} - W_{yy} \right) - 2\omega_{ci} \tau_i b''(2\omega_{ci}) W_{xy} \right],$$

$$\pi_{ixy} = \pi_{iyx} = -n_i T_i \tau_i \left[b'(2\omega_{ci}) W_{xy} - \omega_{ci} \tau_i b''(2\omega_{ci}) (W_{xx} - W_{yy}) \right],$$

$$\pi_{izx} = \pi_{ixz} = -n_i T_i \tau_i \left[b'(\omega_{ci}) W_{xy} - \omega_{ci} \tau_i b''(\omega_{ci}) W_{yz} \right],$$

$$\pi_{iyz} = \pi_{izy} = -n_i T_i \tau_i \left[b'(\omega_{ci}) W_{yz} - \omega_{ci} \tau_i b''(\omega_{ci}) W_{xz} \right].$$

The evaluation of the stress tensor for electrons is similar.

2.3.7 Transport coefficients due to the finite Larmor radius effects

In this subsection, we examine the collisionless limit of the Braginskii transport theory and derive the transport coefficients solo due to the finite Larmor radius effects. With the collision effects included, the kinetic equations have to be solved numerically. Instead, with the collisional effects excluded, the first order kinetic equations can be solved analytically, so that the thermal heat flux and gyroviscous tensor can be obtained.

Excluding the collision effects, the first order kinetic equations for electrons and ions can be obtained from equations (2.127) and (2.133),

$$-f_e^0 \mathbf{v}_e \times \boldsymbol{\omega}_e \cdot \nabla_v \Phi_e = f_e^0 \left(\frac{m_e v_e^2}{2T_e} - \frac{5}{2} \right) \mathbf{v}_e \cdot \nabla \ln T_e, \tag{2.168}$$

$$\begin{aligned} -f_i^0 \mathbf{v}_i \times \boldsymbol{\omega}_i \cdot \nabla_v \Phi_i = {} & f_i^0 \left(\frac{m_i v_i^2}{2T_i} - \frac{5}{2} \right) \mathbf{v}_i \cdot \nabla \ln T_i \\ & + f^0 \frac{m_i}{2T_i} \left(v_{e\alpha} v_{e\beta} - \frac{v_i^2}{3} \delta_{\alpha\beta} \right) W_{i\alpha\beta}. \end{aligned} \tag{2.169}$$

Here, the term related to the electron rate-of-strain tensor in equation (2.168) has been neglected since it is proportional to m_e and is much smaller than the ion one.

Note that the responses to the first terms, respectively, on the right-hand sides of electron and ion equations (2.168) and (2.169) have the same structure. One can use the ion solution to deduce the electron one. Assume the ion solution to be

$$\Phi_i = \Phi_{iv} + \Phi_{iw},$$

where the Φ_{iv} and Φ_{iw} correspond, respectively, to the solutions of the first and second terms on the right-hand side of equation (2.169). One can directly verify that the two parts of the ion solutions are given as follows

$$f_{iv}^{1} = f_{i}^{0}\Phi_{iv} = f_{i}^{0}\frac{2}{\rho\omega_{c}}e_{ljk}v_{\perp j}b_{k}\nabla_{l}\ln T\left(\frac{mv^{2}}{2T} - \frac{5}{2}\right), \tag{2.170}$$

$$f_{iw}^{1} = f_{i}^{0}\Phi_{iv} = f_{i}^{0}\frac{m_{i}}{4T_{i}\omega_{c}}\left(\hat{v}_{\alpha}\hat{v}_{\beta} - \frac{1}{2}\hat{v}_{\perp}^{2}\delta_{\alpha\beta}\right)W_{(3)\alpha\beta}. \tag{2.171}$$

Using equation (2.170), one can evaluate the heat flux for both ions and electrons:

$$\begin{aligned} q_{\delta} &= \frac{1}{\omega_{c}} e_{ljk}b_{k}\nabla_{l}\ln T\frac{\rho}{2}\frac{1}{n}\int d^{3}v v_{\perp j}v_{\delta}v^{2}\left(\frac{mv^{2}}{2T} - \frac{5}{2}\right)f^{0} \\ &= \frac{1}{\omega_{c}} e_{l\delta k}b_{k}\nabla_{l}\ln T\frac{\rho}{2}\frac{1}{n}\int d^{3}v\frac{v_{\perp}^{2}}{2}v^{2}\left(\frac{mv^{2}}{2T} - \frac{5}{2}\right)f^{0} \\ &= \frac{1}{\omega_{c}} e_{l\delta k}b_{k}\nabla_{l}\ln T\frac{\rho}{2}\frac{1}{3n}\int d^{3}v v^{4}\left(\frac{mv^{2}}{2T} - \frac{5}{2}\right)f^{0} \\ &= \frac{1}{\omega_{c}} e_{l\delta k}b_{k}\nabla_{l}\ln T\frac{\rho}{2}\frac{1}{3}\left(\frac{105}{2} - \frac{5}{2}\cdot 15\right)\frac{T^{2}}{m^{2}} \\ &= \frac{5nT}{2m\omega_{c}}e_{l\delta k}b_{k}\nabla_{l}T. \end{aligned} \tag{2.172}$$

Here, the particle species indices are suppressed for brevity.

Using equation (2.171), one can derive the ion gyroviscous tensor as follows

$$\begin{aligned} \pi_{\mu\nu} &= \frac{1}{\hat{n}}\int d^{3}\hat{v}\frac{m}{4T\omega_{c}}\left(v_{\mu}v_{\nu} - \frac{1}{3}v^{3}\delta_{\mu\nu}\right)\left(\hat{v}_{\alpha}\hat{v}_{\beta} - \frac{1}{2}\hat{v}_{\perp}^{2}\delta_{\alpha\beta}\right)W_{(3)\alpha\beta}f^{0} \\ &= \rho\frac{m}{4T\omega_{c}}\frac{1}{n}\int d^{3}v\left(v_{\mu}v_{\nu} - \frac{1}{3}v^{2}\delta_{\mu\nu}\right)(v_{1}v_{1} - v_{2}v_{2})W_{(3)11}f^{0} \\ &+ \rho\frac{m}{2T\omega_{c}}W_{(3)12}\frac{1}{n}\int d^{3}v\left(v_{\mu}v_{\nu} - \frac{1}{3}v^{3}\delta_{\mu\nu}\right)v_{1}v_{2}f^{0}. \end{aligned} \tag{2.173}$$

This can be used to evaluate the specific components as follows

$$\begin{aligned} \pi_{11} = -\pi_{22} &= -\rho\frac{m}{4T\omega_{c}}W_{(3)11}\frac{1}{n}\int d^{3}v v_{2}v_{2}(v_{1}v_{1} - v_{2}v_{2})f^{0} \\ &= \rho\frac{m}{4T\omega_{c}}W_{(3)11}(3 - 1)\frac{T^{2}}{m^{2}} \\ &= -\frac{nT}{2m\omega_{c}}W_{12}, \end{aligned} \tag{2.174}$$

$$\begin{aligned}\pi_{12} = \pi_{21} &= \rho \frac{1}{n} \frac{m}{2T\omega_c} W_{(3)12} \int d^3v v_1^2 v_2^2 f^0 \\ &= \rho \frac{1}{n} \frac{m}{T\omega_c} W_{(3)12} \frac{2T^2}{m^2} \\ &= \frac{nT}{2m\omega_c} \frac{1}{2} (W_{11} - W_{22}).\end{aligned} \tag{2.175}$$

Other components vanish.

These results will be used to compare with the transport coefficients derived in the perpendicular magnetofluid theory in chapter 3.

2.4 Conditions of applicability

In this section, the applicability of the Braginskii transport theory is discussed, including the extension from the classical transport theory underlying the Braginskii framework to the neoclassical and anomalous transport theories [2].

The Braginskii transport theory assumes that there are local equilibria for ion and electron species. They can be specified individually by the lower order moments: the density, velocity, and temperature. The 'fluxes' $\mathbf{R}$, $\boldsymbol{\pi}$, $\mathbf{q}$, and Q are introduced as the transport parameters that lead to the evolution of these lower order moments. This type of framework requires the separation of time scales. The fastest time scale is related to the relaxation process, which forces the distribution function to approach the local Maxwellian so that the density, velocity, and temperature can be defined locally for individual species in time and space. The inhomogeneity of the lowest equilibrium quantities causes the lower order moments to evolve in the next time scale. High collisionality is supposed to lead to the local equilibria. Therefore, the fundamental conditions of applicability for Braginskii transport theory are, the equilibrium time scale is much shorter than the collision time τ and the equilibrium spatial scale L is much larger than the mean free path l in the absence of strong magnetic field, i.e.

$$\frac{d}{dt} \ll \frac{1}{\tau}, \tag{2.176}$$

$$L \gg l. \tag{2.177}$$

In the presence of strong magnetic field ($\omega_c \tau \gg 1$), the time scale requirement in equation (2.176) remains, while for the spatial scale requirement, one has to distinguish the perpendicular and parallel directions:

$$L_\perp \gg r, \quad L_\parallel \gg l, \tag{2.178}$$

where r is the Larmor radius, and $L_\perp$ and $L_\parallel$ are the equilibrium spatial scale lengths perpendicular and parallel to the magnetic field, respectively.

The inhomogeneity of the magnetic field can cause the particle curvature drift. The drift velocity is of order $v_T r / L_B$, where $L_B \sim B/|\nabla B|$ is the scale length of equilibrium magnetic field. The drift motion can give rise to additional heat transport, leading to the deviation of the distribution function from a Maxwellian. In a system with $L_\perp \gg L_B$, the requirement for perpendicular scale changes, so that equation (2.178) becomes

$$L_\perp \gg \tau v_T r / L_B \sim l r / L_B, \quad L_\parallel \gg l.$$

If $L_\perp \sim L_B$, this reduces to

$$L_\perp \gg \sqrt{lr}, \quad L_\parallel \gg l.$$

Otherwise, neoclassical transport needs to be considered [8].

Besides, since the Landau collisional operator is used in the Braginskii transport theory, which does not take into account the effects of magnetic field on collisions, the theory is only applicable when $r \gg \delta_D$. Here, δ_D is the Debye length.

There is another factor that can affect the applicability of the Braginskii transport theory: turbulence transport. Plasma instabilities can generate random fluctuating fields, which in return can produce strong mixing and enhance the transport. The electrostatic modes $\delta\mathbf{E}$ can lead to the velocity across the magnetic field: $V_c = \delta E / B$. If particles move a distance l' across the magnetic field lines in the random fluctuating fields, according to equation (2.72) the turbulence transport coefficient can be estimated as $D_{\rm turb} \sim \delta E l' / B$. From the adiabatic estimate: $\nabla_\parallel (nT) \sim ne\delta E_\parallel$, one has $e\delta E / k_\parallel \sim T$, where $k_\parallel$ is the parallel wave number. Therefore, one has

$$D_{\rm turb} \sim \frac{T}{eB} k_\parallel l'.$$

This is related to the so-called Bohm transport [12, 13]. Besides the electrostatic modes, the electromagnetic instabilities can also cause anomalous transport. On one hand, the electromagnetic modes are usually coupled to the electrostatic modes due to the difference of electron and ion responses [14]. On the other hand, the electromagnetic modes can couple to the tearing modes due to the current interchange effects [15], etc. Tearing modes reconnect the magnetic field line and lead the usually small transport across the field lines to be enhanced by the particle parallel motions.

Actually, the most stringent limitation of the Braginskii transport theory lies in the requirement in the parallel direction: $L_\parallel \gg l$. In the perpendicular direction, the strong magnetic field can play the role of charged particle localization. In the parallel direction, however, there is no particle spatial localization scheme besides the collisions. In the weak collision limit, one cannot even use the parallel fluid velocity as the lowest order description. Some particles are trapped by the inhomogeneous equilibrium magnetic field or by the plasma waves, some others may remain in circulation. Therefore, for magnetically confined fusion plasmas one needs to develop other types of theories. This will be addressed in the following chapters.

References

[1] Braginskii S I 1966 Transport phenomena in a complete ionized two-temperature plasma *ZhETF (J. Exp. Theor. Phys. USSR)* **33** 459
Braginskii S I 1958 *Soviet Phys. JETP* **6** 358–69

[2] Braginskii S I 1966 Transport processes *Reviews of Plasma Physics* vol 1 ed M Leontovich (New York: Consultants Bureau) pp 205–311

[3] Lerner R G and Trigg G L 1991 *Encyclopaedia of Physics* 2nd edn (New York: VHC Publishers)

[4] Bogoliubov N N 1946 Kinetic equations *J. Phys. USSR* **10** 265–74

[5] Klimontovich Y L 1967 *The Statistical Theory of Non-equilibrium Processes in a Plasma* (Oxford: Pergamon)

[6] Landau L D 1937 Kinetic equation for the Coulomb effect *Sov. Phys. JETP* **7** 203–9

[7] Hinton F L and Hazeltine R D 1976 Theory of plasma transport in toroidal confinement systems *Rev. Mod. Phys.* **48** 239308

[8] Helander P and Sigmar D 2002 *Collisional Transport in Magnetized Plasmas* (New York: Cambridge University Press)

[9] Hazeltine R D and Meiss J D 1992 *Plasma Confinement* (New York: Addison-Wesley)

[10] Chapman S and Cowling T G 1953 *Mathematical Theory of Nonuniform Gases* (Cambridge: Cambridge University Press)

[11] Abramowitz M and Stegun I A 1970 *Handbook of Mathematical Functions* (New York: Dover)

[12] Bohm D 1949 Qualitative description of the arc plasma in a magnetic field *The Characteristics of Electrical Discharges in Magnetic Fields* ed A Gutherie and R K Wakerling (New York: McGraw-Hill) pp 1–12

[13] Spitzer L Jr 1960 Particle diffusion across a magnetic field *Phys. Fluids* **3** 659–61

[14] Zheng L-J and Tessarotto M 1994 Collisionless kinetic ballooning equations in the comparable frequency regime *Phys. Plasmas* **1** 2956–62

[15] Zheng L J and Furukawa M 2010 Current-interchange tearing modes: conversion of interchange-type modes to tearing modes *Phys. Plasmas* **17** 052508

Part III

Fluid theory in the collisionless or weak collision limit

IOP Publishing

Magnetically Confined Fusion Plasma Physics, Volume 2
Multifluid theory
Linjin Zheng

Chapter 3

Perpendicular magnetofluid theory in the collisionless limit

The fluid theory is fundamental to magnetically confined fusion researches. Even in the kinetic theory framework, loading particles in Maxwellian relies on the fluid thermodynamic variables: density, velocity, and temperature. The boundary conditions for solving the kinetic equations need to be specified by the fluid variables. Conventionally, the Braginskii two-fluid theory is often employed, assuming the collisionality to be dominant. In reality, however, the magnetically confined plasmas in thermonuclear fusion turn to be collisionless. Some particles are trapped by the inhomogeneous equilibrium field or by the waves, while some others remain in circulation. One cannot expect charged particles to move together as a fluid cell along the magnetic field lines. Consequently, the fluid variables relating to the parallel motion, such as parallel fluid velocity, become inadequate.

In this chapter, the perpendicular magnetofluid theory for magnetically confined plasmas in the collisionless limit is described [1]. It revises Braginskii's transport theory [2; 3] by releasing the collisional dominance assumption. In the perpendicular direction, the particle spatial localization is solo based on the strong magnetic field, while the finite Larmor radius (FLR) effects are taken into account in the next order. In the parallel direction, the particle mobility features are fully retained by constructing only the perpendicular moments. Besides the FLR effects till order $\mathcal{O}(k_\perp^2\rho_i^2)$, the Landau damping, the trapped particle effects, and other parallel kinetic effects are all kept in this formalism. Here, $k_\perp$ is the perpendicular wave number and ρ_i is the ion Larmor radius. The theory can be used both for the analytical and numerical studies of magnetically confined plasmas. The applications will be discussed in the following chapters.

We first describe the kinetic equations for both ion and electron species in the field-line-aligned velocity coordinate system. The perpendicular moments of the kinetic equations are derived next, including the continuity, momentum, and energy transport equations. They govern the perpendicular density n_a, perpendicular velocity V_a, the perpendicular temperature T_a, where the subscript $a = i, e$. The closure of the perpendicular moment equations is facilitated by introducing the transport coefficients. The kinetic equations of first order are then solved to determine the transport coefficients. The conclusions and discussion are given at the end.

3.1 Kinetic equation in the field-line-aligned coordinate system

In this section, the kinetic equations in the field-line-aligned velocity coordinate system are derived. This is necessary for developing the perpendicular magnetofluid theory since the theory relies on constructing the perpendicular moments of the kinetic equations. Note the magnetically confined plasmas are often highly anisotropic. Charged particles gyrate around magnetic field lines in the leading order, while moving rather freely along the field lines. To embrace this feature, a velocity space coordinate system that is aligned to the magnetic field line is favored.

We start with the Vlasov equation [4]:

$$\frac{\partial f}{\partial t} + \mathbf{v} \cdot \frac{\partial f}{\partial \mathbf{x}} + \frac{e}{m} \mathbf{a} \cdot \frac{\partial f}{\partial \mathbf{v}} = 0, \tag{3.1}$$

where f is the distribution function, which depends on the particle coordinate $\mathbf{x}$, velocity $\mathbf{v}$, and time t, and

$$\mathbf{a} = \mathbf{E} + \mathbf{v} \times \mathbf{B}.$$

The subscript a for particle species is suppressed for brevity. The collisional operator $C \equiv (\partial f/\partial t)_c$ is left out in the kinetic equation. The spatial coordinate $\mathbf{x}$ and velocity $\mathbf{v}$ are independent in equation (3.1). However, in order to carry out the perpendicular moments, one needs $\mathbf{x}$ and $(\mathbf{v}_\perp, \mathbf{v}_\parallel)$ to be independent. Because the magnetic field $\mathbf{B}$ depends on $\mathbf{x}$ and t, the velocity projections on the magnetic field $(\mathbf{v}_\perp, \mathbf{v}_\parallel)$ become dependent on $\mathbf{x}$ and t although the complete velocity $\mathbf{v}$ is not. One has to transform equation (3.1) from the $(\mathbf{x}, \mathbf{v})$ coordinates to the field-line-aligned velocity coordinate system $(\mathbf{x}; \mathbf{v}_\perp, \mathbf{v}_\parallel)$.

Two coordinate systems are introduced. The laboratory Cartesian coordinate system is denoted by the basis $(\mathbf{e}_1, \mathbf{e}_2, \mathbf{e}_3)$, where $\mathbf{e}_\alpha$ with $\alpha = 1, 2, 3$ are unit vectors, The field-line-aligned coordinate system is also introduced. It is right-handed and denoted by the basis $(\hat{\mathbf{e}}_1, \hat{\mathbf{e}}_2, \hat{\mathbf{e}}_3)$. Here, $\hat{\mathbf{e}}_\alpha$ with $\alpha = 1, 2, 3$ are unit vectors, $\hat{\mathbf{e}}_3 = \mathbf{B}/B$ is field line unit vector, $\hat{\mathbf{e}}_1$ is perpendicular to the magnetic field, pointing, for example, to the field line curvature direction: $\mathbf{e}_1 \parallel \hat{\mathbf{e}}_3 \cdot \nabla \hat{\mathbf{e}}_3$, and $\hat{\mathbf{e}}_2 = \hat{\mathbf{e}}_3 \times \hat{\mathbf{e}}_1$. Due to the basis $(\hat{\mathbf{e}}_1, \hat{\mathbf{e}}_2, \hat{\mathbf{e}}_3)$ being tied to the magnetic field, one has $\hat{\mathbf{e}}_\alpha = \hat{\mathbf{e}}_\alpha(\mathbf{x}, t)$.

We define the transform matrix, i.e. the metric tensor:

$$g_{\alpha\beta}(\mathbf{x}, t) = \hat{\mathbf{e}}_\alpha \cdot \mathbf{e}_\beta.$$

Therefore, one has

$$\begin{pmatrix} \hat{\mathbf{e}}_1 \\ \hat{\mathbf{e}}_2 \\ \hat{\mathbf{e}}_3 \end{pmatrix} = \begin{pmatrix} \hat{\mathbf{e}}_1 \cdot \mathbf{e}_1 & \hat{\mathbf{e}}_1 \cdot \mathbf{e}_2 & \hat{\mathbf{e}}_1 \cdot \mathbf{e}_3 \\ \hat{\mathbf{e}}_2 \cdot \mathbf{e}_1 & \hat{\mathbf{e}}_2 \cdot \mathbf{e}_2 & \hat{\mathbf{e}}_2 \cdot \mathbf{e}_3 \\ \hat{\mathbf{e}}_3 \cdot \mathbf{e}_1 & \hat{\mathbf{e}}_3 \cdot \mathbf{e}_2 & \hat{\mathbf{e}}_3 \cdot \mathbf{e}_3 \end{pmatrix} \begin{pmatrix} \mathbf{e}_1 \\ \mathbf{e}_2 \\ \mathbf{e}_3 \end{pmatrix},$$

$$\begin{pmatrix} \mathbf{e}_1 \\ \mathbf{e}_2 \\ \mathbf{e}_3 \end{pmatrix} = \begin{pmatrix} \mathbf{e}_1 \cdot \hat{\mathbf{e}}_1 & \mathbf{e}_1 \cdot \hat{\mathbf{e}}_2 & \mathbf{e}_1 \cdot \hat{\mathbf{e}}_3 \\ \mathbf{e}_2 \cdot \hat{\mathbf{e}}_1 & \mathbf{e}_2 \cdot \hat{\mathbf{e}}_2 & \mathbf{e}_2 \cdot \hat{\mathbf{e}}_3 \\ \mathbf{e}_3 \cdot \hat{\mathbf{e}}_1 & \mathbf{e}_3 \cdot \hat{\mathbf{e}}_2 & \mathbf{e}_3 \cdot \hat{\mathbf{e}}_3 \end{pmatrix} \begin{pmatrix} \hat{\mathbf{e}}_1 \\ \hat{\mathbf{e}}_2 \\ \hat{\mathbf{e}}_3 \end{pmatrix},$$

i.e, $\hat{\mathbf{e}}_\alpha = g_{\alpha\beta}\mathbf{e}_\beta$ and $\mathbf{e}_\alpha = g_{\beta\alpha}\hat{\mathbf{e}}_\beta$ with $\alpha, \beta = 1, 2, 3$, where the Einstein summation convention for same indices as given in appendix A is implied. Note here that the definition of the metric tensor is unrelated to the usual covariant and contravariant dual vector bases. One can prove that $g_{\alpha\beta}^{-1} = g_{\alpha\beta}^T$, where superscript '$T$' denotes the matrix transpose. This shows that $g_{\alpha\beta}$ is an orthogonal matrix and has the following properties

$$\begin{aligned} &g_{\alpha\beta}g_{\gamma\beta} = g_{\alpha\beta}g_{\beta\gamma}^T = g_{\alpha\beta}g_{\beta\gamma}^{-1} = \delta_{\alpha\gamma}, \qquad g_{\beta\alpha}g_{\beta\gamma} = \delta_{\alpha\gamma}, \\ &\det|g_{\alpha\beta}| = 1. \end{aligned} \tag{3.2}$$

There are some useful equalities. Using equation (3.2); one has

$$g_{\alpha\beta}\frac{\partial g_{\alpha\beta}}{\partial(x_\nu, t)} = \frac{1}{2}\frac{\partial g_{\alpha\beta}^2}{\partial(x_\nu, t)} = 0, \tag{3.3}$$

$$g_{\alpha\beta}\frac{\partial g_{\alpha\gamma}}{\partial(x_\nu, t)} = -g_{\alpha\gamma}\frac{\partial g_{\alpha\beta}}{\partial(x_\nu, t)}. \tag{3.4}$$

One also has

$$\begin{aligned} g_{3\alpha}g_{3\beta} &= g_{\gamma\alpha}g_{\gamma\beta} - \delta_{\gamma\delta}^\perp g_{\gamma\alpha}g_{\delta\beta} \\ &= \delta_{\alpha\beta} - \delta_{\gamma\delta}^\perp g_{\gamma\alpha}g_{\delta\beta}. \end{aligned} \tag{3.5}$$

Furthermore, for an arbitrary tensor $A_{\alpha\gamma}$ one has

$$A_{\alpha\gamma}g_{\alpha\beta}\frac{\partial g_{\gamma\beta}}{\partial(x_\nu, t)} = -A_{\alpha\gamma}g_{\gamma\beta}\frac{\partial g_{\alpha\beta}}{\partial(x_\nu, t)} = -A_{\alpha\gamma}^T g_{\alpha\beta}\frac{\partial g_{\gamma\beta}}{\partial(x_\nu, t)}.$$

Consequently,

$$A_{\alpha\gamma}g_{\alpha\beta}\frac{\partial g_{\gamma\beta}}{\partial(x_\nu, t)} = 0, \qquad \text{if} \qquad A_{\alpha\gamma} = A_{\alpha\gamma}^T. \tag{3.6}$$

With the basis $(\hat{\mathbf{e}}_1, \hat{\mathbf{e}}_2, \hat{\mathbf{e}}_3)$ introduced, one can project the velocity in the directions parallel and perpendicular to the magnetic field:

$$\begin{aligned} \mathbf{v} &= v_1\mathbf{e}_1 + v_2\mathbf{e}_2 + v_3\mathbf{e}_3 \\ &= \hat{v}_1\hat{\mathbf{e}}_1 + \hat{v}_2\hat{\mathbf{e}}_2 + \hat{v}_3\hat{\mathbf{e}}_3. \end{aligned} \tag{3.7}$$

Here, $\hat{v}_{1,2}$ are the perpendicular components, while $\hat{v}_3$ is the parallel component. From equation (3.7), one obtains

$$\hat{v}_\alpha = g_{\alpha\beta} v_\beta, \qquad v_\alpha = g_{\beta\alpha} \hat{v}_\beta. \tag{3.8}$$

Noting that $\hat{\mathbf{e}}_\alpha$ is a function of time and space and using equation (3.8), one has

$$\left.\frac{\partial \hat{v}_\alpha}{\partial t}\right|_{\mathbf{x},\mathbf{v}} = v_\beta \frac{\partial g_{\alpha\beta}}{\partial t} = g_{\gamma\beta} \hat{v}_\gamma \frac{\partial g_{\alpha\beta}}{\partial t}, \tag{3.9}$$

$$\left.\frac{\partial \hat{v}_\alpha}{\partial x_\delta}\right|_{\mathbf{v},t} = v_\beta \frac{\partial g_{\alpha\beta}}{\partial x_\delta} = g_{\gamma\beta} \hat{v}_\gamma \frac{\partial g_{\alpha\beta}}{\partial x_\delta}, \tag{3.10}$$

$$\left.\frac{\partial \hat{v}_\alpha}{\partial v_\beta}\right|_{\mathbf{x},t} = g_{\alpha\beta}, \tag{3.11}$$

where α, β, γ, $\delta = 1, 2, 3$.

Using equations (3.7)–(3.11), one can transform the kinetic equation, equation (3.1), into the $(\mathbf{x}, \hat{\mathbf{v}})$ coordinate system. The time derivative term becomes

$$\begin{aligned} \left.\frac{\partial f}{\partial t}\right|_{\mathbf{x},\mathbf{v}} &= \left.\frac{\partial f}{\partial t}\right|_{\mathbf{x},\hat{\mathbf{v}}} + \left.\frac{\partial f}{\partial \hat{v}_\mu}\right|_{\mathbf{x},t} \frac{\partial \hat{v}_\mu}{\partial t} \\ &= \left.\frac{\partial f}{\partial t}\right|_{\mathbf{x},\hat{\mathbf{v}}} + g_{\delta\nu} \hat{v}_\delta \frac{\partial g_{\mu\nu}}{\partial t} \left.\frac{\partial f}{\partial \hat{v}_\mu}\right|_{\mathbf{x},t}. \end{aligned} \tag{3.12}$$

The spatial derivative term becomes

$$\begin{aligned} \left.\frac{\partial f}{\partial x_q}\right|_{t,\mathbf{v}} &= \left.\frac{\partial f}{\partial x_q}\right|_{t,\hat{\mathbf{v}}} + \left.\frac{\partial f}{\partial \hat{v}_\mu}\right|_{\mathbf{x},t} \frac{\partial \hat{v}_\mu}{\partial x_q} \\ &= \left.\frac{\partial f}{\partial x_q}\right|_{t,\hat{\mathbf{v}}} + g_{\delta\nu} \hat{v}_\delta \frac{\partial g_{\mu\nu}}{\partial x_q} \left.\frac{\partial f}{\partial \hat{v}_\mu}\right|_{\mathbf{x},t}. \end{aligned} \tag{3.13}$$

The velocity derivative term becomes

$$\left.\frac{\partial f}{\partial v_q}\right|_{t,\mathbf{x}} = \left.\frac{\partial f}{\partial \hat{v}_r}\right|_{t,\mathbf{x}} \frac{\partial \hat{v}_r}{\partial v_q} = g_{rq} \left.\frac{\partial f}{\partial \hat{v}_r}\right|_{t,\mathbf{x}}. \tag{3.14}$$

One can also prove that

$$
\begin{aligned}
e_{\alpha\beta\gamma} v_\beta B_\gamma \frac{\partial}{\partial v_\alpha} &= g_{\nu\beta} e_{\alpha\beta\gamma} \hat{v}_\nu B_\gamma \frac{\partial}{\partial \hat{v}_\mu} \frac{\partial \hat{v}_\mu}{\partial v_\alpha} \\
&= g_{\nu\beta} g_{\mu\alpha} e_{\alpha\beta\gamma} \hat{v}_\nu B_\gamma \frac{\partial}{\partial \hat{v}_\mu} \\
&= \hat{\mathbf{e}}_\mu \times \hat{\mathbf{e}}_\nu \cdot \mathbf{B} \hat{v}_\nu \frac{\partial}{\partial \hat{v}_\mu}.
\end{aligned}
\tag{3.15}
$$

Using these results, the kinetic equation (3.1) is finally expressed in the coordinate system aligned to the magnetic field ($\mathbf{x}$, $\hat{\mathbf{v}}$, t) as follows

$$
\begin{aligned}
&\left.\frac{\partial f}{\partial t}\right|_{\mathbf{x},\hat{\mathbf{v}}} + g_{\beta\alpha} \hat{v}_\beta \left.\frac{\partial f}{\partial x_\alpha}\right|_{\hat{\mathbf{v}},t} \\
&+ \left(\frac{e}{m} a_\alpha g_{\beta\alpha} + g_{\gamma\delta} \hat{v}_\gamma \frac{\partial g_{\beta\delta}}{\partial t} + g_{\delta\alpha} \hat{v}_\delta g_{\gamma\mu} \hat{v}_\gamma \frac{\partial g_{\beta\mu}}{\partial x_\alpha} \right) \left.\frac{\partial f}{\partial \hat{v}_\beta}\right|_{\mathbf{x},t} \\
&= 0.
\end{aligned}
\tag{3.16}
$$

The kinetic equation in the center of fluid motion, $\mathbf{v}' = \mathbf{v} - \mathbf{V}_\perp$, can also be transformed to the field-line-aligned coordinates ($\mathbf{x}$, $\hat{\mathbf{v}}'$, t). The equation in the center of fluid motion has been derived in equation (2.82) in section 2.3. As explained afterwards, the only change needed to be made is the replacement of $\mathbf{V}$ with $\mathbf{V}_\perp$.

$$
\begin{aligned}
&\frac{df}{dt} + v'_\beta \frac{\partial f}{\partial x_\beta} + \left(\frac{e}{m} E^*_\beta - \frac{dV_{\perp\beta}}{dt} \right) \frac{\partial f}{\partial v'_\beta} \\
&= \frac{\partial V_{\perp\gamma}}{\partial x_\beta} v'_\beta \frac{\partial f_\alpha}{\partial v'_\gamma} - \frac{e}{m} e_{\beta\gamma\alpha} v'_\gamma B_\alpha \frac{\partial f}{\partial v'_\beta},
\end{aligned}
\tag{3.17}
$$

where

$$
\begin{aligned}
\frac{d}{dt} &= \left.\frac{\partial}{\partial t}\right|_{\mathbf{x},\mathbf{v}} + V_{\perp\alpha} \left.\frac{\partial}{\partial x_\alpha}\right|_{\mathbf{v},t}, \\
E^*_\alpha &\equiv E_\alpha + e_{\alpha\beta\gamma} V_{\perp\beta} \hat{B}_\gamma.
\end{aligned}
$$

Here, the prime on $\mathbf{v}$ has been dropped for brevity and it has been noted that

$$
\begin{aligned}
\frac{e}{m} e_{\beta\gamma\alpha} v'_\gamma B_\alpha \frac{\partial f}{\partial v'_\delta} \frac{\partial V_{\perp\delta}}{\partial v_\beta} &= \frac{e}{m} e_{\beta\gamma\alpha} v'_\gamma B_\alpha \frac{\partial f}{\partial v'_\delta} \frac{\partial V_{\perp\delta}}{\partial \hat{v}_3} \frac{\partial \hat{v}_3}{\partial v_\beta} \\
&= \frac{e}{m} e_{\beta\gamma\alpha} g_{3\beta} v'_\gamma B_\alpha \frac{\partial f}{\partial v'_\delta} \frac{\partial V_{\perp\delta}}{\partial \hat{v}_3} \\
&= \frac{e}{m} \mathbf{b} \cdot \mathbf{v}' \times \mathbf{B} \frac{\partial f}{\partial v'_\delta} \frac{\partial V_{\perp\delta}}{\partial \hat{v}_3} = 0.
\end{aligned}
\tag{3.18}
$$

Noting also that $V_\perp \ll \hat{v}_3$—the validity condition for fluid description, one can conclude that the $\frac{\partial V_{\perp\delta}}{\partial v_\beta}$ contributions to the last term on the left and the first term on the right-hand side of equation (3.17) can also be neglected.

We can now transform equation (3.17) to the field-line-aligned coordinates term by term as follows.

$$\begin{aligned}\frac{df}{dt} &= \left.\frac{\partial f}{\partial t}\right|_{\mathbf{x},\hat{\mathbf{v}}} + g_{\delta\nu}\hat{v}_\delta \left.\frac{\partial g_{\mu\nu}}{\partial t}\frac{\partial f}{\partial \hat{v}_\mu}\right|_{\mathbf{x},t} \\ &\quad + g_{lq}\hat{V}_{\perp l}\left(\left.\frac{\partial f}{\partial x_q}\right|_{t,\hat{\mathbf{v}}} + g_{\delta\nu}\hat{v}_\delta \left.\frac{\partial g_{\mu\nu}}{\partial x_q}\frac{\partial f}{\partial \hat{v}_\mu}\right|_{\mathbf{x},t}\right), \\ &= \left.\frac{\hat{d}f_{\hat{a}}}{dt}\right|_{\mathbf{x},\hat{\mathbf{v}}} + g_{\delta\nu}\hat{v}_\delta \left.\frac{\hat{d}g_{\mu\nu}}{dt}\frac{\partial f}{\partial \hat{v}_\mu}\right|_{\mathbf{x},t},\end{aligned}$$

where

$$\frac{\hat{d}}{dt} \equiv \left.\frac{\partial}{\partial t}\right|_{\mathbf{x},\hat{\mathbf{v}}} + g_{\beta\alpha}\hat{V}_{\perp\beta}\left.\frac{\partial}{\partial x_\alpha}\right|_{\hat{\mathbf{v}},t}.$$

Because $\mathbf{V}_\perp$ is independent of $\mathbf{v}$, one has

$$\frac{dV_{\perp l}}{dt} = \frac{\hat{d}g_{l\beta}\hat{V}_{\perp l}}{dt}.$$

Furthermore,

$$\begin{aligned}\frac{\partial V_{\perp\gamma}}{\partial x_\beta} v_\beta \frac{\partial f}{\partial v_\gamma} &= \frac{\partial g_{l\gamma}\hat{V}_{\perp l}}{\partial x_\beta} g_{p\beta}\hat{v}_p g_{r\gamma}\left.\frac{\partial f}{\partial \hat{v}_r}\right|_{\mathbf{x},t}, \\ -\frac{e}{m}e_{\beta\gamma\alpha}v_\gamma B_\alpha \frac{\partial f}{\partial v_\beta} &= -\frac{e}{m}e_{\beta\gamma\alpha}g_{l\gamma}\hat{v}_l B_\alpha g_{r\beta}\left.\frac{\partial f}{\partial \hat{v}_r}\right|_{\mathbf{x},t}.\end{aligned}$$

Using these results and equations (3.13) and (3.14), equation (3.17) is transformed to

$$\begin{aligned}&\frac{\hat{d}f}{dt} + g_{\beta\alpha}\hat{v}_\beta \left.\frac{\partial f}{\partial x_\alpha}\right|_{\hat{\mathbf{v}},t} + \left(\frac{e}{m}E^*_\beta - \frac{\hat{d}g_{l\beta}\hat{V}_{\perp l}}{dt}\right) g_{r\beta}\left.\frac{\partial f}{\partial \hat{v}_r}\right|_{\mathbf{x},t} \\ &= g_{p\beta}g_{r\gamma}\hat{v}_p \frac{\partial g_{l\gamma}\hat{V}_{\perp l}}{\partial x_\beta}\left.\frac{\partial f}{\partial \hat{v}_r}\right|_{\mathbf{x},t} - \frac{e}{m}e_{\beta\gamma\alpha}g_{l\gamma}\hat{v}_l B_\alpha g_{r\beta}\left.\frac{\partial f}{\partial \hat{v}_r}\right|_{\mathbf{x},t} \\ &\quad - \left(g_{\gamma\delta}\hat{v}_\gamma \frac{\hat{d}g_{\beta\delta}}{dt} + g_{\delta\alpha}g_{\gamma\mu}\hat{v}_\delta\hat{v}_\gamma \frac{\partial g_{\beta\mu}}{\partial x_\alpha}\right)\left.\frac{\partial f}{\partial \hat{v}_\beta}\right|_{\mathbf{x},t},\end{aligned} \tag{3.19}$$

where

$$E_\alpha^* = E_\alpha + e_{\alpha\gamma\delta} g_{q\gamma} \hat{V}_{\perp q} B_\delta.$$

Note equation (3.19) is given in the $(\mathbf{x}, \hat{\mathbf{v}}, t)$ coordinates. The derivatives with respect to $\mathbf{x}$, $\hat{\mathbf{v}}$, and t in this equation are mutually independent. Consequently, its perpendicular moments can be constructed straightforwardly.

Different from Braginskii's formulation, in which a full Maxwellian is used as the lowest order distribution function, we assume that the distribution function of lowest order is given by the perpendicular Maxwellian

$$f^0 = n\frac{m}{2\pi T} \exp\left\{-\frac{m\hat{v}'^2_\perp/2}{T}\right\}. \tag{3.20}$$

Since we only construct the perpendicular moments, the dependence of parallel velocity can be left to be open. In the gyrokinetic theory, the lowest order equilibrium distribution function is assumed to be a function of the magnetic moment μ [5–9]. Note that

$$\mu = \frac{v_\perp^2}{2B(x)} = \frac{v_\perp^2}{2B(x_0)}\left(1 - \left.\frac{d \ln B}{dx}\right|_{x_0} \delta x + \cdots\right). \tag{3.21}$$

Note that δx is of the order of Larmor radius and in the perpendicular fluid description, the FLR effects are treated in next order, just as with the gyroviscous tensor, etc. Therefore, using the perpendicular Maxwellian in equation (3.20) as the lowest order distribution is consistent with the gyrokinetic ordering. Note also that in equation (3.21) $v_\perp$ is used, instead of $v'_\perp$. Since $\mathbf{V}_\perp$ in leading order is the $\mathbf{E} \times \mathbf{B}$ drift, the difference between $v_\perp$ evaluated at the guiding center and $v'_\perp$ is again of next order in the small Larmor radius ordering. The perpendicular magnetofluid formalism presented in this chapter allows being extended to other types of distribution functions, such as the slowing down distribution function [10]. Note that the distribution in the parallel velocities is close to the Maxwellian, i.e. $n \sim \exp\{-m\hat{v}_3^2/2T\}$, as also employed in the transport theory in the collisionless limit [11]. The $\hat{v}_3$ dependence is mainly absorbed in n so that $\partial \ln n/\partial\hat{v}_3 \gg \partial \ln T/\partial\hat{v}_3$ for equilibrium. This will be further discussed and confirmed when the theory is applied for the equilibrium and stability analyses in the following chapters.

3.2 Moments of kinetic equation

In this section, we derive the general perpendicular moment equation. The difference from the Braginskii formalism described in the previous chapter lies in that the moments are introduced only to the direction perpendicular to the magnetic field, while the particle mobility along the field lines remains unrestricted.

We define the perpendicular density as

$$n(\mathbf{x}, \hat{v}_3, t) = \int\int d\hat{v}_1\, d\hat{v}_2\, f. \tag{3.22}$$

The general perpendicular moment for $\eta(\mathbf{x}, \hat{\mathbf{v}}_\perp)$ is then defined as follows

$$\langle\eta\rangle = \frac{1}{n}\int\int d\hat{v}_1\, d\hat{v}_2 \eta f.$$

To derive the general moment equation, we multiply the kinetic equation (3.19) with a phase-space variable η and integrate it over the velocity space perpendicular to the magnetic field. Note that

$$\int \eta \frac{\partial f}{\partial(x, t)}\bigg|_{\hat{\mathbf{v}}} d\hat{\mathbf{v}}_\perp = \frac{\partial(n\langle\eta\rangle)}{\partial(x, t)} - n\left\langle \frac{\partial \eta}{\partial(x, t)} \right\rangle,$$

$$\begin{aligned}
\int \eta \frac{\partial f}{\partial \hat{v}_\beta} d\hat{\mathbf{v}}_\perp &= -n\left\langle \frac{\partial \eta}{\partial \hat{v}_\beta} \right\rangle + \delta_{\beta 3}\int \frac{\partial \eta f}{\partial \hat{v}_3} d\hat{\mathbf{v}}_\perp \\
&= -n\left\langle \frac{\partial \eta}{\partial \hat{v}_\beta} \right\rangle + \delta_{\beta 3}\frac{\partial n\langle\eta\rangle}{\partial \hat{v}_3}, \quad \int \eta \hat{v}_p \frac{\partial f}{\partial \hat{v}_\beta} d\hat{\mathbf{v}}_\perp \\
&= -n\left(\left\langle \hat{v}_p \frac{\partial \eta}{\partial \hat{v}_\beta} \right\rangle + \langle\eta\rangle\delta_{p\beta}\right) + \delta_{\beta 3}\frac{\partial n\langle\eta\hat{v}_p\rangle}{\partial \hat{v}_3},
\end{aligned}$$

and

$$\begin{aligned}
\int \eta \hat{v}_p \hat{v}_q \frac{\partial f}{\partial \hat{v}_\beta} d\hat{\mathbf{v}}_\perp = &-n\left(\left\langle \hat{v}_p \hat{v}_q \frac{\partial \eta}{\partial \hat{v}_\beta} \right\rangle + \langle\eta\hat{v}_q\rangle\delta_{p\beta} + \langle\eta\hat{v}_p\rangle\delta_{q\beta}\right) \\
&+ \delta_{\beta 3}\frac{\partial n\langle\eta\hat{v}_p\hat{v}_q\rangle}{\partial \hat{v}_3}.
\end{aligned}$$

Here, different from the Braginskii formulation, the terms with derivative with respect to $\hat{v}_3$ appear. We also note that

$$\frac{e}{m}e_{\beta\gamma\alpha}g_{l\gamma}B_\alpha g_{r\beta}\delta_{rl} = \frac{e}{m}e_{\beta\gamma\alpha}g_{r\gamma}B_\alpha g_{r\beta} = \frac{e}{m}e_{\beta\gamma\alpha}B_\alpha\delta_{\gamma\beta} = 0,$$

Using these results and equation (3.13); one can obtain the general moment equation:

$$
\begin{aligned}
&\frac{\partial}{\partial t}(n\langle\eta\rangle) - n\left\langle\frac{\partial\eta}{\partial t}\right\rangle + \frac{\partial}{\partial x_\alpha}\left(n\left\langle g_{\beta\alpha}\hat{V}_{\perp\beta}\eta\right\rangle\right) - n\left\langle\frac{\partial}{\partial x_\alpha}\left(g_{\beta\alpha}\hat{V}_{\perp\beta}\eta\right)\right\rangle \\
&+ \frac{\partial}{\partial x_\alpha}(g_{\beta\alpha}n\langle\eta\hat{v}_\beta\rangle) - n\left\langle\hat{v}_\beta\frac{\partial}{\partial x_\alpha}(\eta g_{\beta\alpha})\right\rangle \\
&- \left(\frac{e}{m}E^*_\beta - \frac{\hat{d}g_{l\beta}\hat{V}_{\perp l}}{dt}\right)g_{r\beta}n\left\langle\frac{\partial\eta}{\partial\hat{v}_r}\right\rangle \\
&= -g_{p\beta}g_{r\gamma}\frac{\partial g_{l\gamma}\hat{V}_{\perp l}}{\partial x_\beta}n\left(\left\langle\hat{v}_p\frac{\partial\eta}{\partial\hat{v}_r}\right\rangle + \langle\eta\rangle\delta_{pr}\right) + \frac{e}{m}e_{\beta\gamma\alpha}g_{l\gamma}B_\alpha g_{r\beta}n\left\langle\frac{\partial\eta}{\partial\hat{v}_r}\hat{v}_l\right\rangle \\
&+ g_{p\delta}\frac{\hat{d}g_{\beta\delta}}{dt}n\left(\left\langle\hat{v}_p\frac{\partial\eta}{\partial\hat{v}_\beta}\right\rangle + \langle\eta\rangle\delta_{p\beta}\right) \\
&+ g_{p\alpha}g_{q\mu}\frac{\partial g_{\beta\mu}}{\partial x_\alpha}n\left(\left\langle\hat{v}_p\hat{v}_q\frac{\partial\eta}{\partial\hat{v}_\beta}\right\rangle + \left\langle\eta\hat{v}_q\right\rangle\delta_{p\beta}\right) \\
&- \left(\frac{e}{m}E^*_\beta - \frac{\hat{d}g_{l\beta}\hat{V}_{\perp l}}{dt}\right)g_{3\beta}\frac{\partial n\langle\eta\rangle}{\partial\hat{v}_3} + g_{p\beta}g_{3\gamma}\frac{\partial g_{l\gamma}\hat{V}_{\perp l}}{\partial x_\beta}\frac{\partial n\left\langle\eta\hat{v}_p\right\rangle}{\partial\hat{v}_3} \\
&- g_{p\delta}\frac{\hat{d}g_{3\delta}}{dt}\frac{\partial n\left\langle\eta\hat{v}_p\right\rangle}{\partial\hat{v}_3} - g_{p\alpha}g_{q\mu}\frac{\partial g_{3\mu}}{\partial x_\alpha}\frac{\partial n\left\langle\eta\hat{v}_p\hat{v}_q\right\rangle}{\partial\hat{v}_3}.
\end{aligned}
\tag{3.23}
$$

It is interesting to compare this moment equation with the conventional one in equation (2.12) in section 2.1.2. Since equation (3.23) is just a perpendicular moment equation, one may integrate over $\hat{v}_3$ to get the conventional moment equation. This integration leads the last four terms on the right-hand side to vanish, using the boundary conditions for $\hat{v}_3 \to \infty$. Comparison for most of the terms is trivial. For simplicity, we limit ourselves to the case with $\hat{\mathbf{V}}_\perp = 0$. Therefore, one just needs to interpret the third and fourth terms on the right. Note here that the time and spatial derivatives are based on holding $\hat{\mathbf{v}}$ constant in equation (3.23). This is different from the conventional one in equation (2.12), in which $\mathbf{v}$ is held constant instead. Therefore, the comparison needs a suitable coordinate transform.

First, let us consider the combination of the second term on the left and the third term on the right in equation (3.23). The second term on the left can be transformed as follows

$$-n\left\langle \frac{\partial \eta}{\partial t}\Big|_{\hat{\mathbf{v}}} \right\rangle = -n\left\langle \frac{\partial \eta}{\partial t}\Big|_{\mathbf{v}} \right\rangle - n\left\langle \frac{\partial \eta}{\partial v_\alpha}\frac{\partial v_\alpha}{\partial t}\Big|_{\hat{\mathbf{v}}} \right\rangle$$
$$= -n\left\langle \frac{\partial \eta}{\partial t}\Big|_{\mathbf{v}} \right\rangle - n\left\langle \frac{\partial \eta}{\partial v_\alpha}\frac{\partial g_{\beta\alpha}}{\partial t}\hat{v}_\beta \right\rangle$$
$$= -n\left\langle \frac{\partial \eta}{\partial t}\Big|_{\mathbf{v}} \right\rangle - g_{\beta\gamma}\frac{\partial g_{\beta\alpha}}{\partial t}n\left\langle v_\gamma\frac{\partial \eta}{\partial v_\alpha} \right\rangle.$$

Meanwhile, the third term on the right becomes

$$g_{p\delta}\frac{\partial g_{\beta\delta}}{\partial t}n\left\langle \hat{v}_p\frac{\partial \eta}{\partial \hat{v}_\beta} \right\rangle = g_{p\delta}\frac{\partial g_{\beta\delta}}{\partial t}n\left\langle g_{p\alpha}v_\alpha\frac{\partial \eta}{\partial v_\gamma}\frac{\partial v_\gamma}{\partial \hat{v}_\beta} \right\rangle$$
$$= g_{\beta\gamma}\frac{\partial g_{\beta\alpha}}{\partial t}n\left\langle v_\alpha\frac{\partial \eta}{\partial v_\gamma} \right\rangle$$
$$= -g_{\beta\alpha}\frac{\partial g_{\beta\gamma}}{\partial t}n\left\langle v_\alpha\frac{\partial \eta}{\partial v_\gamma} \right\rangle.$$

Noting the cancellations, these two terms reduce to the second term in the conventional moment equation (2.12).

Next, let us consider the combination of the sixth term on the left and the fourth term on the right in equation (3.23). The sixth term on the left can be transformed as follows

$$-n\left\langle \hat{v}_\beta\frac{\partial(\eta g_{\beta\alpha})}{\partial x_\alpha}\Big|_{\hat{\mathbf{v}}} \right\rangle$$
$$= -g_{\beta\alpha}n\left\langle \hat{v}_\beta\frac{\partial \eta}{\partial x_\alpha}\Big|_{\hat{\mathbf{v}}} \right\rangle - \frac{\partial g_{\beta\alpha}}{\partial x_\alpha}n\langle \hat{v}_\beta\eta\rangle$$
$$= -n\left\langle v_\alpha\frac{\partial \eta}{\partial x_\alpha}\Big|_{\mathbf{v}} \right\rangle - n\left\langle v_\alpha\frac{\partial \eta}{\partial v_\beta}\frac{\partial v_\beta}{\partial x_\alpha}\Big|_{\hat{\mathbf{v}}} \right\rangle - \frac{\partial g_{\beta\alpha}}{\partial x_\alpha}n\langle \hat{v}_\beta\eta\rangle$$
$$= -n\left\langle v_\alpha\frac{\partial \eta}{\partial x_\alpha}\Big|_{\mathbf{v}} \right\rangle - \frac{\partial g_{\gamma\beta}}{\partial x_\alpha}n\left\langle v_\alpha\hat{v}_\gamma\frac{\partial \eta}{\partial v_\beta} \right\rangle - \frac{\partial g_{\beta\alpha}}{\partial x_\alpha}n\langle \hat{v}_\beta\eta\rangle$$
$$= -n\left\langle v_\alpha\frac{\partial \eta}{\partial x_\alpha}\Big|_{\mathbf{v}} \right\rangle - g_{\gamma\delta}\frac{\partial g_{\gamma\beta}}{\partial x_\alpha}n\left\langle v_\alpha v_\delta\frac{\partial \eta}{\partial v_\beta} \right\rangle - \frac{\partial g_{\beta\alpha}}{\partial x_\alpha}n\langle \hat{v}_\beta\eta\rangle.$$

The fourth term on the right becomes

$$
\begin{aligned}
& g_{p\alpha} g_{q\mu} \frac{\partial g_{\beta\mu}}{\partial x_\alpha} n \left(\left\langle \hat{v}_p \hat{v}_q \frac{\partial \eta}{\partial \hat{v}_\beta} \right\rangle + \left\langle \eta \hat{v}_q \right\rangle \delta_{p\beta} \right) \\
&= \frac{\partial g_{\beta\mu}}{\partial x_\alpha} n \left\langle v_\alpha v_\mu \frac{\partial \eta}{\partial v_\gamma} \frac{\partial v_\gamma}{\partial \hat{v}_\beta} \right\rangle + g_{\beta\alpha} g_{q\mu} \frac{\partial g_{\beta\mu}}{\partial x_\alpha} n \left\langle \eta \hat{v}_q \right\rangle \\
&= g_{\beta\gamma} \frac{\partial g_{\beta\mu}}{\partial x_\alpha} n \left\langle v_\alpha v_\mu \frac{\partial \eta}{\partial v_\gamma} \right\rangle - g_{\beta\mu} g_{q\mu} \frac{\partial g_{\beta\alpha}}{\partial x_\alpha} n \left\langle \eta \hat{v}_q \right\rangle \\
&= - g_{\beta\mu} \frac{\partial g_{\beta\gamma}}{\partial x_\alpha} n \left\langle v_\alpha v_\mu \frac{\partial \eta}{\partial v_\gamma} \right\rangle - \frac{\partial g_{\beta\alpha}}{\partial x_\alpha} n \left\langle \eta \hat{v}_\beta \right\rangle .
\end{aligned}
$$

Noting the cancellations, these two terms reduce to the fourth term in the conventional moment equation (2.12).

3.2.1 Definitions of fluid quantities

Before deriving the individual moment equations, let us first introduce the relevant fluid moments in this subsection.

The density has been defined in equation (3.22). The velocity, temperature, and pressure are defined according to

$$0 = \langle \hat{v}_\alpha \rangle, \tag{3.24}$$

$$T = \frac{m}{2} \left\langle \hat{v}_\alpha^2 \right\rangle, \tag{3.25}$$

$$p = nT. \tag{3.26}$$

Here, equation (3.24) just indicates that the average perpendicular velocity is $\hat{\mathbf{V}}_\perp$. We also introduce the fourth order thermal parameter, referred to as T-square scalar,

$$D = \frac{m_a^2}{4} \langle \hat{v}^4 \rangle. \tag{3.27}$$

The Maxwellian distribution function in equation (3.20) contribute to all of these fluid quantities. Direct proofs are given as follows.

$$
\begin{aligned}
n = \langle 1 \rangle_0 &= \int\!\!\int d\hat{v}_1 \, d\hat{v}_2 \, f^0 \\
&= \frac{m}{2\pi T} \int\!\!\int d\hat{v}_1 \, d\hat{v}_2 n \exp\left\{ -\frac{m\hat{v}_\perp^2/2}{T} \right\} \\
&= n,
\end{aligned} \tag{3.28}
$$

$$0 = \langle v_{\perp\alpha} \rangle_0 = \frac{1}{n} \int\!\!\int d\hat{v}_1 \, d\hat{v}_2 v_{\perp\alpha} f^0, \tag{3.29}$$

$$T = \left\langle \frac{1}{2} m \hat{v}_\perp^2 \right\rangle_0 = \frac{1}{n} \iint d\hat{v}_1 \, d\hat{v}_2 \frac{1}{2} m \hat{v}_\perp^2 f^0 = T, \tag{3.30}$$

$$D = \left\langle \frac{1}{4} m^2 \hat{v}_\perp^4 \right\rangle_0 = \frac{1}{n} \iint d\hat{v}_1 \, d\hat{v}_2 \frac{1}{4} m^2 \hat{v}_\perp^4 f^0 = 2T^2. \tag{3.31}$$

Similar to the constraints introduced in the Braginskii theory in equation (2.135), the following constraints are introduced

$$\langle 1 \rangle_1 = \langle \hat{v}_{\perp\alpha} \rangle_1 = \left\langle \hat{v}_\perp^2 \right\rangle_1 = \left\langle \hat{v}_\perp^4 \right\rangle_1 = 0, \tag{3.32}$$

where $\langle \cdots \rangle_1$ denotes the perpendicular velocity space average with the first order distribution function f^1. They will be confirmed in section 3.4.

We also define the next order fluid quantities, such as viscosity $\pi_{\alpha\beta}$, heat flux q_α, etc.

$$\pi_{\alpha\beta} = \rho \left\langle \hat{v}_{\perp\alpha} \hat{v}_{\perp\beta} - \frac{1}{2} \hat{v}_\perp^2 \delta_{\alpha\beta}^\perp \right\rangle. \tag{3.33}$$

$$q_\alpha = \frac{1}{2} \rho \left\langle \hat{v}_{\perp\nu}^2 \hat{v}_{\perp\alpha} \right\rangle, \tag{3.34}$$

$$\Theta_{\alpha\beta\gamma} = \rho \left\langle \left(\hat{v}_{\perp\alpha} \hat{v}_{\perp\beta} - \frac{1}{2} v'^2_\perp \delta_{\alpha\beta}^\perp \right) \hat{v}_{\perp\gamma} \right\rangle, \tag{3.35}$$

$$\Xi_{\alpha\beta} = \frac{m^2}{2} \left\langle \hat{v}_\perp{}'^2 \left(\hat{v}_{\perp\alpha} \hat{v}_{\perp\beta} - \frac{1}{2} \hat{v}_\perp^2 \delta_{\alpha\beta}^\perp \right) \right\rangle. \tag{3.36}$$

Here, $\Theta_{\alpha\beta\gamma}$ and $\Xi_{\alpha\beta}$ are referred to as the heat flux tensor and T-square tensor, respectively. One can prove that the Maxwellian distribution function in equation (3.20) does not contribute to these quantities, i.e. $\pi_{\alpha\beta 0} = q_{\alpha 0} = \Theta_{\alpha\beta\gamma 0} = \Theta_{\alpha\beta\gamma 0} = 0$

Using these definitions, one has

$$\rho \left\langle \hat{v}_{\perp\alpha} \hat{v}_{\perp\beta} \right\rangle = p \delta_{\alpha\beta}^\perp + \pi_{\alpha\beta}, \tag{3.37}$$

$$\rho \left\langle \hat{v}_{\perp\alpha} \hat{v}_{\perp\beta} \hat{v}_{\perp\gamma} \right\rangle = q_\gamma \delta_{\alpha\beta}^\perp + \Theta_{\alpha\beta\gamma}, \tag{3.38}$$

$$\frac{1}{2} m^2 \left\langle \hat{v}_\perp^2 \hat{v}_{\perp\alpha} \hat{v}_{\perp\beta} \right\rangle = D \delta_{\alpha\beta}^\perp + \Xi_{\alpha\beta}. \tag{3.39}$$

Here, $\delta_{\alpha\beta}^\perp$ $(\alpha, \beta < 3)$ is introduced, which equals $\delta_{\alpha\beta}$ except that α and β do not include the third component 3. This leads to the pressure p in equation (3.37), q_α in equation (3.38), and the parameter D in equation (3.39) appearing only in the perpendicular direction. To make the traces zero in the definitions in equations (3.33), (3.35), and (3.36), the factor 1/2 is introduced. This is different from the isotropic case in the Braginskii two-fluid theory, equation (2.17), where the factor 1/3 is introduced.

We have introduced higher order fluid variables. They are necessary for the perpendicular magnetofluid description. The definitions in this subsection will appear in the individual moment equations to be derived.

3.2.2 Continuity equation

In this subsection, the continuity equation in the perpendicular magnetofluid formalism is derived. This can be obtained by letting $\eta = 1$ in the general moment equation (3.23). Equations (3.2)–(3.6) are used in the simplification.

The term-by-term simplifications are given as follows. Only the terms having actual contributions to the continuity equation are detailed.

$$\frac{\partial}{\partial t}(n\langle\eta\rangle) - n\left\langle\frac{\partial\eta}{\partial t}\right\rangle + \frac{\partial}{\partial x_\alpha}\left(n\left\langle g_{\beta\alpha}\hat{V}_{\perp\beta}\eta\right\rangle\right) - n\left\langle\frac{\partial}{\partial x_\alpha}\left(g_{\beta\alpha}\hat{V}_{\perp\beta}\eta\right)\right\rangle$$
$$= \frac{\partial n}{\partial t} + g_{\beta\alpha}\hat{V}_{\perp\beta}\frac{\partial n}{\partial x_\alpha},$$

$$\frac{\partial}{\partial x_\alpha}\left(g_{\beta\alpha}n\left\langle\eta\hat{v}_\beta\right\rangle\right) - n\left\langle\hat{v}_\beta\frac{\partial}{\partial x_\alpha}(\eta g_{\beta\alpha})\right\rangle$$
$$= \frac{\partial}{\partial x_\alpha}(g_{3\alpha}n\hat{v}_3) - n\hat{v}_3\frac{\partial g_{3\alpha}}{\partial x_\alpha} = g_{3\alpha}\hat{v}_3\frac{\partial n}{\partial x_\alpha},$$

$$-g_{p\beta}g_{r\gamma}\frac{\partial g_{l\gamma}\hat{V}_{\perp l}}{\partial x_\beta}n\left(\left\langle\tilde{v}_p\frac{\partial\eta}{\partial\tilde{v}_r}\right\rangle + \langle\eta\rangle\delta_{pr}\right) = -g_{r\beta}g_{r\gamma}\frac{\partial g_{l\gamma}\hat{V}_{\perp l}}{\partial x_\beta}n = -\frac{\partial g_{l\beta}\hat{V}_{\perp l}}{\partial x_\beta}n,$$

$$g_{p\delta}\frac{\hat{d}g_{\beta\delta}}{dt}n\left(\left\langle\hat{v}_p\frac{\partial\eta}{\partial\hat{v}_\beta}\right\rangle + \langle\eta\rangle\delta_{p\beta}\right) = \frac{n}{2}\frac{\hat{d}g_{\beta\delta}^2}{dt} = 0,$$

$$g_{p\alpha}g_{q\mu}\frac{\partial g_{\beta\mu}}{\partial x_\alpha}n\left(\left\langle\hat{v}_p\hat{v}_q\frac{\partial\eta}{\partial\hat{v}_\beta}\right\rangle + \left\langle\eta\hat{v}_q\right\rangle\delta_{p\beta}\right)$$
$$= ng_{\beta\alpha}g_{3\mu}\hat{v}_3\frac{\partial g_{\beta\mu}}{\partial x_\alpha} = -ng_{\beta\mu}g_{3\mu}\hat{v}_3\frac{\partial g_{\beta\alpha}}{\partial x_\alpha} = -n\hat{v}_3\frac{\partial g_{3\alpha}}{\partial x_\alpha},$$

where $\langle\hat{\mathbf{v}}_\perp\rangle = 0$ in equation (3.24) has been used,

$$-\left(\frac{e}{m}E^*_\beta - \frac{\hat{d}g_{l\beta}\hat{V}_{\perp l}}{dt}\right)g_{3\beta}\frac{\partial n\langle\eta\rangle}{\partial\hat{v}_3} = -\left(\frac{e}{m}E^*_\beta - \frac{\hat{d}g_{l\beta}\hat{V}_{\perp l}}{dt}\right)g_{3\beta}\frac{\partial n}{\partial\hat{v}_3},$$

$$g_{p\beta}g_{3\gamma}\frac{\partial g_{l\gamma}\hat{V}_{\perp l}}{\partial x_\beta}\frac{\partial n\left\langle\hat{v}_p\eta\right\rangle}{\partial\hat{v}_3} = g_{p\beta}g_{3\gamma}\frac{\partial g_{l\gamma}\hat{V}_{\perp l}}{\partial x_\beta}\frac{\partial n\hat{v}_3}{\partial\hat{v}_3}\delta_{p3}$$
$$= g_{3\beta}g_{3\gamma}\frac{\partial g_{l\gamma}\hat{V}_{\perp l}}{\partial x_\beta}\hat{v}_3\frac{\partial n}{\partial\hat{v}_3} + ng_{3\beta}g_{3\gamma}\frac{\partial g_{l\gamma}\hat{V}_{\perp l}}{\partial x_\beta},$$

$$-g_{p\delta}\frac{\hat{d}g_{3\delta}}{dt}\frac{\partial n\langle\eta\hat{v}_p\rangle}{\partial\hat{v}_3} = -g_{3\delta}\frac{\hat{d}g_{3\delta}}{dt}\frac{\partial n\langle\hat{v}_3\rangle}{\partial\hat{v}_3} = 0,$$

and

$$\begin{aligned}
&- g_{p\alpha}g_{q\mu}\frac{\partial g_{3\mu}}{\partial x_\alpha}\frac{\partial n\langle\eta\hat{v}_p\hat{v}_q\rangle}{\partial\hat{v}_3}\\
&= - g_{p\alpha}g_{q\mu}\frac{\partial g_{3\mu}}{\partial x_\alpha}\frac{\partial n\langle\hat{v}_{\perp p}\hat{v}_{\perp q}\rangle}{\partial\hat{v}_3} - g_{3\alpha}g_{3\mu}\frac{\partial g_{3\mu}}{\partial x_\alpha}\frac{\partial n\langle\hat{v}_3\hat{v}_3\rangle}{\partial\hat{v}_3}\\
&= - \frac{1}{m}g_{p\alpha}g_{q\mu}\frac{\partial g_{3\mu}}{\partial x_\alpha}\frac{\partial}{\partial\hat{v}_3}\left(p\delta^{\perp}_{pq} + \pi_{pq}\right)\\
&= - \frac{1}{m}(g_{p\alpha}g_{p\mu} - g_{3\alpha}g_{3\mu})\frac{\partial g_{3\mu}}{\partial x_\alpha}\frac{\partial p}{\partial\hat{v}_3} - \frac{1}{m}g_{p\alpha}g_{q\mu}\frac{\partial g_{3\mu}}{\partial x_\alpha}\frac{\partial\pi_{pq}}{\partial\hat{v}_3}\\
&= - \frac{1}{m}\frac{\partial g_{3\alpha}}{\partial x_\alpha}\frac{\partial nT}{\partial\hat{v}_3} - \frac{1}{m}g_{p\alpha}g_{q\mu}\frac{\partial g_{3\mu}}{\partial x_\alpha}\frac{\partial\pi_{pq}}{\partial\hat{v}_3}.
\end{aligned}$$

Using these results, the continuity equation becomes

$$\begin{aligned}
&\frac{\partial n}{\partial t} + g_{\beta\alpha}\hat{V}_{\perp\beta}\frac{\partial n}{\partial x_\alpha} + g_{3\alpha}\hat{v}_3\frac{\partial n}{\partial x_\alpha} + n\frac{\partial g_{l\gamma}\hat{V}_{\perp l}}{\partial x_\beta}\left(\delta_{\beta\gamma} - g_{3\beta}g_{3\gamma}\right)\\
&+ n\hat{v}_3\frac{\partial g_{3\alpha}}{\partial x_\alpha} + \frac{1}{m}\frac{\partial g_{3\alpha}}{\partial x_\alpha}\frac{\partial nT}{\partial\hat{v}_3} - g_{3\beta}g_{3\gamma}\frac{\partial g_{l\gamma}\hat{V}_{\perp l}}{\partial x_\beta}\hat{v}_3\frac{\partial n}{\partial\hat{v}_3}\\
&= - g_{3\beta}\frac{\partial n}{\partial\hat{v}_3}\left(\frac{e}{m}E^*_\beta - \frac{\hat{d}g_{l\beta}\hat{V}_{\perp l}}{dt}\right) - \frac{1}{m}g_{p\alpha}g_{q\mu}\frac{\partial g_{3\mu}}{\partial x_\alpha}\frac{\partial\pi_{pq}}{\partial\hat{v}_3}.
\end{aligned} \tag{3.40}$$

Note here that, since $\hat{v}_3$ and x_α are independent, one has

$$\begin{aligned}
&n\hat{v}_3\frac{\partial g_{3\alpha}}{\partial x_\alpha} + n\frac{\partial g_{l\gamma}\hat{V}_{\perp l}}{\partial x_\beta}\left(\delta_{\beta\gamma} - g_{3\beta}g_{3\gamma}\right)\\
&= n\frac{\partial g_{l\gamma}(\hat{V}_{\perp l} + \delta_{l3}\hat{v}_3)}{\partial x_\beta}\left(\delta_{\beta\gamma} - g_{3\beta}g_{3\gamma}\right).
\end{aligned}$$

Here, one can see that only the perpendicular fluid compression through $\hat{\mathbf{V}}_\perp$ appears in equation (3.40). The parallel one remains to be described kinetically through the parallel particle velocity $\hat{v}_3$. Because of this hybrid feature, the FLR correction π_{pq} is involved in the continuity equation.

3.2.3 Momentum transport equation

In this subsection, the momentum transport equation in the perpendicular magnetofluid formalism is derived. This can be obtained by letting $\eta = mg_{\nu\lambda}\hat{v}_{\perp\nu} = mg_{\nu\lambda}\hat{v}_l\delta^{\perp}_{l\nu}$ in the general moment equation (3.23). Equations (3.2)–(3.6) are used in the simplification.

The term-by-term simplifications are given as follows. First, we note that using equation (3.24) one can exclude the following terms

$$\frac{\partial}{\partial t}(n\langle\eta\rangle) - n\left\langle\frac{\partial\eta}{\partial t}\right\rangle + \frac{\partial}{\partial x_\alpha}\left(n\left\langle g_{\beta\alpha}\hat{V}_{\perp\beta}\eta\right\rangle\right) - n\left\langle\frac{\partial}{\partial x_\alpha}\left(g_{\beta\alpha}\hat{V}_{\perp\beta}\eta\right)\right\rangle = 0.$$

Next, we reduce each term sequentially. To be specific to determine the signs for mutual cancellations, all terms in equation (3.23) are moved to the left-hand side.

$$\begin{aligned}
&\frac{\partial}{\partial x_\alpha}(g_{\beta\alpha}n\langle\eta\hat{v}_\beta\rangle) - n\left\langle\tilde{v}_\beta\frac{\partial}{\partial x_\alpha}(\eta g_{\beta\alpha})\right\rangle \\
&= \frac{\partial}{\partial x_\alpha}[g_{\nu\lambda}g_{\beta\alpha}\rho\langle\hat{v}_{\perp\nu}\hat{v}_\beta\rangle] - \frac{\partial g_{\nu\lambda}g_{\beta\alpha}}{\partial x_\alpha}\rho\langle\hat{v}_{\perp\nu}\hat{v}_\beta\rangle \\
&= g_{\nu\lambda}g_{\beta\alpha}\frac{\partial}{\partial x_\alpha}\rho\langle\hat{v}_{\perp\nu}\hat{v}_\beta\rangle \\
&= g_{\nu\lambda}g_{\beta\alpha}\frac{\partial}{\partial x_\alpha}\left(p\delta^\perp_{\beta\nu} + \pi_{\beta\nu}\right) \\
&= (\delta_{\lambda\alpha} - g_{3\lambda}g_{3\alpha})\frac{\partial p}{\partial x_\alpha} + g_{\nu\lambda}g_{\beta\alpha}\frac{\partial\pi_{\beta\nu}}{\partial x_\alpha},
\end{aligned}$$

$$-\left(\frac{e}{m}E^*_\beta - \frac{\hat{d}g_{l\beta}\hat{V}_{\perp l}}{dt}\right)g_{r\beta}n\left\langle\frac{\partial\eta}{\partial\hat{v}_r}\right\rangle = -\rho\left(\frac{e}{m}E^*_\beta - \frac{\hat{d}g_{l\beta}\hat{V}_{\perp l}}{dt}\right)g_{r\beta}g_{\nu\lambda}\delta^\perp_{\nu r}$$

$$= -\rho\left(\frac{e}{m}E^*_\beta - \frac{\hat{d}g_{l\beta}\hat{V}_{\perp l}}{dt}\right)\left(\delta_{\beta\lambda} - g_{3\beta}g_{3\lambda}\right),$$

$$\begin{aligned}
g_{p\beta}g_{r\gamma}\frac{\partial g_{l\gamma}\hat{V}_{\perp l}}{\partial x_\beta}n\left(\left\langle\hat{v}_p\frac{\partial\eta}{\partial\hat{v}_r}\right\rangle + \langle\eta\rangle\delta_{pr}\right) &= g_{p\beta}g_{r\gamma}\frac{\partial g_{l\gamma}\hat{V}_{\perp l}}{\partial x_\beta}\rho g_{\nu\lambda}\delta^\perp_{\nu r}\langle\hat{v}_p\rangle \\
&= \left(\delta_{\gamma\lambda} - g_{3\gamma}g_{3\lambda}\right)\rho\hat{v}_3 g_{3\beta}\frac{\partial g_{l\gamma}\hat{V}_{\perp l}}{\partial x_\beta},
\end{aligned}$$

$$-\frac{e}{m}e_{\beta\gamma\alpha}g_{l\gamma}B_\alpha g_{r\beta}n\left\langle\frac{\partial\eta}{\partial\hat{v}_r}\hat{v}_l\right\rangle = 0,$$

$$\begin{aligned}
-g_{p\delta}\frac{\hat{d}g_{\beta\delta}}{dt}n\left(\left\langle\hat{v}_p\frac{\partial\eta}{\partial\hat{v}_\beta}\right\rangle + \langle\eta\rangle\delta_{p\beta}\right) &= -g_{p\delta}\frac{\hat{d}g_{\beta\delta}}{dt}\rho g_{\nu\lambda}\delta^\perp_{\nu\beta}\langle\hat{v}_p\rangle \\
&= g_{\beta\delta}g_{\nu\lambda}\delta^\perp_{\nu\beta}\frac{\hat{d}g_{3\delta}}{dt}\rho\hat{v}_3 \\
&= (\delta_{\delta\lambda} - g_{3\delta}g_{3\lambda})\rho\hat{v}_3\frac{\hat{d}g_{3\delta}}{dt},
\end{aligned}$$

$$
\begin{aligned}
&-g_{p\alpha}g_{q\mu}\frac{\partial g_{\beta\mu}}{\partial x_\alpha}n\left(\left\langle \hat{v}_p\hat{v}_q\frac{\partial\eta}{\partial\hat{v}_\beta}\right\rangle + \left\langle \eta\hat{v}_q\right\rangle\delta_{p\beta}\right)\\
=\;& -g_{p\alpha}g_{q\mu}\frac{\partial g_{\beta\mu}}{\partial x_\alpha}\rho\left(g_{\nu\lambda}\delta^{\perp}_{\beta\nu}\left\langle \hat{v}_p\hat{v}_q\right\rangle + g_{\nu\lambda}\left\langle \hat{v}_q\hat{v}_{\perp\nu}\right\rangle\delta_{p\beta}\right)\\
=\;& -g_{p\alpha}g_{q\mu}\frac{\partial g_{\beta\mu}}{\partial x_\alpha}\rho g_{\nu\lambda}\delta^{\perp}_{\beta\nu}\left\langle \hat{v}_p\hat{v}_q\right\rangle - g_{\beta\alpha}g_{q\mu}\frac{\partial g_{\beta\mu}}{\partial x_\alpha}\rho g_{\nu\lambda}\left\langle \hat{v}_q\hat{v}_{\perp\nu}\right\rangle\\
=\;& -g_{p\alpha}g_{q\mu}\frac{\partial g_{\beta\mu}}{\partial x_\alpha}\rho g_{\nu\lambda}\delta^{\perp}_{\beta\nu}\left\langle \hat{v}_{\perp p}\hat{v}_{\perp q}\right\rangle - g_{3\alpha}g_{3\mu}\hat{v}_3^2\frac{\partial g_{\beta\mu}}{\partial x_\alpha}\rho g_{\nu\lambda}\delta^{\perp}_{\beta\nu}\\
&-g_{\beta\alpha}g_{q\mu}\frac{\partial g_{\beta\mu}}{\partial x_\alpha}\rho g_{p\lambda}\left\langle \hat{v}_{\perp p}\hat{v}_{\perp q}\right\rangle\\
=\;& -g_{p\alpha}g_{q\mu}\frac{\partial g_{\beta\mu}}{\partial x_\alpha}g_{\beta\lambda}\rho\left\langle \hat{v}_{\perp p}\hat{v}_{\perp q}\right\rangle + g_{p\alpha}g_{q\mu}\frac{\partial g_{3\mu}}{\partial x_\alpha}g_{3\lambda}\rho\left\langle \hat{v}_{\perp p}\hat{v}_{\perp q}\right\rangle\\
&-g_{3\alpha}g_{3\mu}\hat{v}_3^2\frac{\partial g_{\beta\mu}}{\partial x_\alpha}\rho g_{\beta\lambda} - g_{\beta\alpha}g_{q\mu}\frac{\partial g_{\beta\mu}}{\partial x_\alpha}g_{p\lambda}\rho\left\langle \hat{v}_{\perp p}\hat{v}_{\perp q}\right\rangle\\
=\;& g_{p\alpha}g_{q\mu}g_{\beta\mu}\frac{\partial g_{\beta\lambda}}{\partial x_\alpha}\rho\left\langle \hat{v}_{\perp p}\hat{v}_{\perp q}\right\rangle + g_{p\alpha}g_{q\mu}\frac{\partial g_{3\mu}}{\partial x_\alpha}g_{3\lambda}\rho\left\langle \hat{v}_{\perp p}\hat{v}_{\perp q}\right\rangle\\
&+g_{3\alpha}g_{\beta\mu}g_{\beta\lambda}\hat{v}_3^2\frac{\partial g_{3\mu}}{\partial x_\alpha}\rho + g_{\beta\alpha}g_{\beta\mu}\frac{\partial g_{q\mu}}{\partial x_\alpha}g_{p\lambda}\rho\left\langle \hat{v}_{\perp p}\hat{v}_{\perp q}\right\rangle\\
=\;& g_{p\alpha}\frac{\partial g_{q\lambda}}{\partial x_\alpha}\rho\left\langle \hat{v}_{\perp p}\hat{v}_{\perp q}\right\rangle + g_{p\lambda}\frac{\partial g_{q\alpha}}{\partial x_\alpha}\rho\left\langle \hat{v}_{\perp p}\hat{v}_{\perp q}\right\rangle\\
&+g_{p\alpha}g_{q\mu}\frac{\partial g_{3\mu}}{\partial x_\alpha}g_{3\lambda}\rho\left\langle \hat{v}_{\perp p}\hat{v}_{\perp q}\right\rangle + g_{3\alpha}\hat{v}_3^2\frac{\partial g_{3\lambda}}{\partial x_\alpha}\rho\\
=\;& \frac{\partial g_{p\alpha}g_{q\lambda}}{\partial x_\alpha}\rho\left\langle \hat{v}_{\perp p}\hat{v}_{\perp q}\right\rangle + g_{p\alpha}g_{q\mu}g_{3\lambda}\frac{\partial g_{3\mu}}{\partial x_\alpha}\rho\left\langle \hat{v}_{\perp p}\hat{v}_{\perp q}\right\rangle + g_{3\alpha}\hat{v}_3^2\frac{\partial g_{3\lambda}}{\partial x_\alpha}\rho\\
=\;& \frac{\partial g_{p\alpha}g_{q\lambda}}{\partial x_\alpha}\left(p\delta^{\perp}_{pq} + \pi_{pq}\right) + g_{p\alpha}g_{q\mu}g_{3\lambda}\frac{\partial g_{3\mu}}{\partial x_\alpha}\left(p\delta^{\perp}_{pq} + \pi_{pq}\right) + g_{3\alpha}\hat{v}_3^2\frac{\partial g_{3\lambda}}{\partial x_\alpha}\rho\\
=\;& -\frac{\partial g_{3\alpha}g_{3\lambda}}{\partial x_\alpha}p + g_{3\lambda}\frac{\partial g_{3\alpha}}{\partial x_\alpha}p + g_{3\alpha}\hat{v}_3^2\frac{\partial g_{3\lambda}}{\partial x_\alpha}\rho + \left(\frac{\partial g_{p\alpha}g_{q\lambda}}{\partial x_\alpha} + g_{p\alpha}g_{q\mu}g_{3\lambda}\frac{\partial g_{3\mu}}{\partial x_\alpha}\right)\pi_{pq}\\
=\;& \underbrace{-g_{3\alpha}\frac{\partial g_{3\lambda}}{\partial x_\alpha}p}_{x1} + g_{3\alpha}\hat{v}_3^2\frac{\partial g_{3\lambda}}{\partial x_\alpha}\rho + \left(\frac{\partial g_{p\alpha}g_{q\lambda}}{\partial x_\alpha} + g_{p\alpha}g_{q\mu}g_{3\lambda}\frac{\partial g_{3\mu}}{\partial x_\alpha}\right)\pi_{pq},
\end{aligned}
$$

$$
\begin{aligned}
\left(\frac{e}{m}E^*_\beta - \frac{\hat{d}g_{l\beta}\hat{V}_{\perp l}}{dt}\right)g_{3\beta}\frac{\partial n\langle\eta\rangle}{\partial\hat{v}_3} &= \left(\frac{e}{m}E^*_\beta - \frac{\hat{d}g_{l\beta}\hat{V}_{\perp l}}{dt}\right)g_{3\beta}g_{\nu\lambda}\frac{\partial\rho\langle\hat{v}_{\perp\nu}\rangle}{\partial\hat{v}_3}\\
&= 0,
\end{aligned}
$$

$$
\begin{aligned}
&\left(g_{p\delta}\frac{\hat{d}g_{3\delta}}{dt} - g_{p\beta}g_{3\gamma}\frac{\partial g_{l\gamma}\hat{V}_{\perp l}}{\partial x_\beta}\right)\frac{\partial n\left\langle \hat{v}_p\eta\right\rangle}{\partial \hat{v}_3} \\
&= \left(g_{p\delta}\frac{\hat{d}g_{3\delta}}{dt} - g_{p\beta}g_{3\gamma}\frac{\partial g_{l\gamma}\hat{V}_{\perp l}}{\partial x_\beta}\right)g_{\nu\lambda}\frac{\partial \rho\left\langle \hat{v}_p v_{\perp\nu}\right\rangle}{\partial \hat{v}_3} \\
&= \left(g_{p\delta}\frac{\hat{d}g_{3\delta}}{dt} - g_{p\beta}g_{3\gamma}\frac{\partial g_{l\gamma}\hat{V}_{\perp l}}{\partial x_\beta}\right)g_{\nu\lambda}\frac{\partial \rho\left\langle \hat{v}_{\perp p} v_{\perp\nu}\right\rangle}{\partial \hat{v}_3} \\
&= \left(g_{p\delta}\frac{\hat{d}g_{3\delta}}{dt} - g_{p\beta}g_{3\gamma}\frac{\partial g_{l\gamma}\hat{V}_{\perp l}}{\partial x_\beta}\right)g_{\nu\lambda}\frac{\partial}{\partial \hat{v}_3}\left(p\delta_{p\nu}^{\perp} + \pi_{p\nu}\right) \\
&= \frac{\hat{d}g_{3\lambda}}{dt}\frac{\partial p}{\partial \hat{v}_3} - \left(\delta_{\beta\lambda} - g_{3\beta}g_{3\lambda}\right)g_{3\gamma}\frac{\partial g_{l\gamma}\hat{V}_{\perp l}}{\partial x_\beta}\frac{\partial p}{\partial \hat{v}_3} \\
&\quad + \left(g_{p\delta}\frac{\hat{d}g_{3\delta}}{dt} - g_{p\beta}g_{3\gamma}\frac{\partial g_{l\gamma}\hat{V}_{\perp l}}{\partial x_\beta}\right)g_{\nu\lambda}\frac{\partial \pi_{p\nu}}{\partial \hat{v}_3},
\end{aligned}
$$

and

$$
g_{p\alpha}g_{q\mu}\frac{\partial g_{3\mu}}{\partial x_\alpha}\frac{\partial n\left\langle \eta\hat{v}_p\hat{v}_q\right\rangle}{\partial \hat{v}_3} = mg_{p\alpha}g_{q\mu}\frac{\partial g_{3\mu}}{\partial x_\alpha}g_{\nu\lambda}\frac{\partial n\left\langle \hat{v}_p\hat{v}_q\hat{v}_{\perp\nu}\right\rangle}{\partial \hat{v}_3}
$$

$$
\begin{aligned}
&= mg_{p\alpha}g_{q\mu}\frac{\partial g_{3\mu}}{\partial x_\alpha}g_{\nu\lambda}\left(\frac{\partial n\left\langle \hat{v}_{\perp p}\hat{v}_{\perp q}\hat{v}_{\perp\nu}\right\rangle}{\partial \hat{v}_3} + \frac{\partial n\left\langle \hat{v}_3\hat{v}_{\perp q}\hat{v}_{\perp\nu}\right\rangle}{\partial \hat{v}_3}\delta_{p3} + \frac{\partial n\left\langle \hat{v}_p\hat{v}_3\hat{v}_{\perp\nu}\right\rangle}{\partial \hat{v}_3}\delta_{q3}\right) \\
&= g_{p\alpha}g_{q\mu}g_{\nu\lambda}\frac{\partial g_{3\mu}}{\partial x_\alpha}\left(\frac{\partial \rho\left\langle \hat{v}_{\perp p}\hat{v}_{\perp q}\hat{v}_{\perp\nu}\right\rangle}{\partial \hat{v}_3} + \rho\left\langle \hat{v}_{\perp q}\hat{v}_{\perp\nu}\right\rangle\delta_{p3} + \hat{v}_3\frac{\partial \rho\left\langle \hat{v}_{\perp q}\hat{v}_{\perp\nu}\right\rangle}{\partial \hat{v}_3}\delta_{p3}\right) \\
&= g_{p\alpha}g_{q\mu}g_{\nu\lambda}\frac{\partial g_{3\mu}}{\partial x_\alpha}\frac{\partial}{\partial \hat{v}_3}\left(q_\nu\delta_{pq}^{\perp} + \Theta_{pq\nu}\right) + g_{3\alpha}g_{q\mu}g_{\nu\lambda}\frac{\partial g_{3\mu}}{\partial x_\alpha}\left(p\delta_{q\nu}^{\perp} + \pi_{q\nu}\right) \\
&\quad + \hat{v}_3g_{3\alpha}g_{q\mu}g_{\nu\lambda}\frac{\partial g_{3\mu}}{\partial x_\alpha}\frac{\partial\left(p\delta_{q\nu}^{\perp} + \pi_{q\nu}\right)}{\partial \hat{v}_3} \\
&= (g_{p\alpha}g_{p\mu} - g_{3\alpha}g_{3\mu})g_{\nu\lambda}\frac{\partial g_{3\mu}}{\partial x_\alpha}\frac{\partial q_\nu}{\partial \hat{v}_3} + g_{p\alpha}g_{q\mu}g_{\nu\lambda}\frac{\partial g_{3\mu}}{\partial x_\alpha}\frac{\partial \Theta_{pq\nu}}{\partial \hat{v}_3} \\
&\quad + g_{3\alpha}(g_{q\mu}g_{q\lambda} - g_{3\mu}g_{3\lambda})\frac{\partial g_{3\mu}}{\partial x_\alpha}p + g_{3\alpha}g_{q\mu}g_{\nu\lambda}\frac{\partial g_{3\mu}}{\partial x_\alpha}\pi_{q\nu} \\
&\quad + \hat{v}_3g_{3\alpha}(g_{q\mu}g_{q\lambda} - g_{3\mu}g_{3\lambda})\frac{\partial g_{3\mu}}{\partial x_\alpha}\frac{\partial p}{\partial \hat{v}_3} + \hat{v}_3g_{3\alpha}g_{q\mu}g_{\nu\lambda}\frac{\partial g_{3\mu}}{\partial x_\alpha}\frac{\partial \pi_{q\nu}}{\partial \hat{v}_3} \\
&= g_{\nu\lambda}\frac{\partial g_{3\alpha}}{\partial x_\alpha}\frac{\partial q_\nu}{\partial \hat{v}_3} + g_{p\alpha}g_{q\mu}g_{\nu\lambda}\frac{\partial g_{3\mu}}{\partial x_\alpha}\frac{\partial \Theta_{pq\nu}}{\partial \hat{v}_3} + \underbrace{g_{3\alpha}\frac{\partial g_{3\lambda}}{\partial x_\alpha}p}_{x1} + g_{3\alpha}g_{q\mu}g_{\nu\lambda}\frac{\partial g_{3\mu}}{\partial x_\alpha}\pi_{q\nu} \\
&\quad + \hat{v}_3g_{3\alpha}\frac{\partial g_{3\lambda}}{\partial x_\alpha}\frac{\partial p}{\partial \hat{v}_3} + \hat{v}_3g_{3\alpha}g_{q\mu}g_{\nu\lambda}\frac{\partial g_{3\mu}}{\partial x_\alpha}\frac{\partial \pi_{q\nu}}{\partial \hat{v}_3}.
\end{aligned}
$$

Using these results and noting the cancellations with the same indices '*xi*', one obtains the momentum equation:

$$\begin{aligned}
&(\delta_{\lambda\alpha} - g_{3\alpha}g_{3\lambda})\left(neE_\alpha^* - \rho\frac{\hat{d}g_{l\alpha}\hat{V}_{\perp l}}{dt} - \frac{\partial p}{\partial x_\alpha} - \hat{v}_3 g_{3\beta}\frac{\partial g_{3\alpha}}{\partial x_\beta}\frac{\partial p}{\partial \hat{v}_3}\right.\\
&- g_{3\beta}\hat{v}_3^2\frac{\partial g_{3\alpha}}{\partial x_\beta}\rho + g_{3\gamma}\frac{\partial g_{l\gamma}\hat{V}_{\perp l}}{\partial x_\alpha}\frac{\partial p}{\partial \hat{v}_3} - \rho\hat{v}_3 g_{3\beta}\frac{\partial g_{l\alpha}\hat{V}_{\perp l}}{\partial x_\beta} - \rho\hat{v}_3\frac{\hat{d}g_{3\alpha}}{dt}\\
&\left.- \frac{\hat{d}g_{3\alpha}}{dt}\frac{\partial p}{\partial \hat{v}_3}\right)\\
&= g_{\nu\lambda}g_{\beta\alpha}\frac{\partial \pi_{\beta\nu}}{\partial x_\alpha} + g_{\nu\lambda}\left(g_{p\delta}\frac{\hat{d}g_{3\delta}}{dt} - g_{p\beta}g_{3\gamma}\frac{\partial g_{l\gamma}\hat{V}_{\perp l}}{\partial x_\beta} + \hat{v}_3 g_{3\alpha}g_{p\mu}\frac{\partial g_{3\mu}}{\partial x_\alpha}\right)\frac{\partial \pi_{p\nu}}{\partial \hat{v}_3}\\
&+ \left(\frac{\partial g_{p\alpha}g_{q\lambda}}{\partial x_\alpha} + g_{p\alpha}g_{q\mu}g_{3\lambda}\frac{\partial g_{3\mu}}{\partial x_\alpha} + g_{3\alpha}g_{q\mu}g_{p\lambda}\frac{\partial g_{3\mu}}{\partial x_\alpha}\right)\pi_{pq}\\
&+ g_{\nu\lambda}\frac{\partial g_{3\alpha}}{\partial x_\alpha}\frac{\partial q_\nu}{\partial \hat{v}_3} + g_{p\alpha}g_{q\mu}g_{\nu\lambda}\frac{\partial g_{3\mu}}{\partial x_\alpha}\frac{\partial \Theta_{pq\nu}}{\partial \hat{v}_3}.
\end{aligned} \tag{3.41}$$

Here, $E_\alpha^* = E_\alpha + e_{\alpha\gamma\delta}g_{q\gamma}\hat{V}_q B_\delta$.

The momentum equation (3.41) is a vector equation. It can be proved that (3.41) contains only two perpendicular components by applying the parallel-projection operator $g_{3\lambda}$. For the left-hand side, one has

$$g_{3\lambda}(\delta_{\lambda\alpha} - g_{3\alpha}g_{3\lambda}) = 0,$$

For the right-hand side, it is noted that $\pi_{p3} = q_3 = \Theta_{pq3} = 0$ as will be proved in section 3.4. Therefore, one just needs to prove that

$$\begin{aligned}
&g_{3\lambda}\left(\frac{\partial g_{p\alpha}g_{q\lambda}}{\partial x_\alpha} + g_{p\alpha}g_{q\mu}g_{3\lambda}\frac{\partial g_{3\mu}}{\partial x_\alpha}\right)\pi_{pq}\\
&= \left(g_{p\alpha}g_{3\lambda}\frac{\partial g_{q\lambda}}{\partial x_\alpha} + g_{p\alpha}g_{q\mu}\frac{\partial g_{3\mu}}{\partial x_\alpha}\right)\pi_{pq}\\
&= \left(- g_{p\alpha}g_{q\lambda}\frac{\partial g_{3\lambda}}{\partial x_\alpha} + g_{p\alpha}g_{q\mu}\frac{\partial g_{3\mu}}{\partial x_\alpha}\right)\pi_{pq} = 0.
\end{aligned}$$

These show that the parallel projection of equation (3.41) is trivial equality.

3.2.4 Energy transport equation

In this subsection, the energy transport equation in the perpendicular magnetofluid formalism is derived. This can be obtained by letting $\eta = m\hat{v}_{\perp\nu}\hat{v}_{\perp\nu}/2$ in the general moment equation (3.23). Equations (3.2)–(3.6) are used in the simplification.

The term-by-term simplifications are given as follows. Only the terms having actual contributions to the energy transport equation are detailed.

$$\begin{aligned}
&\frac{\partial}{\partial t}(n\langle\eta\rangle) - n\left\langle\frac{\partial\eta}{\partial t}\right\rangle + \frac{\partial}{\partial x_\alpha}\left(n\left\langle g_{\beta\alpha}\hat{V}_{\perp\beta}\eta\right\rangle\right) - n\left\langle\frac{\partial}{\partial x_\alpha}\left(g_{\beta\alpha}\hat{V}_{\perp\beta}\eta\right)\right\rangle\\
&= \frac{\partial}{\partial t}(nT) + \frac{\partial}{\partial x_\alpha}\left(g_{\beta\alpha}\hat{V}_{\perp\beta}nT\right) - nT\frac{\partial}{\partial x_\alpha}\left(g_{\beta\alpha}\hat{V}_{\perp\beta}\right)\\
&= \frac{\partial}{\partial t}(nT) + g_{\beta\alpha}\hat{V}_{\perp\beta}\frac{\partial}{\partial x_\alpha}(nT),
\end{aligned}$$

$$\begin{aligned}
&\frac{\partial}{\partial x_\alpha}\left(g_{\beta\alpha}n\langle\eta\hat{v}_\beta\rangle\right) - n\left\langle\hat{v}_\beta\frac{\partial}{\partial x_\alpha}(\eta g_{\beta\alpha})\right\rangle\\
&= \frac{\partial}{\partial x_\alpha}\left(g_{\beta\alpha}\rho\left\langle\frac{1}{2}\hat{v}_{\perp\nu}^2\hat{v}_\beta\right\rangle\right) - \rho\left\langle\frac{1}{2}\hat{v}_{\perp\nu}^2\hat{v}_\beta\right\rangle\frac{\partial g_{\beta\alpha}}{\partial x_\alpha}\\
&= \frac{\partial}{\partial x_\alpha}\left(g_{\beta\alpha}\rho\left\langle\frac{1}{2}\hat{v}_{\perp\nu}^2\hat{v}_{\perp\beta}\right\rangle\right) + \frac{\partial}{\partial x_\alpha}\left(g_{3\alpha}\rho\left\langle\frac{1}{2}\hat{v}_{\perp\nu}^2\hat{v}_3\right\rangle\right) - \rho\left\langle\frac{1}{2}\hat{v}_{\perp\nu}^2\hat{v}_\beta\right\rangle\frac{\partial g_{\beta\alpha}}{\partial x_\alpha}\\
&= \frac{\partial}{\partial x_\alpha}(g_{\beta\alpha}q_\beta) + \underbrace{\hat{v}_3\frac{\partial}{\partial x_\alpha}(g_{3\alpha}p)}_{o1} - \underbrace{\rho\left\langle\frac{1}{2}\hat{v}_{\perp\nu}^2\hat{v}_\beta\right\rangle\frac{\partial g_{\beta\alpha}}{\partial x_\alpha}}_{x2},
\end{aligned}$$

$$\begin{aligned}
&g_{q\alpha}g_{p\mu}\frac{\partial g_{\beta\mu}}{\partial x_\alpha}n\left(\left\langle\hat{v}_p\hat{v}_q\frac{\partial\eta}{\partial\hat{v}_\beta}\right\rangle + \langle\eta\hat{v}_q\rangle\delta_{p\beta}\right)\\
&= g_{p\alpha}g_{q\mu}\frac{\partial g_{\beta\mu}}{\partial x_\alpha}\rho\left(\langle\hat{v}_p\hat{v}_q\hat{v}_{\perp\beta}\rangle + \left\langle\frac{1}{2}\hat{v}_{\perp\nu}^2\hat{v}_q\right\rangle\delta_{p\beta}\right)\\
&= g_{p\alpha}g_{q\mu}\frac{\partial g_{\beta\mu}}{\partial x_\alpha}\rho\langle\hat{v}_p\hat{v}_q\hat{v}_\beta\rangle - g_{p\alpha}g_{q\mu}\frac{\partial g_{3\mu}}{\partial x_\alpha}\rho\langle\hat{v}_p\hat{v}_q\hat{v}_3\rangle + g_{p\alpha}g_{q\mu}\frac{\partial g_{\beta\mu}}{\partial x_\alpha}\rho\left(\left\langle\frac{1}{2}\hat{v}_{\perp\nu}^2\hat{v}_q\right\rangle\delta_{p\beta}\right)\\
&= -g_{p\alpha}g_{q\mu}\frac{\partial g_{3\mu}}{\partial x_\alpha}\rho\langle\hat{v}_p\hat{v}_q\hat{v}_3\rangle + g_{p\alpha}g_{q\mu}\frac{\partial g_{\beta\mu}}{\partial x_\alpha}\rho\left(\left\langle\frac{1}{2}\hat{v}_{\perp\nu}^2\hat{v}_q\right\rangle\delta_{p\beta}\right)\\
&= -g_{p\alpha}g_{q\mu}\hat{v}_3\frac{\partial g_{3\mu}}{\partial x_\alpha}\rho\langle\hat{v}_p\hat{v}_q\rangle + g_{p\alpha}g_{q\mu}\frac{\partial g_{p\mu}}{\partial x_\alpha}\rho\left(\left\langle\frac{1}{2}\hat{v}_{\perp\nu}^2\hat{v}_q\right\rangle\right)\\
&= -\hat{v}_3g_{p\alpha}g_{q\mu}\frac{\partial g_{3\mu}}{\partial x_\alpha}\left(p\delta_{pq}^{\perp} + \pi_{pq}\right) - g_{p\alpha}g_{p\mu}\frac{\partial g_{q\mu}}{\partial x_\alpha}\rho\left(\left\langle\frac{1}{2}\hat{v}_{\perp\nu}^2\hat{v}_q\right\rangle\right)\\
&= -\underbrace{\hat{v}_3\frac{\partial g_{3\alpha}}{\partial x_\alpha}p}_{o1} - \hat{v}_3g_{p\alpha}g_{q\mu}\frac{\partial g_{3\mu}}{\partial x_\alpha}\pi_{pq} - \underbrace{\frac{\partial g_{p\alpha}}{\partial x_\alpha}\rho\left(\left\langle\frac{1}{2}\hat{v}_{\perp\nu}^2\hat{v}_p\right\rangle\right)}_{x2},
\end{aligned}$$

where equation (3.6) has been used in the third step,

$$-\left(\frac{e}{m}E_{\beta}^{*} - \frac{\hat{d}g_{l\beta}\hat{V}_{\perp l}}{dt}\right)g_{r\beta}n\left\langle\frac{\partial\eta}{\partial\hat{v}_r}\right\rangle = -\left(\frac{e}{m}E_{\beta}^{*} - \frac{\hat{d}g_{l\beta}\hat{V}_{\perp l}}{dt}\right)g_{r\beta}\rho\langle\hat{v}_{\perp r}\rangle = 0,$$

where equation (3.24) has been used,

$$\begin{aligned}
&- g_{p\beta}g_{r\gamma}\frac{\partial g_{l\gamma}\hat{V}_{\perp l}}{\partial x_{\beta}}n\left(\left\langle\hat{v}_p\frac{\partial\eta}{\partial\hat{v}_r}\right\rangle + \langle\eta\rangle\delta_{pr}\right)\\
&= - g_{p\beta}g_{r\gamma}\frac{\partial g_{l\gamma}\hat{V}_{\perp l}}{\partial x_{\beta}}\rho\left(\left\langle\hat{v}_p\hat{v}_{\perp r}\right\rangle + \frac{1}{2}\left\langle v_{\perp\nu}^2\right\rangle\delta_{pr}\right)\\
&= - g_{p\beta}g_{r\gamma}\frac{\partial g_{l\gamma}\hat{V}_{\perp l}}{\partial x_{\beta}}\rho\left(\left\langle\hat{v}_{\perp p}\hat{v}_{\perp r}\right\rangle + \frac{1}{2}\left\langle v_{\perp\nu}^2\right\rangle\delta_{pr}\right)\\
&= - g_{p\beta}g_{r\gamma}\frac{\partial g_{l\gamma}\hat{V}_{\perp l}}{\partial x_{\beta}}\left(p\delta_{pr}^{\perp} + \pi_{pr}\right) - \frac{\partial g_{l\beta}\hat{V}_{\perp l}}{\partial x_{\beta}}p\\
&= - 2\frac{\partial g_{l\beta}\hat{V}_{\perp l}}{\partial x_{\beta}}p + \underbrace{g_{3\beta}g_{3\gamma}\frac{\partial g_{l\gamma}\hat{V}_{\perp l}}{\partial x_{\beta}}p}_{o3} - g_{p\beta}g_{r\gamma}\frac{\partial g_{l\gamma}\hat{V}_{\perp l}}{\partial x_{\beta}}\pi_{pr},
\end{aligned}$$

$$\frac{e}{m}e_{\beta\gamma\alpha}g_{l\gamma}B_{\alpha}g_{r\beta}n\left\langle\frac{\partial\eta}{\partial\hat{v}_r}\hat{v}_l\right\rangle = ee_{\beta\gamma\alpha}g_{l\gamma}B_{\alpha}g_{r\beta}n\langle\hat{v}_{\perp r}\hat{v}_{\perp l}\rangle = 0,$$

$$\begin{aligned}
g_{p\delta}\frac{\hat{d}g_{\beta\delta}}{dt}n\left(\left\langle\hat{v}_p\frac{\partial\eta}{\partial\hat{v}_{\beta}}\right\rangle + \langle\eta\rangle\delta_{p\beta}\right) &= g_{p\delta}\frac{\hat{d}g_{\beta\delta}}{dt}nm\left\langle\hat{v}_p\hat{v}_{\perp\beta}\right\rangle\\
&= g_{p\delta}\frac{\hat{d}g_{\beta\delta}}{dt}nm\left\langle\hat{v}_{\perp p}\hat{v}_{\perp\beta}\right\rangle + g_{3\delta}\frac{\hat{d}g_{3\delta}}{dt}nm\left\langle\hat{v}_3\hat{v}_{\perp\beta}\right\rangle\\
&= 0,
\end{aligned}$$

$$\begin{aligned}
&-\left(\frac{e}{m}E_{\beta}^{*} - \frac{\hat{d}g_{l\beta}\hat{V}_{\perp l}}{dt}\right)g_{3\beta}\frac{\partial n\langle\eta\rangle}{\partial\hat{v}_3}\\
&= -\left(\frac{e}{m}E_{\beta}^{*} - \frac{\hat{d}g_{l\beta}\hat{V}_{\perp l}}{dt}\right)g_{3\beta}\frac{m}{2}\frac{\partial n\left\langle\hat{v}_{\perp\nu}^2\right\rangle}{\partial\hat{v}_3}\\
&= -\left(\frac{e}{m}E_{\beta}^{*} - \frac{\hat{d}g_{l\beta}\hat{V}_{\perp l}}{dt}\right)g_{3\beta}\frac{\partial p}{\partial\hat{v}_3},
\end{aligned}$$

$$
\begin{aligned}
&g_{p\beta}g_{3\gamma}\frac{\partial g_{l\gamma}\hat{V}_{\perp l}}{\partial x_\beta}\frac{\partial n\left\langle \hat{v}_p\eta\right\rangle}{\partial \hat{v}_3} \\
&= g_{p\beta}g_{3\gamma}\frac{\partial g_{l\gamma}\hat{V}_{\perp l}}{\partial x_\beta}\frac{m}{2}\frac{\partial n\left\langle \hat{v}_{\perp\nu}^2\hat{v}_p\right\rangle}{\partial \hat{v}_3} \\
&= g_{p\beta}g_{3\gamma}\frac{\partial g_{l\gamma}\hat{V}_{\perp l}}{\partial x_\beta}\frac{m}{2}\frac{\partial n\left\langle \hat{v}_{\perp\nu}^2\hat{v}_{\perp p}\right\rangle}{\partial \hat{v}_3} + g_{3\beta}g_{3\gamma}\frac{\partial g_{l\gamma}\hat{V}_{\perp l}}{\partial x_\beta}\frac{m}{2}\frac{\partial n\left\langle \hat{v}_{\perp\nu}^2\hat{v}_3\right\rangle}{\partial \hat{v}_3} \\
&= g_{p\beta}g_{3\gamma}\frac{\partial g_{l\gamma}\hat{V}_{\perp l}}{\partial x_\beta}\frac{\partial q_p}{\partial \hat{v}_3} + \hat{v}_3 g_{3\beta}g_{3\gamma}\frac{\partial g_{l\gamma}\hat{V}_{\perp l}}{\partial x_\beta}\frac{\partial p}{\partial \hat{v}_3} + \underbrace{g_{3\beta}g_{3\gamma}\frac{\partial g_{l\gamma}\hat{V}_{\perp l}}{\partial x_\beta}p}_{o3},
\end{aligned}
$$

$$
\begin{aligned}
-g_{p\delta}\frac{\hat{d}g_{3\delta}}{dt}\frac{\partial n\left\langle \hat{v}_p\eta\right\rangle}{\partial \hat{v}_3} &= -g_{p\delta}\frac{\hat{d}g_{3\delta}}{dt}\frac{1}{2}\frac{\partial \rho\left\langle \hat{v}_p\hat{v}_{\perp\nu}^2\right\rangle}{\partial \hat{v}_3} \\
&= -g_{p\delta}\frac{\hat{d}g_{3\delta}}{dt}\frac{1}{2}\frac{\partial \rho\left\langle \hat{v}_{\perp p}\hat{v}_{\perp\nu}^2\right\rangle}{\partial \hat{v}_3} \\
&= -g_{p\delta}\frac{\hat{d}g_{3\delta}}{dt}\frac{\partial q_p}{\partial \hat{v}_3},
\end{aligned}
$$

and

$$
\begin{aligned}
&-g_{p\alpha}g_{q\mu}\frac{\partial g_{3\mu}}{\partial x_\alpha}\frac{\partial n\left\langle \eta\hat{v}_p\hat{v}_q\right\rangle}{\partial \hat{v}_3} \\
&= -g_{p\alpha}g_{q\mu}\frac{\partial g_{3\mu}}{\partial x_\alpha}\frac{m}{2}\frac{\partial n\left\langle \hat{v}_{\perp\nu}^2\hat{v}_p\hat{v}_q\right\rangle}{\partial \hat{v}_3} \\
&= -g_{p\alpha}g_{q\mu}\frac{\partial g_{3\mu}}{\partial x_\alpha}\frac{m}{2}\frac{\partial n\left\langle \hat{v}_{\perp\nu}^2\hat{v}_p\hat{v}_{\perp q}\right\rangle}{\partial \hat{v}_3} \\
&= -g_{p\alpha}g_{q\mu}\frac{\partial g_{3\mu}}{\partial x_\alpha}\frac{m}{2}\frac{\partial n\left\langle \hat{v}_{\perp\nu}^2\hat{v}_{\perp p}\hat{v}_{\perp q}\right\rangle}{\partial \hat{v}_3} - g_{3\alpha}g_{q\mu}\frac{\partial g_{3\mu}}{\partial x_\alpha}\frac{m}{2}\frac{\partial n\left\langle \hat{v}_{\perp\nu}^2\hat{v}_3\hat{v}_{\perp q}\right\rangle}{\partial \hat{v}_3} \\
&= -g_{p\alpha}g_{q\mu}\frac{\partial g_{3\mu}}{\partial x_\alpha}\frac{1}{m}\frac{\partial}{\partial \hat{v}_3}\left(nD\delta_{pq}^{\perp} + n\Xi_{pq}\right) \\
&\quad -\hat{v}_3 g_{3\alpha}g_{q\mu}\frac{\partial g_{3\mu}}{\partial x_\alpha}\frac{\partial q_q}{\partial \hat{v}_3} - g_{3\alpha}g_{q\mu}\frac{\partial g_{3\mu}}{\partial x_\alpha}q_q \\
&= -\frac{\partial g_{3\alpha}}{\partial x_\alpha}\frac{1}{m}\frac{\partial nD}{\partial \hat{v}_3} - g_{p\alpha}g_{q\mu}\frac{\partial g_{3\mu}}{\partial x_\alpha}\frac{1}{m}\frac{\partial n\Xi_{pq}}{\partial \hat{v}_3} \\
&\quad -\hat{v}_3 g_{3\alpha}g_{q\mu}\frac{\partial g_{3\mu}}{\partial x_\alpha}\frac{\partial q_q}{\partial \hat{v}_3} - g_{3\alpha}g_{q\mu}\frac{\partial g_{3\mu}}{\partial x_\alpha}q_q.
\end{aligned}
$$

Using these results and noting the cancellations with the same indices '*xi*' and the combinations with indices '*oi*', the energy transport equation becomes

$$\begin{aligned}
&\frac{\partial p}{\partial t} + g_{\beta\alpha}\hat{V}_{\perp\beta}\frac{\partial p}{\partial x_\alpha} + g_{3\alpha}\hat{v}_3\frac{\partial p}{\partial x_\alpha} + 2p\frac{\partial g_{l\gamma}\hat{V}_{\perp l}}{\partial x_\beta}(\delta_{\beta\gamma} - g_{3\beta}g_{3\gamma})\\
&+ 2\hat{v}_3\frac{\partial g_{3\alpha}}{\partial x_\alpha}p + \frac{\partial g_{3\alpha}}{\partial x_\alpha}\frac{1}{m}\frac{\partial nD}{\partial \hat{v}_3} - \hat{v}_3 g_{3\beta}g_{3\gamma}\frac{\partial g_{l\gamma}\hat{V}_{\perp l}}{\partial x_\beta}\frac{\partial p}{\partial \hat{v}_3}\\
&= - g_{3\alpha}\left(\frac{e}{m}E_\alpha^* - \frac{\hat{d}g_{l\alpha}\hat{V}_{\perp l}}{dt}\right)\frac{\partial p}{\partial \hat{v}_3} - g_{p\alpha}g_{q\mu}\left(\hat{v}_3\frac{\partial g_{3\mu}}{\partial x_\alpha} + \frac{\partial g_{l\mu}\hat{V}_{\perp l}}{\partial x_\alpha}\right)\pi_{pq} \qquad (3.42)\\
&+ \left(g_{p\beta}g_{3\gamma}\frac{\partial g_{l\gamma}\hat{V}_{\perp l}}{\partial x_\beta} - g_{p\delta}\frac{\hat{d}g_{3\delta}}{dt}\right)\frac{\partial q_p}{\partial \hat{v}_3} - \frac{\partial}{\partial x_\alpha}(g_{\beta\alpha}q_\beta)\\
&- \hat{v}_3 g_{3\alpha}g_{q\mu}\frac{\partial g_{3\mu}}{\partial x_\alpha}\frac{\partial q_q}{\partial \hat{v}_3} - g_{3\alpha}g_{q\mu}\frac{\partial g_{3\mu}}{\partial x_\alpha}q_q - g_{p\alpha}g_{q\mu}\frac{\partial g_{3\mu}}{\partial x_\alpha}\frac{1}{m}\frac{\partial n\Xi_{pq}}{\partial \hat{v}_3}.
\end{aligned}$$

Using the continuity equation in equation (3.40), one can reduce equation (3.42) to

$$\begin{aligned}
&n\frac{\partial T}{\partial t} + ng_{\beta\alpha}\hat{V}_{\perp\beta}\frac{\partial T}{\partial x_\alpha} + ng_{3\alpha}\hat{v}_3\frac{\partial T}{\partial x_\alpha} + p\frac{\partial g_{l\gamma}\hat{V}_{\perp l}}{\partial x_\beta}(\delta_{\gamma\beta} - g_{3\beta}g_{3\gamma})\\
&+ \hat{v}_3\frac{\partial g_{3\alpha}}{\partial x_\alpha}p + \frac{\partial g_{3\alpha}}{\partial x_\alpha}\frac{T^2}{m}\frac{\partial n}{\partial \hat{v}_3} + \frac{\partial g_{3\alpha}}{\partial x_\alpha}\frac{3p}{m}\frac{\partial T}{\partial \hat{v}_3} - \hat{v}_3 g_{3\beta}g_{3\gamma}\frac{\partial g_{l\gamma}\hat{V}_{\perp l}}{\partial x_\beta}n\frac{\partial T}{\partial \hat{v}_3}\\
&= - g_{3\alpha}\left(\frac{e}{m}E_\alpha^* - \frac{\hat{d}g_{l\alpha}\hat{V}_{\perp l}}{dt}\right)n\frac{\partial T}{\partial \hat{v}_3} + \frac{T}{m}g_{p\alpha}g_{q\mu}\frac{\partial g_{3\mu}}{\partial x_\alpha}\frac{\partial \pi_{pq}}{\partial \hat{v}_3} \qquad (3.43)\\
&- g_{p\alpha}g_{q\mu}\left(\hat{v}_3\frac{\partial g_{3\mu}}{\partial x_\alpha} + \frac{\partial g_{l\mu}\hat{V}_{\perp l}}{\partial x_\alpha}\right)\pi_{pq} + \left(g_{p\beta}g_{3\gamma}\frac{\partial g_{l\gamma}\hat{V}_{\perp l}}{\partial x_\beta} - g_{p\delta}\frac{\hat{d}g_{3\delta}}{dt}\right)\frac{\partial q_p}{\partial \hat{v}_3}\\
&- \frac{\partial}{\partial x_\alpha}(g_{\beta\alpha}q_\beta) - \hat{v}_3 g_{3\alpha}g_{q\mu}\frac{\partial g_{3\mu}}{\partial x_\alpha}\frac{\partial q_q}{\partial \hat{v}_3} - g_{3\alpha}g_{q\mu}\frac{\partial g_{3\mu}}{\partial x_\alpha}q_q\\
&- g_{p\alpha}g_{q\mu}\frac{\partial g_{3\mu}}{\partial x_\alpha}\frac{1}{m}\frac{\partial n\Xi_{pq}}{\partial \hat{v}_3}.
\end{aligned}$$

3.3 Kinetic closure of perpendicular magnetofluid equations

In this section, we describe the kinetic closure procedure for the perpendicular magnetofluid equations. We first derive the first order kinetic equations for various particle species. The moment equations of lowest order will be used to simplify the equations. The solutions of the first order kinetic equations will be described afterwards. In the end, these solutions will be used to evaluate the transport coefficients, so that the kinetic closure of the perpendicular magnetofluid equations is materialized.

3.3.1 The first order kinetic equations

In this subsection, the kinetic equations of first order are derived using the moment equations of lowest order for simplification as in the Braginskii two-fluid theory. The derivations apply to each particle species. For brevity, the particle species indices are suppressed.

The gyrofrequency ω_c is assumed to be large and the distribution functions are expanded using the inverse gyrofrequency ordering:

$$f = f^0 + f^1 + \cdots. \tag{3.44}$$

with

$$f^0 = n\frac{m}{2\pi T}\exp\left\{-\frac{m\hat{v}_\perp^2/2}{T}\right\}.$$

For the lowest order distribution function, one has

$$\frac{\partial f^0}{\partial \hat{v}_r} = \left(\frac{\partial \ln n}{\partial \hat{v}_3}\delta_{r3} - \frac{m}{T}\hat{v}_{\perp r}\right)f^0. \tag{3.45}$$

Here, it has been noted that $\partial \ln n/\partial \hat{v}_3 \gg \partial \ln T/\partial \hat{v}_3$, as explained after introducing the perpendicular Maxwellian distribution in equation (3.20).

The continuity, momentum, and energy transport equations in lowest order are obtained from equations (3.40), (3.41), and (3.43), respectively, as follows.

$$\begin{aligned}
&\frac{\partial n}{\partial t} + g_{\beta\alpha}\hat{V}_{\perp\beta}\frac{\partial n}{\partial x_\alpha} + g_{3\alpha}\hat{v}_3\frac{\partial n}{\partial x_\alpha} + n\frac{\partial g_{l\gamma}\hat{V}_{\perp l}}{\partial x_\beta}\left(\delta_{\beta\gamma} - g_{3\beta}g_{3\gamma}\right)\\
&+ n\hat{v}_3\frac{\partial g_{3\alpha}}{\partial x_\alpha} + \frac{1}{m}m\frac{\partial g_{3\alpha}}{\partial x_\alpha}\frac{\partial nT}{\partial \hat{v}_3} - g_{3\beta}g_{3\gamma}\frac{\partial g_{l\gamma}\hat{V}_{\perp l}}{\partial x_\beta}\hat{v}_3\frac{\partial n}{\partial \hat{v}_3}\\
&= - g_{3\beta}\frac{\partial n}{\partial \hat{v}_3}\left(\frac{e}{m}E_\beta^* - \frac{\hat{d}g_{l\beta}\hat{V}_{\perp l}}{dt}\right),
\end{aligned} \tag{3.46}$$

$$\begin{aligned}
&\left(\delta_{\lambda\alpha} - g_{3\alpha}g_{3\lambda}\right)\left(neE_\alpha^* - \rho\frac{\hat{d}g_{l\alpha}\hat{V}_{\perp l}}{dt} - \frac{\partial p}{\partial x_\alpha} - \hat{v}_3 g_{3\beta}\frac{\partial g_{3\alpha}}{\partial x_\beta}\frac{\partial p}{\partial \hat{v}_3}\right.\\
&- g_{3\beta}\hat{v}_3^2\frac{\partial g_{3\alpha}}{\partial x_\beta}\rho + g_{3\gamma}\frac{\partial g_{l\gamma}\hat{V}_{\perp l}}{\partial x_\alpha}\frac{\partial p}{\partial \hat{v}_3} - \rho\hat{v}_3 g_{3\beta}\frac{\partial g_{l\alpha}\hat{V}_{\perp l}}{\partial x_\beta} - \rho\hat{v}_3\frac{\hat{d}g_{3\alpha}}{dt}\\
&\left. - \frac{\hat{d}g_{3\alpha}}{dt}\frac{\partial p}{\partial \hat{v}_3}\right) = 0,
\end{aligned} \tag{3.47}$$

and

$$
\begin{aligned}
&n\frac{\partial T}{\partial t} + ng_{\beta\alpha}\hat{V}_{\perp\beta}\frac{\partial T}{\partial x_\alpha} + ng_{3\alpha}\hat{v}_3\frac{\partial T}{\partial x_\alpha} + p\frac{\partial g_{l\gamma}\hat{V}_{\perp l}}{\partial x_\beta}\left(\delta_{\gamma\beta} - g_{3\beta}g_{3\gamma}\right)\\
&+\hat{v}_3\frac{\partial g_{3\alpha}}{\partial x_\alpha}p + \frac{\partial g_{3\alpha}}{\partial x_\alpha}\frac{T^2}{m}\frac{\partial n}{\partial \hat{v}_3} = 0.
\end{aligned}
\tag{3.48}
$$

Note that the particle species indices are suppressed in these equations.

Noting that

$$
\begin{aligned}
&\frac{e}{m}e_{\beta\gamma\alpha}g_{l\gamma}\hat{v}_{\perp l}B_\alpha g_{r\beta}\frac{\partial f^0}{\partial \hat{v}_r}\bigg|_{\mathbf{x},t}\\
&= \frac{e}{m}e_{\beta\gamma\alpha}g_{l\gamma}\hat{v}_{\perp l}B_\alpha g_{r\beta}\left(\frac{\partial \ln n}{\partial \hat{v}_3}\delta_{r3} - \frac{m}{T}\hat{v}_{\perp r}\right)f^0 = 0,
\end{aligned}
\tag{3.49}
$$

where equation (3.45) has been used, the first order kinetic equation can be derived from equation (3.19)

$$
\begin{aligned}
&\frac{e}{m}e_{\beta\gamma\alpha}g_{l\gamma}\hat{v}_{\perp l}B_\alpha g_{r\beta}\frac{\partial f^1}{\partial \hat{v}_r}\bigg|_{\mathbf{x},t}\\
&= -\frac{\hat{d}f^0}{dt} - g_{\beta\alpha}\hat{v}_\beta\frac{\partial f^0}{\partial x_\alpha}\bigg|_{\hat{\mathbf{v}},t} - \left(\frac{e}{m}E^*_\beta - \frac{\hat{d}g_{l\beta}\hat{V}_{\perp l}}{dt}\right)g_{r\beta}\frac{\partial f^0}{\partial \hat{v}_r}\bigg|_{\mathbf{x},t}\\
&+ g_{p\beta}g_{r\gamma}\hat{v}_p\frac{\partial g_{l\gamma}\hat{V}_{\perp l}}{\partial x_\beta}\frac{\partial f^0}{\partial v_r'}\bigg|_{\mathbf{x},t} - \left(g_{\gamma\delta}\hat{v}_\gamma\frac{\hat{d}g_{\beta\delta}}{dt} + g_{\delta\alpha}g_{\gamma\mu}\hat{v}_\delta\hat{v}_\gamma\frac{\partial g_{\beta\mu}}{\partial x_\alpha}\right)\frac{\partial f^0}{\partial \hat{v}_\beta}\bigg|_{\mathbf{x},t},
\end{aligned}
\tag{3.50}
$$

where

$$
\begin{aligned}
&\frac{\hat{d}}{dt} \equiv \frac{\partial_{o3}}{\partial t}\bigg|_{\mathbf{x},\hat{\mathbf{v}}} + g_{\beta\alpha}\hat{V}_{\perp\beta}\frac{\partial}{\partial x_\alpha}\bigg|_{\hat{\mathbf{v}},t},\\
&E^*_\alpha = E_\alpha + e_{\alpha\gamma\delta}g_{q\gamma}\hat{V}_{\perp q}B_\delta.
\end{aligned}
$$

Equation (3.50) can be simplified as follows using the moment equations of lowest order in equations (3.46)–(3.48). The first term on the right-hand side of equation (3.50) becomes

$$
\begin{aligned}
-\frac{\hat{d}f^0}{dt}\bigg|_{\mathbf{x},\hat{v}} &= -f^0\left(-1+\frac{m\hat{v}_\perp^2}{2T}\right)\frac{\hat{d}\ln T}{dt} - f^0\frac{\hat{d}\ln n}{dt}\\
&= -f^0\frac{m\hat{v}_\perp^2}{2T}\frac{\hat{d}\ln T}{dt} + \left(f^0\frac{\hat{d}\ln T}{dt} - f^0\frac{\hat{d}\ln n}{dt}\right)\\
&= f^0\frac{m\hat{v}_\perp^2}{2T}\left(\underbrace{g_{3\alpha}\hat{v}_3\frac{\partial \ln T}{\partial x_\alpha}}_{x1} + \frac{\partial g_{l\gamma}\hat{V}_{\perp l}}{\partial x_\beta}\left(\delta_{\gamma\beta} - g_{3\beta}g_{3\gamma}\right)\right.\\
&\quad\left. + \underbrace{\hat{v}_3\frac{\partial g_{3\alpha}}{\partial x_\alpha}}_{o2} + \underbrace{\frac{\partial g_{3\alpha}}{\partial x_\alpha}\frac{T}{m}\frac{\partial \ln n}{\partial \hat{v}_3}}_{o3}\right)\underbrace{-g_{3\alpha}\hat{v}_3\frac{\partial \ln T}{\partial x_\alpha}f^0}_{x4}\ \underbrace{+g_{3\alpha}\hat{v}_3\frac{\partial \ln n}{\partial x_\alpha}f^0}_{x5}\\
&\quad \underbrace{-f^0 g_{3\beta}g_{3\gamma}\frac{\partial g_{l\gamma}\hat{V}_{\perp l}}{\partial x_\beta}\hat{v}_3\frac{\partial n}{\partial \hat{v}_3}}_{x6} + f^0 g_{3\alpha}\frac{\partial \ln n}{\partial \hat{v}_3}\left(\underbrace{\frac{e}{m}E_\alpha^*}_{x7} - \underbrace{\frac{\hat{d}g_{l\alpha}\hat{V}_{\perp l}}{dt}}_{x8}\right),
\end{aligned}
\tag{3.51}
$$

where the continuity and energy transport equations of lowest order, equations (3.46) and (3.48), have been used.

The second term on the right-hand side of (3.50) becomes

$$
\begin{aligned}
-g_{\beta\alpha}\hat{v}_\beta\frac{\partial f^0}{\partial x_\alpha} &= -g_{\beta\alpha}\hat{v}_{\perp\beta}\left(\underbrace{-1}_{o9} + \frac{m\hat{v}_\mu^2/2}{T}\right)\frac{\partial \ln T}{\partial x_\alpha}f^0 \underbrace{-g_{\beta\alpha}\hat{v}_{\perp\beta}\frac{\partial \ln n}{\partial x_\alpha}f^0}_{x10}\\
&\quad - g_{3\alpha}\hat{v}_3\left(\underbrace{-1}_{x4} + \underbrace{\frac{m\hat{v}_\mu^2/2}{T}}_{x1}\right)\frac{\partial \ln T}{\partial x_\alpha}f^0 \underbrace{-g_{3\alpha}\hat{v}_3\frac{\partial \ln n}{\partial x_\alpha}f^0}_{x5}.
\end{aligned}
\tag{3.52}
$$

The third term on the right-hand side of (3.50) is reduced as follows. The momentum equation of lowest order, equation (3.47), can be used for simplification. For this, one needs to apply the factor $g_{3\alpha}g_{3\beta} + (\delta_{\alpha\beta} - g_{3\alpha}g_{3\beta})$ on the third term on the right. For the $g_{3\alpha}g_{3\beta}$ part, using equation (3.45), one obtains

$$
\begin{aligned}
-g_{3\alpha}g_{3\beta}\frac{e}{m}E_\alpha^* g_{r\beta}\frac{\partial f^0}{\partial \hat{v}_r}\bigg|_{t,\mathbf{x}} &= g_{3\alpha}g_{3\beta}\frac{e}{m}E_\alpha^*\left(\frac{m}{T}g_{r\beta}\hat{v}_{\perp r} - g_{3\beta}\frac{\partial \ln n}{\partial \hat{v}_3}\right)f^0\\
&= \underbrace{-g_{3\alpha}\frac{e}{m}E_\alpha^*\frac{\partial \ln n}{\partial \hat{v}_3}f^0}_{x7},
\end{aligned}
\tag{3.53}
$$

and

$$g_{3\alpha}g_{3\beta}\frac{\hat{d}g_{l\alpha}\hat{V}_{\perp l}}{dt}g_{r\beta}\frac{\partial f^0}{\partial \hat{v}_r}\bigg|_{t,\mathbf{x}} = g_{3\alpha}g_{3\beta}\frac{\hat{d}g_{l\alpha}\hat{V}_{\perp l}}{dt}\left(\frac{m}{T}g_{r\beta}\hat{v}_{\perp r} - g_{3\beta}\frac{\partial \ln n}{\partial \hat{v}_3}\right)f^0$$
$$= \underbrace{g_{3\alpha}\frac{\partial \ln n}{\partial \hat{v}_3}\frac{\hat{d}g_{l\alpha}\hat{V}_{\perp l}}{dt}f^0}_{x8}. \tag{3.54}$$

For the $(\delta_{\alpha\beta} - g_{3\alpha}g_{3\beta})$ part, using the lowest order momentum equation (3.47) and equation (3.45), one obtains

$$-\left(\delta_{\alpha\beta} - g_{3\alpha}g_{3\beta}\right)\left(\frac{e}{m}E^*_\beta - \frac{\hat{d}g_{l\beta}\hat{V}_{\perp l}}{dt}\right)g_{r\beta}\frac{\partial f^0}{\partial \hat{v}_r}\bigg|_{t,\mathbf{x}}$$
$$= -\left(\delta_{\alpha\beta} - g_{3\alpha}g_{3\beta}\right)\frac{1}{\rho}\left(-\frac{\partial p}{\partial x_\alpha} - \hat{v}_3 g_{3\gamma}\frac{\partial g_{3\alpha}}{\partial x_\gamma}\frac{\partial p}{\partial \hat{v}_3} - g_{3\gamma}\hat{v}_3^2\frac{\partial g_{3\alpha}}{\partial x_\gamma}\rho\right.$$
$$+ g_{3\gamma}\frac{\partial g_{l\gamma}\hat{V}_{\perp l}}{\partial x_\alpha}\frac{\partial p}{\partial \hat{v}_3}$$
$$\left.- \rho\hat{v}_3 g_{3\gamma}\frac{\partial g_{l\alpha}\hat{V}_{\perp l}}{\partial x_\gamma} - \rho\hat{v}_3\frac{\hat{d}g_{3\alpha}}{dt} - \frac{\hat{d}g_{3\alpha}}{dt}\frac{\partial p}{\partial \hat{v}_3}\right)\left(\frac{m}{T}g_{r\beta}\hat{v}_{\perp r} - g_{3\beta}\frac{\partial \ln n}{\partial \hat{v}_3}\right)f^0$$
$$= \frac{1}{\rho}\left(\frac{\partial p}{\partial x_\alpha} + \hat{v}_3 g_{3\beta}\frac{\partial g_{3\alpha}}{\partial x_\beta}\frac{\partial p}{\partial \hat{v}_3} + g_{3\beta}\hat{v}_3^2\frac{\partial g_{3\alpha}}{\partial x_\beta}\rho - g_{3\beta}\frac{\partial g_{l\beta}\hat{V}_{\perp l}}{\partial x_\alpha}\frac{\partial p}{\partial \hat{v}_3}\right. \tag{3.55}$$
$$\left.+ \rho\hat{v}_3 g_{3\beta}\frac{\partial g_{l\alpha}\hat{V}_{\perp l}}{\partial x_\beta} + \rho\hat{v}_3\frac{\hat{d}g_{3\alpha}}{dt} + \frac{\hat{d}g_{3\alpha}}{dt}\frac{\partial p}{\partial \hat{v}_3}\right)\frac{m}{T}g_{r\alpha}\hat{v}_{\perp r}f^0$$
$$= \frac{1}{\rho}\left(\underbrace{T\frac{\partial n}{\partial x_\alpha}}_{x10} \underbrace{+n\frac{\partial T}{\partial x_\alpha}}_{o9} + \underbrace{\hat{v}_3 g_{3\beta}\frac{\partial g_{3\alpha}}{\partial x_\beta}\frac{\partial p}{\partial \hat{v}_3}}_{x11} + \underbrace{g_{3\beta}\hat{v}_3^2\frac{\partial g_{3\alpha}}{\partial x_\beta}\rho}_{x12} \underbrace{-g_{3\beta}\frac{\partial g_{l\beta}\hat{V}_{\perp l}}{\partial x_\alpha}\frac{\partial p}{\partial \hat{v}_3}}_{x13}\right.$$
$$\left.+ \underbrace{\rho\hat{v}_3 g_{3\beta}\frac{\partial g_{l\alpha}\hat{V}_{\perp l}}{\partial x_\beta}}_{x14} + \underbrace{\rho\hat{v}_3\frac{\hat{d}g_{3\alpha}}{dt}}_{x15} + \underbrace{\frac{\hat{d}g_{3\alpha}}{dt}\frac{\partial p}{\partial \hat{v}_3}}_{x16}\right)\frac{m}{T}g_{r\alpha}\hat{v}_{\perp r}f^0.$$

The fourth term on the right-hand side of equation (3.50) becomes

$$
\begin{aligned}
&g_{\nu\alpha}g_{r\beta}\hat{v}_\nu\frac{\partial g_{l\beta}\hat{V}_{\perp l}}{\partial x_\alpha}\frac{\partial f^0}{\partial \hat{v}_r}\bigg|_{t,\mathbf{x}}\\
&= -g_{\nu\alpha}\hat{v}_{\perp\nu}\frac{\partial g_{l\beta}\hat{V}_{\perp l}}{\partial x_\alpha}\left(\frac{m}{T}g_{r\beta}\hat{v}_{\perp r} - g_{3\beta}\frac{\partial \ln n}{\partial \hat{v}_3}\right)f^0\\
&\quad - g_{3\alpha}\hat{v}_3\frac{\partial g_{l\beta}\hat{V}_{\perp l}}{\partial x_\alpha}\left(\frac{m}{T}g_{r\beta}\hat{v}_{\perp r} - g_{3\beta}\frac{\partial \ln n}{\partial \hat{v}_3}\right)f^0\\
&= -\frac{m}{T}g_{\nu\alpha}\hat{v}_{\perp\nu}\frac{\partial g_{l\beta}\hat{V}_{\perp l}}{\partial x_\alpha}g_{r\beta}\hat{v}_{\perp r}f^0 \underbrace{+g_{\nu\alpha}\hat{v}_{\perp\nu}\frac{\partial g_{l\beta}\hat{V}_{\perp l}}{\partial x_\alpha}g_{3\beta}\frac{\partial \ln n}{\partial \hat{v}_3}f^0}_{x13}\\
&\quad - g_{3\alpha}\hat{v}_3\frac{\partial g_{l\beta}\hat{V}_{\perp l}}{\partial x_\alpha}\left(\underbrace{\frac{m}{T}g_{r\beta}\hat{v}_{\perp r}}_{x14} - \underbrace{g_{3\beta}\frac{\partial \ln n}{\partial \hat{v}_3}}_{x6}\right)f^0.
\end{aligned}
\tag{3.56}
$$

The first part of the last term on the right-hand side of equation (3.50) becomes

$$
\begin{aligned}
-g_{\gamma\delta}\hat{v}_\gamma\frac{\hat{d}g_{\beta\delta}}{dt}\frac{\partial f^0}{\partial \hat{v}_\beta} &= g_{\gamma\delta}\hat{v}_\gamma\frac{\hat{d}g_{\beta\delta}}{dt}\left(\frac{m}{T}\hat{v}_{\perp\beta} - \delta_{3\beta}\frac{\partial \ln n}{\partial \hat{v}_3}\right)f^0\\
&= \frac{m}{T}\hat{v}_{\perp\beta}g_{3\delta}\hat{v}_3\frac{\hat{d}g_{\beta\delta}}{dt}f^0 - g_{\gamma\delta}\hat{v}_\gamma\frac{\hat{d}g_{3\delta}}{dt}\frac{\partial \ln n}{\partial \hat{v}_3}f^0\\
&= \underbrace{-\frac{m}{T}g_{\beta\delta}\hat{v}_{\perp\beta}\hat{v}_3\frac{\hat{d}g_{3\delta}}{dt}f^0}_{x15}\ \underbrace{-g_{\gamma\delta}\hat{v}_{\perp\gamma}\frac{\hat{d}g_{3\delta}}{dt}\frac{\partial \ln n}{\partial \hat{v}_3}f^0}_{x16},
\end{aligned}
\tag{3.57}
$$

where equation (3.2) has been used. The second part of the last term on the right-hand side of equation (3.50) becomes

$$
\begin{aligned}
&-g_{\delta\alpha}g_{\gamma\mu}\hat{v}_\delta\hat{v}_\gamma\frac{\partial g_{\beta\mu}}{\partial x_\alpha}\frac{\partial f}{\partial \hat{v}_\beta}\\
&= g_{\delta\alpha}g_{\gamma\mu}\hat{v}_\delta\hat{v}_\gamma\frac{\partial g_{\beta\mu}}{\partial x_\alpha}\left(\frac{m}{T}\hat{v}_{\perp\beta} - \delta_{3\beta}\frac{\partial \ln n}{\partial \hat{v}_3}\right)f^0\\
&= \frac{m}{T}\hat{v}_{\perp\beta}g_{\delta\alpha}g_{3\mu}\hat{v}_\delta\hat{v}_3\frac{\partial g_{\beta\mu}}{\partial x_\alpha}f^0 - g_{\delta\alpha}g_{\gamma\mu}\hat{v}_\delta\hat{v}_\gamma\frac{\partial g_{3\mu}}{\partial x_\alpha}\frac{\partial \ln n}{\partial \hat{v}_3}f^0\\
&= -\frac{m}{T}\hat{v}_{\perp\beta}g_{\delta\alpha}g_{\beta\mu}\hat{v}_\delta\hat{v}_3\frac{\partial g_{3\mu}}{\partial x_\alpha}f^0 - g_{\delta\alpha}g_{\gamma\mu}\hat{v}_\delta\hat{v}_{\perp\gamma}\frac{\partial g_{3\mu}}{\partial x_\alpha}\frac{\partial \ln n}{\partial \hat{v}_3}f^0\\
&= \underbrace{-\frac{m}{T}\hat{v}_{\perp\beta}g_{\delta\alpha}g_{\beta\mu}\hat{v}_{\perp\delta}\hat{v}_3\frac{\partial g_{3\mu}}{\partial x_\alpha}f^0}_{o2}\ \underbrace{-\frac{m}{T}\hat{v}_{\perp\beta}g_{3\alpha}g_{\beta\mu}\hat{v}_3^2\frac{\partial g_{3\mu}}{\partial x_\alpha}f^0}_{x12}\\
&\quad \underbrace{-g_{\delta\alpha}g_{\gamma\mu}\hat{v}_{\perp\delta}\hat{v}_{\perp\gamma}\frac{\partial g_{3\mu}}{\partial x_\alpha}\frac{\partial \ln n}{\partial \hat{v}_3}f^0}_{o3}\ \underbrace{-g_{3\alpha}g_{\gamma\mu}\hat{v}_3\hat{v}_{\perp\gamma}\frac{\partial g_{3\mu}}{\partial x_\alpha}\frac{\partial \ln n}{\partial \hat{v}_3}f^0}_{x11},
\end{aligned}
\tag{3.58}
$$

where equation (3.2) has been used.

Using these results and noting the cancellations with the same indices '*xi*' and combinations with same indices '*oi*', the first order kinetic equation (3.50) is reduced to

$$\begin{aligned} &-\frac{e}{m}e_{\beta\gamma\alpha}g_{l\gamma}\hat{v}_{\perp l}B_{\alpha}g_{r\beta}\frac{\partial f^{1}}{\partial \hat{v}_{r}}\Bigg|_{t,\mathbf{x}} \\ &= g_{\beta\alpha}\hat{v}_{\perp\beta}\left(\frac{m\hat{v}_{\mu}^{2}}{2T}-2\right)\frac{\partial \ln T}{\partial x_{\alpha}}f^{0}+\frac{m}{T}g_{\nu\alpha}g_{\mu\beta}\hat{v}_{\perp\mu}\hat{v}_{\perp\nu}\frac{\partial g_{l\beta}\hat{V}_{\perp l}}{\partial x_{\alpha}}f^{0} \\ &-f^{0}\frac{m\hat{v}_{\perp}^{2}}{2T}\frac{\partial g_{l\beta}\hat{V}_{\perp l}}{\partial x_{\alpha}}\left(\delta_{\alpha\beta}-g_{3\alpha}g_{3\beta}\right) \\ &+\frac{m}{T}g_{\delta\alpha}g_{\gamma\mu}\frac{\partial g_{3\mu}}{\partial x_{\alpha}}\left(\hat{v}_{\perp\delta}\hat{v}_{\perp\gamma}-\frac{1}{2}\hat{v}_{\perp\nu}^{2}\delta_{\delta\gamma}^{\perp}\right)\left(\hat{v}_{3}+\frac{T}{m}\frac{\partial \ln n}{\partial \hat{v}_{3}}\right)f^{0}. \end{aligned} \tag{3.59}$$

Here, the second and third terms on the right-hand side can be further reduced as follows.

$$\begin{aligned} &\frac{m}{T}g_{\nu\alpha}g_{\mu\beta}\hat{v}_{\perp\mu}\hat{v}_{\perp\nu}\frac{\partial g_{l\beta}\hat{V}_{\perp l}}{\partial x_{\alpha}}f^{0}-f^{0}\frac{m\hat{v}_{\perp}^{2}}{2T}\frac{\partial g_{l\beta}\hat{V}_{\perp l}}{\partial x_{\alpha}}\left(\delta_{\alpha\beta}-g_{3\alpha}g_{3\beta}\right) \\ &= f^{0}\frac{m}{T}\left(\hat{v}_{\perp\mu}\hat{v}_{\perp\nu}-\frac{1}{2}\hat{v}_{\perp}^{2}\delta_{\mu\nu}^{\perp}\right)g_{\mu\alpha}g_{\nu\beta}\frac{\partial g_{l\beta}\hat{V}_{\perp l}}{\partial x_{\alpha}} \\ &= f^{0}\frac{m}{2T}\left(\hat{v}_{\perp\mu}\hat{v}_{\perp\nu}-\frac{1}{2}\hat{v}_{\perp}^{2}\delta_{\mu\nu}^{\perp}\right)\left(g_{\mu\alpha}g_{\nu\beta}\frac{\partial g_{l\beta}\hat{V}_{\perp l}}{\partial x_{\alpha}}+g_{\nu\alpha}g_{\mu\beta}\frac{\partial g_{l\beta}\hat{V}_{\perp l}}{\partial x_{\alpha}}\right) \\ &= f^{0}\frac{m}{2T}\left(\hat{v}_{\perp\mu}\hat{v}_{\perp\nu}-\frac{1}{2}\hat{v}_{\perp}^{2}\delta_{\mu\nu}^{\perp}\right) \\ &\times\left(g_{\mu\alpha}g_{\nu\beta}\frac{\partial g_{l\beta}\hat{V}_{\perp l}}{\partial x_{\alpha}}+g_{\nu\alpha}g_{\mu\beta}\frac{\partial g_{l\beta}\hat{V}_{\perp l}}{\partial x_{\alpha}}-\delta_{\mu\nu}^{\perp}\frac{\partial g_{l\alpha}\hat{V}_{\perp l}}{\partial x_{\alpha}}\right). \end{aligned}$$

Similarly, the last term on the right-hand side of equation (3.59) can be reduced as follows.

$$\begin{aligned} &g_{\delta\alpha}g_{\gamma\mu}\frac{\partial g_{3\mu}}{\partial x_{\alpha}}\left(\hat{v}_{\perp\delta}\hat{v}_{\perp\gamma}-\frac{1}{2}\hat{v}_{\perp\nu}^{2}\delta_{\delta\gamma}^{\perp}\right)g_{\delta\alpha}g_{\gamma\mu}\frac{\partial g_{3\mu}}{\partial x_{\alpha}} \\ &= \frac{1}{2}\left(\hat{v}_{\perp\delta}\hat{v}_{\perp\gamma}-\frac{1}{2}\hat{v}_{\perp\nu}^{2}\delta_{\delta\gamma}^{\perp}\right)\left(g_{\delta\alpha}g_{\gamma\mu}\frac{\partial g_{3\mu}}{\partial x_{\alpha}}+g_{\gamma\alpha}g_{\delta\mu}\frac{\partial g_{3\mu}}{\partial x_{\alpha}}\right) \\ &= \frac{1}{2}\left(\hat{v}_{\perp\delta}\hat{v}_{\perp\gamma}-\frac{1}{2}\hat{v}_{\perp\nu}^{2}\delta_{\delta\gamma}^{\perp}\right) \\ &\times\left(g_{\delta\alpha}g_{\gamma\mu}\frac{\partial g_{3\mu}}{\partial x_{\alpha}}+g_{\gamma\alpha}g_{\delta\mu}\frac{\partial g_{3\mu}}{\partial x_{\alpha}}-\delta_{\delta\gamma}^{\perp}\frac{\partial g_{3\alpha}}{\partial x_{\alpha}}\right) \end{aligned}$$

Therefore, the first order kinetic equation (3.59) is further reduced to

$$-\frac{e}{m}e_{\beta\gamma\alpha}g_{l\gamma}\hat{v}_{\perp l}B_\alpha g_{r\beta}\frac{\partial f^1}{\partial\hat{v}_r}\Bigg|_{t,\mathbf{x}} = g_{\beta\alpha}\hat{v}_{\perp\beta}\left(\frac{m\hat{v}_\mu^2}{2T}-2\right)\frac{\partial\ln T}{\partial x_\alpha}f^0 + \frac{m}{2T}\left(\hat{v}_{\perp\mu}\hat{v}_{\perp\nu}-\frac{1}{2}\hat{v}_\perp^2\delta_{\mu\nu}^\perp\right)f^0 W_{\mu\nu}, \tag{3.60}$$

where the rate-of-strain tensor is

$$W_{\mu\nu} = W_{\mu\nu}^\perp + W_{\mu\nu}^\parallel, \tag{3.61}$$

$$W_{\mu\nu}^\perp = g_{\mu\alpha}g_{\nu\beta}\frac{\partial g_{l\beta}\hat{V}_{\perp l}}{\partial x_\alpha} + g_{\nu\alpha}g_{\mu\beta}\frac{\partial g_{l\beta}\hat{V}_{\perp l}}{\partial x_\alpha} - \delta_{\mu\nu}^\perp\frac{\partial g_{l\alpha}\hat{V}_{\perp l}}{\partial x_\alpha},$$

$$W_{\mu\nu}^\parallel = \left(g_{\mu\alpha}g_{\nu\beta}\frac{\partial g_{3\beta}}{\partial x_\alpha} + g_{\nu\alpha}g_{\mu\beta}\frac{\partial g_{3\beta}}{\partial x_\alpha} - \delta_{\mu\nu}^\perp\frac{\partial g_{3\alpha}}{\partial x_\alpha}\right)\left(\hat{v}_3 + \frac{T}{m}\frac{\partial\ln n}{\partial\hat{v}_3}\right).$$

Note that this definition is different from the Braginskii one in equation (2.39). The parallel fluid velocity is absent, but $W_{\mu\nu}^\parallel$ appears. The species indices are suppressed in equations (3.60) and (3.61).

3.3.2 Solutions of the first order kinetic equations

In this subsection, we proceed to solve the first order kinetic equation (3.60). The solution procedure is similar to the Braginskii theory in section 2.3.6. For brevity, the species indices are suppressed.

Since the first term on the right-hand side equation (3.60) is a vector and the second one is a tensor in velocity space, their solutions can be split as follows

$$f^1 = f_v^1 + f_w^1 \equiv f^0\Phi_v + f^0\Phi_w,$$

where the first and second terms on the right contain the subscripts v and w, which represent, respectively, the solutions to the first and second terms on the right-hand side of equation equation (3.60).

We first solve the contribution from the first term on the right-hand side of (3.60). Using equation (3.49), one obtains the equation for Φ_v:

$$-\frac{e}{m}e_{\beta\gamma\alpha}g_{\mu\gamma}\hat{v}_\mu B_\alpha g_{\delta\beta}\frac{\partial\Phi_v}{\partial\hat{v}_\delta}\Bigg|_{t,\mathbf{x}} = g_{\beta\alpha}\hat{v}_{\perp\beta}\left(\frac{m\hat{v}_\mu^2}{2T}-2\right)\frac{\partial\ln T}{\partial x_\alpha}. \tag{3.62}$$

Similar to the Braginskii theory in section 2.3.6, one can solve this equation. Note that

$$e_{\beta\gamma\alpha}g_{\mu\gamma}\hat{v}_{\perp\mu}\omega_{c\alpha}g_{\delta\beta}\frac{\partial}{\partial\hat{v}_\delta}A(v^2) = 0$$

and

$$
\begin{aligned}
&e_{\beta\gamma\alpha}g_{\mu\gamma}\hat{v}_{\perp\mu}\omega_{c\alpha}g_{\delta\beta}\frac{\partial}{\partial\hat{v}_\delta}\left(e_{pqr}g_{lq}\hat{v}_{\perp l}\omega_{cr}\frac{\partial \ln T}{\partial x_p}\right)\\
&= e_{\beta\gamma\alpha}g_{\mu\gamma}\hat{v}_{\perp\mu}\omega_{c\alpha}e_{p\beta r}\omega_{cr}\frac{\partial \ln T}{\partial x_p}\\
&= g_{\mu\gamma}\hat{v}_{\perp\mu}\omega_{c\alpha}\omega_{c\gamma}\frac{\partial \ln T}{\partial x_\alpha} - g_{\mu\gamma}\hat{v}_{\perp\mu}\omega_{c\alpha}\omega_{c\alpha}\frac{\partial \ln T}{\partial x_\gamma}\\
&= -\,\omega_{c\alpha}^2 g_{\mu\gamma}\hat{v}_{\perp\mu}\frac{\partial \ln T}{\partial x_\gamma},
\end{aligned}
$$

where ω_c is gyrofrequency, A is an arbitrary function of v^2, and the identity $e_{\beta\gamma\alpha}e_{\beta rp} = \delta_{\gamma r}\delta_{\alpha p} - \delta_{\gamma p}\delta_{\alpha r}$ as given in appendix A has been used. Therefore, one obtains the vector part of the solution to equation (3.62) as

$$
f_v^1 \equiv f^0\Phi_v = f^0\frac{1}{\omega_c^2}\left(\frac{m\hat{v}_\mu^2}{2T} - 2\right)e_{pqr}g_{lq}\hat{v}_{\perp l}\omega_{cr}\frac{\partial \ln T}{\partial x_p}. \tag{3.63}
$$

We next solve the contribution from the second term on the right-hand side of equation (3.60). Using equation (3.49), one obtains the equation for Φ_w:

$$
-\frac{e}{m}e_{\beta\gamma\alpha}g_{l\gamma}\hat{v}_{\perp l}B_\alpha g_{r\beta}\frac{\partial\Phi_w^1}{\partial\hat{v}_r} = \frac{m}{2T}\hat{v}_{\mu\nu}W_{\mu\nu}, \tag{3.64}
$$

where

$$
\hat{v}_{\mu\nu} = \hat{v}_\mu\hat{v}_\nu - \frac{1}{2}\hat{v}_\perp^2\delta_{\mu\nu}.
$$

Similar to the Braginskii theory in section 2.3.6, to solve equation (3.64), we decompose $W_{\mu\nu}$ as follows

$$
W = W_{(0)} + W_{(1)}, \tag{3.65}
$$

where

$$
W_{(0)} = \begin{pmatrix} \frac{1}{2}(W_{11} + W_{22}) & 0 & 0 \\ 0 & \frac{1}{2}(W_{11} + W_{22}) & 0 \\ 0 & 0 & 0 \end{pmatrix}, \tag{3.66}
$$

$$
W_{(1)} = \begin{pmatrix} \frac{1}{2}(W_{11} - W_{22}) & W_{12} & 0 \\ W_{21} & -\frac{1}{2}(W_{11} - W_{22}) & 0 \\ 0 & 0 & 0 \end{pmatrix}. \tag{3.67}
$$

We also introduce the following accompanying matrix to $W_{(1)}$:

$$W_{(3)} = \begin{pmatrix} -W_{12} & \frac{1}{2}(W_{11} - W_{22}) & 0 \\ \frac{1}{2}(W_{11} - W_{22}) & W_{12} & 0 \\ 0 & 0 & 0 \end{pmatrix}. \tag{3.68}$$

Noting that the trace of matrix sum $W_{11} + W_{22}$ vanishes and $W_{\mu\nu} = W_{\nu\mu}$, one can see that these three matrices $W_{(0,\,1,\,3)}$ are symmetric and their traces vanish.

One can prove that

$$e_{\beta\gamma\alpha}g_{l\gamma}\hat{v}_{\perp l}b_\alpha g_{r\beta}\frac{\partial\hat{v}_{\mu\nu}}{\partial\hat{v}_r} = e_{\hat{3}\beta\gamma}g_{l\gamma}\hat{v}_{\perp l}g_{\mu\beta}\hat{v}_\nu + e_{\hat{3}\beta\gamma}g_{l\gamma}\hat{v}_{\perp l}g_{\nu\beta}\hat{v}_\mu. \tag{3.69}$$

Noting that $e_{\hat{3}\beta\gamma}g_{l\gamma}\hat{v}_{\perp l}g_{\mu\beta}\hat{v}_\mu = 0$ and using equation (3.69), one has

$$\begin{aligned} e_{\beta\gamma\alpha}g_{l\gamma}\hat{v}_{\perp l}b_\alpha g_{r\beta}\frac{\partial\hat{v}_{\mu\nu}}{\partial\hat{v}_r}W_{(0)\mu\nu} &= e_{\hat{3}\beta\gamma}g_{l\gamma}\hat{v}_{\perp l}g_{\mu\beta}\hat{v}_\nu\delta^{\perp}_{\mu\nu}W_{(0)11} \\ &= e_{\hat{3}\beta\gamma}g_{l\gamma}\hat{v}_{\perp l}g_{\nu\beta}\hat{v}_{\perp\nu}W_{(0)11} = 0. \end{aligned} \tag{3.70}$$

where $b_\alpha = B_\alpha/B$ and the diagonal feature of $W_{(0)\mu\nu}$ has been exploited. One can also prove that

$$\hat{v}_{\mu\nu}W_{(0)\mu\nu} = \left(\hat{v}_1\hat{v}_1 - \frac{1}{2}\hat{v}^2\right)W_{(0)11} + \left(\hat{v}_2\hat{v}_2 - \frac{1}{2}\hat{v}^2\right)W_{(0)22} = 0. \tag{3.71}$$

Here, it has been noted that $W_{(0)11} = W_{(0)22}$. Furthermore, taking into account equation (3.70) one obtains

$$\begin{aligned} e_{\beta\gamma\alpha}g_{l\gamma}\hat{v}_{\perp l}b_\alpha g_{r\beta}\frac{\partial\hat{v}_{11}}{\partial\hat{v}_r}W_{(3)11} &= \left(e_{\hat{3}\beta\gamma}g_{l\gamma}\hat{v}_{\perp l}g_{1\beta}\hat{v}_1 + e_{\hat{3}\beta\gamma}g_{l\gamma}\hat{v}_{\perp l}g_{1\beta}\hat{v}_1\right)W_{(3)11} \\ &= 2e_{\hat{3}\beta\gamma}g_{2\gamma}g_{1\beta}\hat{v}_1\hat{v}_2W_{(3)11} \\ &= 2\mathbf{b}\cdot\mathbf{e}_\beta(\hat{\mathbf{v}}_1\cdot\mathbf{e}_\beta)\times\ \mathbf{e}_\gamma(\hat{\mathbf{v}}_2\cdot\mathbf{e}_\gamma)W_{(3)11} \\ &= 2\mathbf{b}\cdot\hat{\mathbf{v}}_1\ \times\ \hat{\mathbf{v}}_2 \\ &= 2\hat{v}_1\hat{v}_2W_{(3)11} \\ &= -\,2\hat{v}_{12}W_{(1)12}, \end{aligned}$$

$$\begin{aligned} e_{\beta\gamma\alpha}g_{l\gamma}\hat{v}_{\perp l}b_\alpha g_{r\beta}\frac{\partial\hat{v}_{22}}{\partial\hat{v}_r}W_{(3)22} &= \left(e_{\hat{3}\beta\gamma}g_{l\gamma}\hat{v}_{\perp l}g_{2\beta}\hat{v}_2 + e_{\hat{3}\beta\gamma}g_{l\gamma}\hat{v}_{\perp l}g_{2\beta}\hat{v}_2\right)W_{(3)22} \\ &= \left(e_{\hat{3}\beta\gamma}g_{1\gamma}\hat{v}_1g_{2\beta}\hat{v}_2 + e_{\hat{3}\beta\gamma}g_{1\gamma}\hat{v}_1g_{2\beta}\hat{v}_2\right)W_{(3)22} \\ &= -\,2\hat{v}_1\hat{v}_2W_{(3)22} \\ &= -\,2\hat{v}_{21}W_{(1)21}, \end{aligned}$$

and

$$
\begin{aligned}
& e_{\beta\gamma\alpha} g_{l\gamma} \hat{v}_{\perp l} b_\alpha g_{r\beta} \frac{\partial \hat{v}_{12}}{\partial \hat{v}_r} W_{(3)12} + e_{\beta\gamma\alpha} g_{l\gamma} \hat{v}_{\perp l} b_\alpha g_{r\beta} \frac{\partial \hat{v}_{21}}{\partial \hat{v}_r} W_{(3)21} \\
&= \left(e_{\hat{3}\beta\gamma} g_{l\gamma} \hat{v}_{\perp l} g_{1\beta} \hat{v}_2 + e_{\hat{3}\beta\gamma} g_{l\gamma} \hat{v}_{\perp l} g_{2\beta} \hat{v}_1 \right) W_{(3)12} \\
&\quad + \left(e_{\hat{3}\beta\gamma} g_{l\gamma} \hat{v}_{\perp l} g_{1\beta} \hat{v}_2 + e_{\hat{3}\beta\gamma} g_{l\gamma} \hat{v}_{\perp l} g_{2\beta} \hat{v}_1 \right) W_{(3)21} \\
&= \left(e_{\hat{3}\beta\gamma} g_{2\gamma} \hat{v}_2 g_{1\beta} \hat{v}_2 + e_{\hat{3}\beta\gamma} g_{1\gamma} \hat{v}_1 g_{2\beta} \hat{v}_1 \right) W_{(3)12} \\
&\quad + \left(e_{\hat{3}\beta\gamma} g_{2\gamma} \hat{v}_2 g_{1\beta} \hat{v}_2 + e_{\hat{3}\beta\gamma} g_{1\gamma} \hat{v}_1 g_{2\beta} \hat{v}_1 \right) W_{(3)21} \\
&= (\hat{v}_2 \hat{v}_2 - \hat{v}_1 \hat{v}_1) W_{(3)12} + (\hat{v}_2 \hat{v}_2 - \hat{v}_1 \hat{v}_1) W_{(3)21} \\
&= - 2\hat{v}_{11} W_{(1)11} - 2\hat{v}_{22} W_{(1)22}.
\end{aligned}
$$

Here, it has been noted that $e_{\hat{3}\beta\gamma} g_{1\beta} g_{2\gamma} = \hat{\mathbf{e}}_3 \cdot \hat{\mathbf{e}}_1 \times \hat{\mathbf{e}}_2 = 1$, $e_{\hat{3}\beta\gamma} g_{1\beta} g_{1\gamma} = e_{\hat{3}\beta\gamma} g_{2\beta} g_{2\gamma} = 0$, and $W_{(3)21} = W_{(3)12}$. In summary, one has

$$
e_{\beta\gamma\alpha} g_{l\gamma} \hat{v}_{\perp l} b_\alpha g_{r\beta} \frac{\partial \hat{v}_{\alpha\beta}}{\partial \hat{v}_r} W_{(3)\alpha\beta} = -2 v_{\alpha\beta} W_{(1)\alpha\beta}, \tag{3.72}
$$

According to equation (3.71), $W_{(0)\alpha\beta}$ vanishes on the right-hand side of equation (3.64). Therefore, using equations (3.70) and (3.72), the tensor part of the solution to equation (3.71) becomes

$$
\Phi = \frac{m}{4T\omega} \hat{v}_{\alpha\beta} W_{(3)\alpha\beta}. \tag{3.73}
$$

Here, it is pointed out again that the species indices in the solutions in equations (3.63) and (3.73) are suppressed.

3.4 Transport coefficients

In this section, the solutions of the first order kinetic equations (3.63) and (3.73) are used to compute the transport coefficients. This provides the kinetic closure of perpendicular magnetofluid equations.

For the convenience of reference, the solutions of the first order kinetic equations are recited here

$$
f^1 = f_v^1 + f_w^1, \tag{3.74}
$$

where

$$
f_v^1 = f^0 \frac{1}{\omega_{ca}^2 g} \left(\frac{m \hat{v}_\mu^2}{2T} - 2 \right) e_{pqr} g_{lq} \hat{v}_{\perp l} \omega_{cr} \frac{\partial \ln T}{\partial x_p}, \tag{3.75}
$$

$$
f_w^1 = f^0 \frac{m}{4T\omega} \hat{v}_{\alpha\beta} W_{(3)\alpha\beta}, \tag{3.76}
$$

and $\hat{v}_{\alpha\beta} = \hat{v}_\alpha \hat{v}_\beta - \frac{1}{2}\hat{v}_\perp^2 \delta_{\alpha\beta}^\perp$. The moments from f_v^1 and f_w^1 can be generally computed as follows:

$$\langle\eta\rangle_v = \frac{1}{n}\int\int d\hat{v}_1\, d\hat{v}_2 \eta f^0 \frac{1}{\omega_{ca}^2}\left(\frac{m\hat{v}_\mu^2}{2T} - 2\right) e_{pqr} g_{lq} \hat{v}_{\perp l} \omega_{cr} \frac{\partial \ln T}{\partial x_p} \tag{3.77}$$

and

$$\begin{aligned}
\langle\eta\rangle_w &= \frac{1}{n}\int\int d\hat{v}_1\, d\hat{v}_2 \eta \frac{m}{4T\omega}\left(\hat{v}_\alpha \hat{v}_\beta - \frac{1}{2}\hat{v}_\perp^2 \delta_{\alpha\beta}\right) W_{(3)\alpha\beta} f^0 \\
&= \frac{1}{n}\int\int d\hat{v}_1\, d\hat{v}_2 \eta \frac{m}{4T\omega}(\hat{v}_1\hat{v}_1 - \hat{v}_2\hat{v}_2) W_{(3)11} f^0 \\
&\quad + \frac{1}{n}\int\int d\hat{v}_1\, d\hat{v}_2 \eta \frac{m}{T\omega}(\hat{v}_1\hat{v}_2 + \hat{v}_2\hat{v}_1) W_{(3)12} f^0 \\
&= \frac{m}{4T\omega} W_{(3)11} \frac{1}{n}\int\int d\hat{v}_1\, d\hat{v}_2 \eta(\hat{v}_1\hat{v}_1 - \hat{v}_2\hat{v}_2) f^0 \\
&\quad + \frac{m}{2T\omega} W_{(3)12} \frac{1}{n}\int\int d\hat{v}_1\, d\hat{v}_2 \eta \hat{v}_1 \hat{v}_2 f^0 .
\end{aligned} \tag{3.78}$$

First, we check the constraints in equation (3.32) for the definitions of n, $\mathbf{V}$, T, and D. We carry it out one by one for f_v^1 and f_w^1 parts in equations (3.75) and (3.76). Because f_v^1 is odd in $\hat{v}_\perp$, its contributions to the even moments n, T, and D vanish. Therefore, we only need to check the $\hat{v}_\perp$ contribution to $\mathbf{V}$. Using equation (3.77), one obtains

$$\begin{aligned}
V_\delta &= \frac{1}{n}\int\int d\hat{v}_1\, d\hat{v}_2 \hat{v}_\delta f^0 \frac{1}{\omega_{ca}^2}\left(\frac{m\hat{v}_\mu^2}{2T} - 2\right) e_{pqr} g_{lq} \hat{v}_{\perp l} \omega_{cr} \frac{\partial \ln T}{\partial x_p} \\
&= \frac{1}{n}\delta_{l\delta}^\perp e_{pqr} g_{lq} \omega_{cr} \frac{1}{\omega_{ca}^2}\frac{\partial \ln T}{\partial x_p} \int\int d\hat{v}_1\, d\hat{v}_2 \hat{v}_k^2 \left(\frac{m\hat{v}_\mu^2}{2T} - 2\right) f^0 \\
&= 0.
\end{aligned}$$

Here, the integration table in appendix B.2 has been used. As for f_w^1 contributions, equation (3.78) shows that the moments to n, $\mathbf{V}$, T, and D with $\eta = 1$, $\hat{v}_\perp$, $\hat{v}_\perp^2$, and $\hat{v}_\perp^4$, respectively, vanish. Therefore, the constraints in equation (3.32) are proved.

Next, we compute the transport coefficients one by one with the first order solution in equation (3.74). For the heat flux, using equation (3.78) one can prove that f_w^1 contribution to the heat flux vanishes. Therefore, one only needs to take into account the f_v^1 contribution. Using equation (3.77), one obtains

$$
\begin{aligned}
q_\delta &= \frac{\rho}{2}\frac{1}{\hat{n}}\int\int d\hat{v}_1\, d\hat{v}_2 \hat{v}_\perp^2 \hat{v}_\delta f^0 \frac{1}{\omega_{ca}^2}\left(\frac{m\hat{v}_\mu^2}{2T} - 2\right) e_{pqr} g_{lq} \hat{v}_{\perp l} \omega_{cr} \frac{\partial \ln T}{\partial x_p} \\
&= \frac{\rho}{2}\frac{1}{\hat{n}}\int\int d\hat{v}_1\, d\hat{v}_2 \frac{1}{2}\hat{v}_\perp^4 \delta_{\delta l}^{\perp} f^0 \frac{1}{\omega_{ca}^2}\left(\frac{m\hat{v}_\mu^2}{2T} - 2\right) e_{pqr} g_{lq} \omega_{cr} \frac{\partial \ln T}{\partial x_p} \\
&= \frac{\rho}{4}\frac{T^2}{m^2\omega_c^2} e_{pqr} g_{\delta q} \omega_{cr}\left(\frac{48}{2} - 2\cdot 8\right)\frac{\partial \ln T}{\partial x_p} \\
&= \frac{2nT}{m\omega_c}\hat{\mathbf{e}}_\delta \cdot \mathbf{b}\times\nabla T.
\end{aligned}
\tag{3.79}
$$

For the pressure tensor, using equation (3.77), one can prove that f_v^1 contribution vanishes. Therefore, one only needs to take into account the f_w^1 contribution. Using equation (3.78), one obtains

$$
\begin{aligned}
\pi_{\mu\nu} &= \frac{m}{4T\omega} W_{(3)11}\rho\frac{1}{n}\int\int d\hat{v}_1\, d\hat{v}_2\left(\hat{v}_\mu\hat{v}_\nu - \frac{1}{2}\hat{v}_\perp^2\delta_{\mu\nu}\right)(\hat{v}_1\hat{v}_1 - \hat{v}_2\hat{v}_2) f^0 \\
&+ \frac{m}{2T\omega} W_{(3)12}\rho\frac{1}{n}\int\int d\hat{v}_1\, d\hat{v}_2\left(\hat{v}_\mu\hat{v}_\nu - \frac{1}{2}\hat{v}_\perp^2\delta_{\mu\nu}\right)\hat{v}_1\hat{v}_2 f^0.
\end{aligned}
\tag{3.80}
$$

Specifically, for each component, one has

$$
\begin{aligned}
\pi_{11} = -\pi_{22} &= -\frac{m}{4T\omega_c} W_{(3)11}\rho\frac{1}{n}\int\int d\hat{v}_1\, d\hat{v}_2 \hat{v}_2\hat{v}_2(\hat{v}_1\hat{v}_1 - \hat{v}_2\hat{v}_2) f^0 \\
&= \rho\frac{m}{4T\omega_c} W_{(3)11}(3 - 1)\frac{T^2}{m^2} \\
&= -\frac{nT}{2\omega_c} W_{12}
\end{aligned}
\tag{3.81}
$$

and

$$
\begin{aligned}
\pi_{12} = \pi_{21} &= \frac{m}{2T\omega_c} W_{(3)12}\rho\frac{1}{n}\int\int d\hat{v}_1\, d\hat{v}_2 \hat{v}_1^2\hat{v}_2^2 f^0 \\
&= \rho\frac{m}{2T\omega_c} W_{(3)12}\frac{T^2}{m^2} \\
&= \frac{nT}{2\omega_c}\frac{1}{2}(W_{11} - W_{22}).
\end{aligned}
\tag{3.82}
$$

The rest of the components vanish.

In passing, one can also express the gyroviscous tensor π in the index notation as in the Braginskii transport theory [3].

$$
\pi_{\alpha\beta} = \eta_3 W_{3\alpha\beta},
\tag{3.83}
$$

where

$$\eta_3 = nT/2\omega_c,$$
$$W_{3\alpha\beta} = \frac{1}{2}\left(\delta^{\perp}_{\alpha\mu}e_{\beta\gamma\nu} + \delta^{\perp}_{\beta\nu}e_{\alpha\gamma\mu}\right)b_\gamma W_{\mu\nu}.$$

This can be directly verified as follows.

$$W_{311} = \frac{1}{2}\left(\delta^{\perp}_{1\mu}e_{13\nu} + \delta^{\perp}_{1\nu}e_{13\mu}\right)b_3 W_{\mu\nu} = \frac{1}{2}(-W_{12} - W_{21}) = -W_{12},$$
$$W_{322} = \frac{1}{2}\left(\delta^{\perp}_{2\mu}e_{23\nu} + \delta^{\perp}_{2\nu}e_{23\mu}\right)b_3 W_{\mu\nu} = \frac{1}{2}(W_{21} + W_{12}) = W_{12} = -W_{311},$$
$$W_{312} = \frac{1}{2}\left(\delta^{\perp}_{1\mu}e_{23\nu} + \delta^{\perp}_{2\nu}e_{13\mu}\right)b_3 W_{\mu\nu} = \frac{1}{2}(W_{11} - W_{22}),$$
$$W_{321} = \frac{1}{2}\left(\delta^{\perp}_{2\mu}e_{13\nu} + \delta^{\perp}_{1\nu}e_{23\mu}\right)b_3 W_{\mu\nu} = \frac{1}{2}(-W_{22} + W_{11}) = W_{312}.$$

For the heat flux tensor, using equation (3.78) one can prove that f_w^1 contribution vanishes. Therefore, one only needs to take into account the f_v^1 contribution. Using equation (3.77); one obtains

$$\Theta_{\alpha\beta\gamma} = \rho\frac{1}{n}\int\int d\hat{v}_1\, d\hat{v}_2\left(\hat{v}_\alpha\hat{v}_\beta - \frac{1}{2}\hat{v}_\nu^2\delta^{\perp}_{\alpha\beta}\right)\hat{v}_\gamma$$
$$\times f^0\frac{1}{\omega_c^2}\left(\frac{m\hat{v}_\mu^2}{2T} - 2\right)e_{pqr}g_{lq}\hat{v}_{\perp l}\omega_{cr}\frac{\partial \ln T}{\partial x_p}. \tag{3.84}$$

Specifically, for each component, one has

$$\begin{aligned}
\Theta_{111} &= \rho\frac{1}{2n}\int\int d\hat{v}_1\, d\hat{v}_2(\hat{v}_1\hat{v}_1 - \hat{v}_2\hat{v}_2)\hat{v}_1^2 \\
&\quad\times f^0\frac{1}{\omega_c^2}\left(\frac{m\hat{v}_\mu^2}{2T} - 2\right)e_{pqr}g_{1q}\omega_{cr}\frac{\partial \ln T}{\partial x_p} \\
&= \rho\frac{1}{2\omega_c^2}e_{pqr}g_{1q}\omega_{cr}\frac{\partial \ln T}{\partial x_p}\frac{1}{n}\int\int d\hat{v}_1\, d\hat{v}_2 \\
&\quad\times \hat{v}_1^2\left[\frac{m}{2T}(\hat{v}_1^4 - \hat{v}_2^4) - 2(\hat{v}_1^2 - \hat{v}_2^2)\right] \\
&= \rho\frac{1}{2\omega_c^2}e_{pqr}g_{1q}\omega_{cr}\frac{\partial \ln T}{\partial x_p}\frac{T^2}{m^2}\left(\frac{15}{2} - \frac{3}{2} - 6 + 2\right) \\
&= \frac{1}{\omega_c^2}e_{pqr}g_{1q}\omega_{cr}\frac{nT^2}{m}\frac{\partial \ln T}{\partial x_p} \\
&= \frac{nT^2}{m\omega_c^2}\hat{\mathbf{e}}_1 \times \boldsymbol{\omega}_c \cdot \nabla \ln T,
\end{aligned} \tag{3.85}$$

$$\begin{aligned}\Theta_{112} &= \rho\frac{1}{2n}\int\int d\hat{v}_1\, d\hat{v}_2(\hat{v}_1\hat{v}_1 - \hat{v}_2\hat{v}_2)\hat{v}_2^2 \\ &\quad \times f^0\frac{1}{\omega_c^2}\left(\frac{m\hat{v}_\mu^2}{2T} - 2\right)e_{pqr}g_{2q}\hat{\omega}_{cr}\frac{\partial \ln T}{\partial x_p} \\ &= -\frac{1}{\omega_c^2}e_{pqr}g_{2q}\hat{\omega}_{cr}\frac{nT^2}{m}\frac{\partial \ln T}{\partial x_p} \\ &= -\frac{nT^2}{m\omega_c^2}\hat{\mathbf{e}}_2 \times \boldsymbol{\omega}_c \cdot \nabla \ln T,\end{aligned} \tag{3.86}$$

$$\begin{aligned}\Theta_{121} = \Theta_{211} &= \rho\frac{1}{n}\int\int d\hat{v}_1\, d\hat{v}_2\hat{v}_1^2\hat{v}_2^2 \\ &\quad \times f^0\frac{1}{\omega_c^2}\left(\frac{m\hat{v}_\mu^2}{2T} - 2\right)e_{pqr}g_{2q}\hat{\omega}_{cr}\frac{\partial \ln T}{\partial x_p} \\ &= \frac{1}{\omega_c^2}e_{pqr}g_{2q}\hat{\omega}_{cr}\frac{nT^2}{m}\frac{\partial \ln T}{\partial x_p}\left[\frac{1}{2}(3 + 3) - 2\right] \\ &= \omega_c^2 e_{pqr}g_{2q}\hat{\omega}_{cr}\frac{nT^2}{m}\frac{\partial \ln T}{\partial x_p} \\ &= -\Theta_{112},\end{aligned} \tag{3.87}$$

$$\begin{aligned}\Theta_{122} = \Theta_{212} &= \rho\frac{1}{n}\int\int d\hat{v}_1\, d\hat{v}_2\hat{v}_1^2\hat{v}_2^2 \\ &\quad \times f^0\frac{1}{\omega_c^2}\left(\frac{m\hat{v}_\mu^2}{2T} - 2\right)e_{pqr}g_{1q}\hat{\omega}_{cr}\frac{\partial \ln T}{\partial x_p} \\ &= \frac{1}{\omega_c^2}e_{pqr}g_{1q}\hat{\omega}_{cr}\frac{nT^2}{m}\frac{\partial \ln T}{\partial x_p}\left[\frac{1}{2}(3 + 3) - 2\right] \\ &= \omega_c^2 e_{pqr}g_{1q}\hat{\omega}_{cr}\frac{nT^2}{m}\frac{\partial \ln T}{\partial x_p} \\ &= \Theta_{111},\end{aligned} \tag{3.88}$$

$$\begin{aligned}\Theta_{221} &= -\rho\frac{1}{2n}\int\int d\hat{v}_1\, d\hat{v}_2(\hat{v}_1\hat{v}_1 - \hat{v}_2\hat{v}_2)\hat{v}_1^2 \\ &\quad \times f^0\frac{1}{\omega_c^2}\left(\frac{m\hat{v}_\mu^2}{2T} - 2\right)e_{pqr}g_{1q}\hat{\omega}_{cr}\frac{\partial \ln T}{\partial x_p} \\ &= -\Theta_{111},\end{aligned} \tag{3.89}$$

and

$$\begin{aligned}\Theta_{222} &= -\rho\frac{1}{2n}\int\int d\hat{v}_1\, d\hat{v}_2(\hat{v}_1\hat{v}_1 - \hat{v}_2\hat{v}_2)\hat{v}_2^2 \\ &\quad\times f^0\frac{1}{\omega_c^2}\left(\frac{m\hat{v}_\mu^2}{2T} - 2\right)e_{pqr}g_{2q}\hat{\omega}_{cr}\frac{\partial \ln T}{\partial x_p} \\ &= -\Theta_{112}.\end{aligned} \tag{3.90}$$

The rest of the components vanish.

For the T-square tensor, using equation (3.77), one can prove that f_v^1 contribution vanishes. Therefore, one only needs to take into account the f_w^1 contribution. Using equation (3.78), one obtains

$$\begin{aligned}\Xi_{\mu\nu} &= \frac{m}{4T\omega_c}W_{(3)11}\frac{m^2}{2}\int\int d\hat{v}_1 d\hat{v}_2\hat{v}_\perp^2\left(\hat{v}_\mu\hat{v}_\nu - \frac{1}{2}\hat{v}_\perp^2\delta_{\mu\nu}\right)(\hat{v}_1\hat{v}_1 - \hat{v}_2\hat{v}_2)f^0 \\ &\quad+ \frac{m^2}{2}\frac{m}{2T\omega_c}W_{(3)12}\frac{1}{n}\int\int d\hat{v}_1 d\hat{v}_2\hat{v}_\perp^2\left(\hat{v}_\mu\hat{v}_\nu - \frac{1}{2}\hat{v}_\perp^2\delta_{\mu\nu}\right)\hat{v}_1\hat{v}_2 f^0.\end{aligned} \tag{3.91}$$

Specifically, for each components one has

$$\begin{aligned}\Xi_{11} = -\Xi_{22} &= -\frac{m^2}{2}\frac{m}{4T\omega_c}W_{(3)11}\int\int d\hat{v}_1\, d\hat{v}_2\hat{v}_\perp^2\hat{v}_2\hat{v}_2(\hat{v}_1\hat{v}_1 - \hat{v}_2\hat{v}_2)f^0 \\ &= -\frac{m^2}{2}\frac{m}{4T\omega_c}W_{(3)11}\frac{1}{n}\int\int d\hat{v}_1\, d\hat{v}_2\hat{v}_2^2(\hat{v}_1^4 - \hat{v}_2^4)f^0 \\ &= \frac{m^2}{2}\frac{m}{4T\omega_c}W_{(3)11}(3 - 15)\frac{T^3}{m^3} \\ &= \frac{3T^2}{2\omega_c}W_{12},\end{aligned} \tag{3.92}$$

and

$$\begin{aligned}\Xi_{12} = \Xi_{21} &= \frac{m^2}{2}\frac{m}{2T\omega_c}W_{(3)12}\frac{1}{n}\int\int d\hat{v}_1\, d\hat{v}_2\hat{v}_\perp^2\hat{v}_1^2\hat{v}_2^2 f^0 \\ &= \frac{m^2}{2}\frac{1}{n}\frac{m}{2T\omega_c}W_{(3)12}\frac{6T^3}{m^3} \\ &= \frac{3T^2}{2\omega_c}\frac{1}{2}(W_{11} - W_{22}).\end{aligned} \tag{3.93}$$

The rest of the components vanish.

It is interesting to compare the transport coefficients in the perpendicular magnetofluid theory with the Braginskii ones in the collisionless limit in section 2.3.7. In the Braginskii transport theory, there are only two collisionless transport parameters: the thermal heat flux and gyroviscous tensor.

Comparing equation (2.172) in the Braginskii theory with equation (3.79) here, one can see that the thermal heat flux assumes the same structure, except for the constant coefficients: 5/2 versus 2. This is because only two perpendicular velocity components are taken into account in the perpendicular magnetofluid theory, while the Braginskii one involves three components.

Comparing equations (2.174)–(2.175) in the Braginskii theory with equations (3.81)–(3.82) here, one can see that the gyroviscous tensors also have the same structure, except for the difference in the definitions of the rate-of-strain tensors. The parallel fluid velocity does not appear in the perpendicular magnetofluid theory.

Also, we note that there are higher order transport coefficients in the perpendicular magnetofluid description: the thermal heat flux tensor $\Theta_{\alpha\beta\gamma}$ and the T-square tensor $\Xi_{\alpha\beta}$.

3.5 Equation summary

The basic set of equations is summarized as follows. The continuity equation is collected from equation (3.40),

$$\begin{aligned}&\frac{\partial n}{\partial t} + g_{\beta\alpha}\hat{V}_{\perp\beta}\frac{\partial n}{\partial x_\alpha} + g_{3\alpha}\hat{v}_3\frac{\partial n}{\partial x_\alpha} + n\frac{\partial g_{l\gamma}\hat{V}_{\perp l}}{\partial x_\beta}\left(\delta_{\beta\gamma} - g_{3\beta}g_{3\gamma}\right)\\&+ n\hat{v}_3\frac{\partial g_{3\alpha}}{\partial x_\alpha} + \frac{1}{m}\frac{\partial g_{3\alpha}}{\partial x_\alpha}\frac{\partial nT}{\partial\hat{v}_3} - g_{3\beta}g_{3\gamma}\frac{\partial g_{l\gamma}\hat{V}_{\perp l}}{\partial x_\beta}\hat{v}_3\frac{\partial n}{\partial\hat{v}_3}\\&= - g_{3\beta}\frac{\partial n}{\partial\hat{v}_3}\left(\frac{e}{m}E^*_\beta - \frac{\hat{d}g_{l\beta}\hat{V}_{\perp l}}{dt}\right) - \frac{1}{m}g_{p\alpha}g_{q\mu}\frac{\partial g_{3\mu}}{\partial x_\alpha}\frac{\partial\pi_{pq}}{\partial\hat{v}_3}.\end{aligned} \tag{3.94}$$

The momentum is collected from equation (3.41),

$$\begin{aligned}&(\delta_{\lambda\alpha} - g_{3\alpha}g_{3\lambda})\Bigg(neE^*_\alpha - \rho\frac{\hat{d}g_{l\alpha}\hat{V}_{\perp l}}{dt} - \frac{\partial p}{\partial x_\alpha} - \hat{v}_3 g_{3\beta}\frac{\partial g_{3\alpha}}{\partial x_\beta}\frac{\partial p}{\partial\hat{v}_3}\\&- g_{3\beta}\hat{v}_3^2\frac{\partial g_{3\alpha}}{\partial x_\beta}\rho + g_{3\gamma}\frac{\partial g_{l\gamma}\hat{V}_{\perp l}}{\partial x_\alpha}\frac{\partial p}{\partial\hat{v}_3} - \rho\hat{v}_3 g_{3\beta}\frac{\partial g_{l\alpha}\hat{V}_{\perp l}}{\partial x_\beta} - \rho\hat{v}_3\frac{\hat{d}g_{3\alpha}}{dt}\\&- \frac{\hat{d}g_{3\alpha}}{dt}\frac{\partial p}{\partial\hat{v}_3}\Bigg)\\&= g_{\nu\lambda}g_{\beta\alpha}\frac{\partial\pi_{\beta\nu}}{\partial x_\alpha} + g_{\nu\lambda}\left(g_{p\delta}\frac{\hat{d}g_{3\delta}}{dt} - g_{p\beta}g_{3\gamma}\frac{\partial g_{l\gamma}\hat{V}_{\perp l}}{\partial x_\beta} + \hat{v}_3 g_{3\alpha}g_{p\mu}\frac{\partial g_{3\mu}}{\partial x_\alpha}\right)\frac{\partial\pi_{p\nu}}{\partial\hat{v}_3}\\&+ \left(\frac{\partial g_{p\alpha}g_{q\lambda}}{\partial x_\alpha} + g_{p\alpha}g_{q\mu}g_{3\lambda}\frac{\partial g_{3\mu}}{\partial x_\alpha} + g_{3\alpha}g_{q\mu}g_{p\lambda}\frac{\partial g_{3\mu}}{\partial x_\alpha}\right)\pi_{pq}\\&+ g_{\nu\lambda}\frac{\partial g_{3\alpha}}{\partial x_\alpha}\frac{\partial q_\nu}{\partial\hat{v}_3} + g_{p\alpha}g_{q\mu}g_{\nu\lambda}\frac{\partial g_{3\mu}}{\partial x_\alpha}\frac{\partial\Theta_{pq\nu}}{\partial\hat{v}_3}.\end{aligned} \tag{3.95}$$

The energy transport equation is collected from equations (3.42) and (3.43),

$$\begin{aligned}
&\frac{\partial p}{\partial t} + g_{\beta\alpha}\hat{V}_{\perp\beta}\frac{\partial p}{\partial x_\alpha} + g_{3\alpha}\hat{v}_3\frac{\partial p}{\partial x_\alpha} + 2p\frac{\partial g_{l\gamma}\hat{V}_{\perp l}}{\partial x_\beta}\left(\delta_{\beta\gamma} - g_{3\beta}g_{3\gamma}\right)\\
&+ 2\hat{v}_3\frac{\partial g_{3\alpha}}{\partial x_\alpha}p + \frac{\partial g_{3\alpha}}{\partial x_\alpha}\frac{1}{m}\frac{\partial nD}{\partial \hat{v}_3} - \hat{v}_3 g_{3\beta}g_{3\gamma}\frac{\partial g_{l\gamma}\hat{V}_{\perp l}}{\partial x_\beta}\frac{\partial p}{\partial \hat{v}_3}\\
&= - g_{3\alpha}\left(\frac{e}{m}E^*_\alpha - \frac{\hat{d}g_{l\alpha}\hat{V}_{\perp l}}{dt}\right)\frac{\partial p}{\partial \hat{v}_3} - g_{p\alpha}g_{q\mu}\left(\hat{v}_3\frac{\partial g_{3\mu}}{\partial x_\alpha} + \frac{\partial g_{l\mu}\hat{V}_{\perp l}}{\partial x_\alpha}\right)\pi_{pq}\\
&+ \left(g_{p\beta}g_{3\gamma}\frac{\partial g_{l\gamma}\hat{V}_{\perp l}}{\partial x_\beta} - g_{p\delta}\frac{\hat{d}g_{3\delta}}{dt}\right)\frac{\partial q_p}{\partial \hat{v}_3} - \frac{\partial}{\partial x_\alpha}(g_{\beta\alpha}q_\beta)\\
&- \hat{v}_3 g_{3\alpha}g_{q\mu}\frac{\partial g_{3\mu}}{\partial x_\alpha}\frac{\partial q_q}{\partial \hat{v}_3} - g_{3\alpha}g_{q\mu}\frac{\partial g_{3\mu}}{\partial x_\alpha}q_q - g_{p\alpha}g_{q\mu}\frac{\partial g_{3\mu}}{\partial x_\alpha}\frac{1}{m}\frac{\partial n\Xi_{pq}}{\partial \hat{v}_3}
\end{aligned} \tag{3.96}$$

or

$$\begin{aligned}
&n\frac{\partial T}{\partial t} + ng_{\beta\alpha}\hat{V}_{\perp\beta}\frac{\partial T}{\partial x_\alpha} + ng_{3\alpha}\hat{v}_3\frac{\partial T}{\partial x_\alpha} + p\frac{\partial g_{l\gamma}\hat{V}_{\perp l}}{\partial x_\beta}\left(\delta_{\gamma\beta} - g_{3\beta}g_{3\gamma}\right)\\
&+ \hat{v}_3\frac{\partial g_{3\alpha}}{\partial x_\alpha}p + \frac{\partial g_{3\alpha}}{\partial x_\alpha}\frac{T^2}{m}\frac{\partial n}{\partial \hat{v}_3} + \frac{\partial g_{3\alpha}}{\partial x_\alpha}\frac{3p}{m}\frac{\partial T}{\partial \hat{v}_3} - \hat{v}_3 g_{3\beta}g_{3\gamma}\frac{\partial g_{l\gamma}\hat{V}_{\perp l}}{\partial x_\beta}n\frac{\partial T}{\partial \hat{v}_3}\\
&= - g_{3\alpha}\left(\frac{e}{m}E^*_\alpha - \frac{\hat{d}g_{l\alpha}\hat{V}_{\perp l}}{dt}\right)n\frac{\partial T}{\partial \hat{v}_3} + \frac{T}{m}g_{p\alpha}g_{q\mu}\frac{\partial g_{3\mu}}{\partial x_\alpha}\frac{\partial \pi_{pq}}{\partial \hat{v}_3}\\
&- g_{p\alpha}g_{q\mu}\left(\hat{v}_3\frac{\partial g_{3\mu}}{\partial x_\alpha} + \frac{\partial g_{l\mu}\hat{V}_{\perp l}}{\partial x_\alpha}\right)\pi_{pq} + \left(g_{p\beta}g_{3\gamma}\frac{\partial g_{l\gamma}\hat{V}_{\perp l}}{\partial x_\beta} - g_{p\delta}\frac{\hat{d}g_{3\delta}}{dt}\right)\frac{\partial q_p}{\partial \hat{v}_3}\\
&- \frac{\partial}{\partial x_\alpha}(g_{\beta\alpha}q_\beta) - \hat{v}_3 g_{3\alpha}g_{q\mu}\frac{\partial g_{3\mu}}{\partial x_\alpha}\frac{\partial q_q}{\partial \hat{v}_3} - g_{3\alpha}g_{q\mu}\frac{\partial g_{3\mu}}{\partial x_\alpha}q_q\\
&- g_{p\alpha}g_{q\mu}\frac{\partial g_{3\mu}}{\partial x_\alpha}\frac{1}{m}\frac{\partial n\Xi_{pq}}{\partial \hat{v}_3}.
\end{aligned} \tag{3.97}$$

Here, one has

$$E^*_\alpha = E_\alpha + e_{\alpha\gamma\delta}g_{q\gamma}\hat{V}_q B_\delta,$$
$$p = nT.$$

The transport coefficients are summarized as follows. The pressure tensor is collected from equations (3.81) and (3.82)

$$\pi_{11} = -\pi_{22} = -\frac{nT}{2\omega_c}W_{12},$$

$$\pi_{12} = \pi_{21} = \frac{nT}{2\omega_c}\frac{1}{2}(W_{11} - W_{22}),$$

and the rest of the components tied to the parallel direction vanish. Here, the rate-of-strain tensor is collected from equation (3.61).

$$W_{\mu\nu} = W_{\mu\nu}^{\perp} + W_{\mu\nu}^{\parallel},$$

$$W_{\mu\nu}^{\perp} = g_{\mu\alpha}g_{\nu\beta}\frac{\partial g_{l\beta}\hat{V}_{\perp l}}{\partial x_\alpha} + g_{\nu\alpha}g_{\mu\beta}\frac{\partial g_{l\beta}\hat{V}_{\perp l}}{\partial x_\alpha} - \delta_{\mu\nu}^{\perp}\frac{\partial g_{l\alpha}\hat{V}_{\perp l}}{\partial x_\alpha},$$

$$W_{\mu\nu}^{\parallel} = \left(g_{\mu\alpha}g_{\nu\beta}\frac{\partial g_{3\beta}}{\partial x_\alpha} + g_{\nu\alpha}g_{\mu\beta}\frac{\partial g_{3\beta}}{\partial x_\alpha} - \delta_{\mu\nu}^{\perp}\frac{\partial g_{3\alpha}}{\partial x_\alpha}\right)\left(\hat{v}_3 + \frac{T}{m}\frac{\partial \ln n}{\partial \hat{v}_3}\right).$$

From equation (3.83), the gyroviscous tensor π can also be expressed in the index notation:

$$\pi_{\alpha\beta} = \eta_3 W_{3\alpha\beta}, \tag{3.98}$$

where

$$\eta_3 = nT/2\omega_c,$$

$$W_{3\alpha\beta} = \frac{1}{2}\left(\delta_{\alpha\mu}^{\perp}e_{\beta\gamma\nu} + \delta_{\beta\nu}^{\perp}e_{\alpha\gamma\mu}\right)b_\gamma W_{\mu\nu}.$$

The heat flux vector is collected from equation (3.79)

$$\mathbf{q} = \frac{2nT}{m\omega_c}\mathbf{b} \times \nabla T.$$

The heat flux tensor is collected from equations (3.85)–3.90

$$\Theta_{111} = \frac{nT^2}{m\omega_c^2}\hat{\mathbf{e}}_1 \times \boldsymbol{\omega}_c \cdot \nabla \ln T,$$

$$\Theta_{112} = -\frac{nT^2}{m\omega_c^2}\hat{\mathbf{e}}_2 \times \boldsymbol{\omega}_c \cdot \nabla \ln T,$$

$$\Theta_{121} = -\Theta_{112},$$
$$\Theta_{122} = \Theta_{111},$$
$$\Theta_{211} = -\Theta_{112},$$
$$\Theta_{212} = \Theta_{111},$$
$$\Theta_{221} = -\Theta_{111},$$
$$\Theta_{222} = -\Theta_{112},$$

and the rest of the components tied to the parallel direction vanish. The T-square tensor is collected from equations (3.92) and (3.93)

$$\Xi_{11} = -\,\Xi_{22} = \frac{3T^2}{2\omega_c} W_{12},$$
$$\Xi_{12} = \Xi_{21} = \frac{3T^2}{2\omega_c}\frac{1}{2}(W_{11} - W_{22}),$$

and the rest of the components tied to the parallel direction vanish.

To constitute the complete set of equations, the Maxwell equations are also required.

$$\nabla \cdot \mathbf{E} = \frac{1}{\epsilon_0}\sum_a n_a e_a,$$
$$\nabla \times \mathbf{E} = -\frac{\partial \mathbf{B}}{\partial t},$$
$$\nabla \cdot \mathbf{B} = 0,$$
$$\nabla \times \mathbf{B} = \mu_0\left(\mathbf{J} + \epsilon_0\frac{\partial \mathbf{E}}{\partial t}\right),$$

where ϵ_0 and μ_0 are, respectively, the permittivity and the permeability of free space. Here, the Gauss law is often replaced by the quasineutrality condition. The set of the equations replace the Braginskii two-fluid equations for describing the collisionless or weakly collisional plasmas.

3.6 Conclusions and discussion

In conclusion, the perpendicular magnetofluid theory in reference [1] is described in this chapter. The new set of transport equations release the assumption of particle spatial localization along the magnetic field lines in the Braginskii transport theory [2; 3]. This is particularly relevant noting that the magnetically confined plasmas in thermonuclear fusion turn out to be collisionless. One cannot expect the charged particles to be collectively localized in space by collisions along the magnetic field lines. Some particles are trapped by the inhomogeneous equilibrium field or by the waves, while some others keep in circulation along the field lines. Consequently, the fluid variables based on the parallel motion in the conventional fluid theory, such as the parallel fluid velocity, become inadequate. The particle motion along the magnetic field lines does not tie together to move together as a fluid cell.

The perpendicular magnetofluid theory described in this chapter relies only on the assumption of the existence of strong magnetic field, which causes the charged particles to localize in the perpendicular direction, while the finite Larmor radius effects are taken into account in the next order. In the parallel direction, the particle mobility feature is fully retained by taking only the perpendicular moments. The current theory is based on the systematic and rigorous perturbation procedure, It is valid till the second order for FLR effects, i.e. $\mathcal{O}(k_\perp^2\rho_i^2)$. The results extend the gyroviscous tensor, thermal heat flux, etc, in the Braginskii transport theory to the collisionless limit. Besides the FLR effects, the wave-particle resonance (the Landau damping), the trapped particle effects, and other kinetic effects are all retained. This addresses the major concern about the applicability of Braginskii's fluid theory discussed in section 2.4.

The perpendicular magnetofluid theory can be used both in the analytical and numerical studies of magnetically confined plasmas in parallel with the gyrokinetic theory. The applications will be detailed in the following chapters.

References

[1] Zheng L 2020 Perpendicular magnetofluid theory for magnetically confined plasmas in the collisionless limit *Phys. Plasmas* **27** 052503

[2] Braginskii S I 1958 Transport phenomena in a complete ionized two-temperature plasma *ZhETF (J. Exp. Theor. Phys. USSR)* **33** 459
Braginskii S I 1958 *Soviet Phys. JETP* **6** 358–69

[3] Braginskii S I 1966 *Reviews of Plasma Physics* 1 ed M Leontovich (New York: Consultants Bureau) 205–311

[4] Vlasov A A 1968 The vibrational properties of an electron gas *Sov. Phys. Uspekhi* **10** 721–33

[5] Rutherford P H and Frieman E A 1968 Drift instabilities in general magnetic field configurations *Phys. Fluids* **11** 569–85

[6] Taylor J B and Hastie R J 1968 Stability of general plasma equilibria: I. formal theory *Plasma Phys.* **10** 479

[7] Antonsen T M and Lane B 1980 Kinetic equations for low frequency instabilities in inhomogeneous plasmas *Phys. Fluids* **23** 1205–14

[8] Catto P J, Tang W M and Baldwin D E 1981 Generalized gyrokinetics *Plasma Phys.* **23** 639–50

[9] Zheng L J, Kotschenreuther M T and Van Dam J W 2007 Revisiting linear gyrokinetics to recover ideal magnetohydrodynamics and missing finite Larmor radius effects *Phys. Plasmas* **14** 072505

[10] Helander P and Sigmar D 2002 *Collisional Transport in Magnetized Plasmas* (New York: Cambridge University Press)

[11] Hazeltine R D 1998 Transport theory in the collisionless limit *Phys. Plasmas* **5** 3282–86

IOP Publishing

Magnetically Confined Fusion Plasma Physics, Volume 2

Multifluid theory

Linjin Zheng

Chapter 4

Equilibrium analyses with the perpendicular magnetofluid theory

In this chapter, we describe the equilibrium analyses based on the perpendicular magnetofluid theory in the collisionless limit [1], which is reviewed in chapter 3.

The continuity, momentum, and energy transport equations have been given in equations (3.94), (3.95), and (3.97), respectively. For equilibrium calculation, however, the next order corrections from the finite Larmor radius effects can be neglected and also the time dependence can be suppressed. Consequently, the basic set of equilibrium equations becomes

$$\begin{aligned}
&g_{\beta\alpha}\hat{V}_{\perp\beta}\frac{\partial n}{\partial x_\alpha} + g_{3\alpha}\hat{v}_3\frac{\partial n}{\partial x_\alpha} + n\frac{\partial g_{l\gamma}\hat{V}_{\perp l}}{\partial x_\beta}\left(\delta_{\beta\gamma} - g_{3\beta}g_{3\gamma}\right)\\
&+ n\hat{v}_3\frac{\partial g_{3\alpha}}{\partial x_\alpha} + \frac{1}{m}\frac{\partial g_{3\alpha}}{\partial x_\alpha}\frac{\partial nT}{\partial \hat{v}_3} - g_{3\beta}g_{3\gamma}\frac{\partial g_{l\gamma}\hat{V}_{\perp l}}{\partial x_\beta}\hat{v}_3\frac{\partial n}{\partial \hat{v}_3}\\
&= -g_{3\beta}\frac{\partial n}{\partial \hat{v}_3}\left(\frac{e}{m}E^*_\beta - g_{\gamma\alpha}\hat{V}_{\perp\gamma}\frac{\partial g_{l\beta}\hat{V}_{\perp l}}{\partial x_\alpha}\right),
\end{aligned} \tag{4.1}$$

$$\begin{aligned}
&(\delta_{\lambda\alpha} - g_{3\alpha}g_{3\lambda})\left(neE^*_\alpha - \rho g_{\mu\nu}\hat{V}_{\perp\mu}\frac{\partial g_{l\alpha}\hat{V}_{\perp l}}{\partial x_\nu} - \frac{\partial p}{\partial x_\alpha} - \hat{v}_3 g_{3\beta}\frac{\partial g_{3\alpha}}{\partial x_\beta}\frac{\partial p}{\partial \hat{v}_3}\right.\\
&- g_{3\beta}\hat{v}_3^2\frac{\partial g_{3\alpha}}{\partial x_\beta}\rho + g_{3\gamma}\frac{\partial g_{l\gamma}\hat{V}_{\perp l}}{\partial x_\alpha}\frac{\partial p}{\partial \hat{v}_3} - \rho\hat{v}_3 g_{3\beta}\frac{\partial g_{l\alpha}\hat{V}_{\perp l}}{\partial x_\beta}\\
&\left.- \rho\hat{v}_3 g_{\mu\nu}\hat{V}_{\perp\mu}\frac{\partial g_{3\alpha}}{\partial x_\nu} - g_{\mu\nu}\hat{V}_{\perp\mu}\frac{\partial g_{3\alpha}}{\partial x_\nu}\frac{\partial p}{\partial \hat{v}_3}\right) = 0,
\end{aligned} \tag{4.2}$$

and

$$
\begin{aligned}
& n g_{\beta\alpha} \hat{V}_{\perp\beta} \frac{\partial T}{\partial x_\alpha} + n g_{3\alpha} \hat{v}_3 \frac{\partial T}{\partial x_\alpha} + p \frac{\partial g_{l\gamma} \hat{V}_{\perp l}}{\partial x_\beta} \left(\delta_{\gamma\beta} - g_{3\beta} g_{3\gamma} \right) \\
& + \hat{v}_3 \frac{\partial g_{3\alpha}}{\partial x_\alpha} p + \frac{\partial g_{3\alpha}}{\partial x_\alpha} \frac{T^2}{m} \frac{\partial n}{\partial \hat{v}_3} + \frac{\partial g_{3\alpha}}{\partial x_\alpha} \frac{3p}{m} \frac{\partial T}{\partial \hat{v}_3} - \hat{v}_3 g_{3\beta} g_{3\gamma} \frac{\partial g_{l\gamma} \hat{V}_{\perp l}}{\partial x_\beta} n \frac{\partial T}{\partial \hat{v}_3} \\
& = -g_{3\alpha} \left(\frac{e}{m} E_\alpha^* - \frac{\hat{d} g_{l\alpha} \hat{V}_{\perp l}}{dt} \right) n \frac{\partial T}{\partial \hat{v}_3}.
\end{aligned} \tag{4.3}
$$

We can seek for the solution with $\partial \ln n / \partial \hat{v}_3 \gg \partial \ln T / \partial \hat{v}_3$, which can be proved *a posteriori*. It is also assumed that $\hat{v}_3 \gg \hat{V}_\perp$. Therefore, the continuity and energy transport equations (4.1) and (4.3) are reduced, respectively, to

$$
g_{3\alpha} \hat{v}_3 \frac{\partial n}{\partial x_\alpha} + n \hat{v}_3 \frac{\partial g_{3\alpha}}{\partial x_\alpha} + \frac{T}{m} \frac{\partial g_{3\alpha}}{\partial x_\alpha} \frac{\partial n}{\partial \hat{v}_3} = -g_{3\beta} \frac{\partial n}{\partial \hat{v}_3} \frac{e}{m} E_\beta^*, \tag{4.4}
$$

and

$$
n g_{3\alpha} \hat{v}_3 \frac{\partial T}{\partial x_\alpha} + T \frac{\partial g_{3\alpha}}{\partial x_\alpha} \left(n \hat{v}_3 + \frac{T}{m} \frac{\partial n}{\partial \hat{v}_3} \right) = 0. \tag{4.5}
$$

As in reference [2, 3], the equilibrium electromagnetic field is decomposed as $\mathbf{E} = -\nabla \Phi - \nabla \varphi$ with $-\nabla \Phi + \mathbf{V} \times \mathbf{B} = 0$. The electric potential Φ corresponds to the plasma rotation, while φ represents the potential in the rotating frame. Therefore, (4.4) is reduced to

$$
g_{3\alpha} \left(\hat{v}_3 \frac{\partial n}{\partial x_\alpha} + \frac{\partial n}{\partial \hat{v}_3} \frac{e}{m} \frac{\partial \varphi}{\partial x_\alpha} \right) + \frac{\partial g_{3\alpha}}{\partial x_\alpha} \left(n \hat{v}_3 + \frac{T}{m} \frac{\partial n}{\partial \hat{v}_3} \right) = 0. \tag{4.6}
$$

The solutions of equations of (4.5) and (4.6) are

$$
n = n_0 \exp \left\{ -\frac{m \hat{v}_3^2 / 2 + e\varphi}{T} \right\} \quad \text{and} \quad \mathbf{b} \cdot \nabla (n_0, T) = \frac{\partial T}{\partial \hat{v}_3} = 0. \tag{4.7}
$$

The solution is consistent with the equilibrium distribution function used in the transport theory in the collisionless limit in reference [4]. It is interesting to point out that for the solution in (4.7), one has

$$
\mathbf{b} \cdot (-n e \nabla \varphi - \nabla p) = 0
$$

noting that $\mathbf{b} \cdot \nabla \Phi_0 = 0$. Similar to reference [3], one can determine φ in the rotating plasmas with finite Φ using the quasineutrality condition. This is not detailed here.

Next, we consider the force balance (4.2). Noting the solution in (4.7), one has

$$
-\hat{v}_3 g_{3\beta} \frac{\partial g_{3\alpha}}{\partial x_\beta} \frac{\partial p}{\partial \hat{v}_3} - g_{3\beta} \hat{v}_3^2 \frac{\partial g_{3\alpha}}{\partial x_\beta} \rho = 0.
$$

Therefore, the momentum (4.2) of lowest order becomes

$$[ne(-\nabla\varphi + \mathbf{V}_\perp \times \mathbf{B}) - \nabla p]_\perp = 0. \tag{4.8}$$

Note further that n and p in (4.8) depend on the parallel velocity $\hat{v}_3$ and species charges. Integrating over $\hat{v}_3$, summing over charged particle species, and using the quasineutrality condition

$$\sum_{i,e} \int d\hat{v}_3 ne = 0,$$

one obtains the equilibrium equation in the single fluid description,

$$\mathbf{J}_s \times \mathbf{B} = \nabla_\perp p_s, \tag{4.9}$$

where the total pressure and current are given as follows

$$p_s = \sum_{i,e} \int d\hat{v}_3 p, \qquad \mathbf{J}_s = \sum_{i,e} \int d\hat{v}_3 ne\mathbf{V}_\perp.$$

Equation (4.9) is the equilibrium equation of lowest order with the perpendicular magnetofluid description in the collisionless limit. The difference from the conventional MHD theory [5, 6] lies in that the pressure is determined only by the particle velocities perpendicular to the magnetic field. These results will be used in the stability analyses in the following chapters.

One can also work out the equilibrium solutions in the next order. The equilibrium difference between the perpendicular magnetofluid and the ideal MHD descriptions in the next order can be extensive. It is not further exploited here.

References

[1] Zheng L 2020 Perpendicular magnetofluid theory for magnetically confined plasmas in the collisionless limit *Phys. Plasmas* **27** 052503

[2] Helander P and Sigmar D 2002 *Collisional Transport in Magnetized Plasmas* (New York: Cambridge University Press)

[3] Artun M and Tang W M 1994 Nonlinear electromagnetic gyrokinetic equations for rotating axisymmetric plasmas *Phys. Plasmas* **1** 2682–92

[4] Hazeltine R D 1998 Transport theory in the collisionless limit *Phys. Plasmas* **5** 3282–6

[5] Kulsrud R M M 1983 MHD Description of Plasma *Basic Plasma Physics I (Handbook of Plasma Physics)* vol 1 ed A A Galeev and R N Sudan (Amsterdam: North-Holland) pp 1–31

[6] Freidberg J P 1987 *Ideal Magnetohydrodynamics* (New York: Plenum)

IOP Publishing

Magnetically Confined Fusion Plasma Physics, Volume 2
Multifluid theory
Linjin Zheng

Chapter 5

Electrostatic modes in the perpendicular magnetofluid description

In this chapter, the perpendicular magnetofluid theory [1], as reviewed in chapter 3, will be used to study the electrostatic modes in the collisionless limit. The linear theory is considered. The equilibrium solution in chapter 4 is used. It is mainly aimed at demonstrating how to use the perpendicular magnetofluid theory to study the electrostatic modes. The basic features are exploited so that they can be compared with the gyrokinetic description [2–6], as well as the Braginskii one [7, 8]. The subtle toroidal structure of electrostatic modes, such as the slab and toroid branches, have been reviewed in the author's earlier book [9]. Similar analyses for the mode structure can be applied to the perpendicular magnetofluid formulation to be described in this chapter and, therefore, will not be further detailed.

We first describe the basic set of equations. They will be reduced in order to explain the key features of electrostatic modes in the perpendicular magnetofluid description. Next, the dispersion relation of electrostatic modes will be derived to show how the kinetic effects, such as the wave-particle resonance, the trapped-particle effects, the magnetic and diamagnetic drift effects, can be taken into account in the perpendicular magnetofluid description.

5.1 Basic set of equations

In this section, the basic set of equations in the perpendicular magnetofluid theory in the collisionless limit is reduced in order to unveil the basic features of the electrostatic modes in this formulation.

The continuity, the momentum, and the energy transport equations have been given in equations (3.94), (3.95), and (3.97), respectively. They are recited here with terms labelled for the convenience of reference.

doi:10.1088/978-0-7503-3575-1ch5

$$
\begin{aligned}
&\underbrace{\frac{\partial n}{\partial t}}_{n_1} + \underbrace{g_{\beta\alpha}\hat{V}_{\perp\beta}\frac{\partial n}{\partial x_\alpha}}_{n_2} + \underbrace{g_{3\alpha}\hat{v}_3\frac{\partial n}{\partial x_\alpha}}_{n_3} + \underbrace{n\frac{\partial g_{l\gamma}\hat{V}_{\perp l}}{\partial x_\beta}(\delta_{\beta\gamma} - g_{3\beta}g_{3\gamma})}_{n_4} \\
&+ \underbrace{n\hat{v}_3\frac{\partial g_{3\alpha}}{\partial x_\alpha}}_{n_5} + \underbrace{\frac{1}{m}\frac{\partial g_{3\alpha}}{\partial x_\alpha}\frac{\partial nT}{\partial \hat{v}_3}}_{n_6} - \underbrace{g_{3\beta}g_{3\gamma}\frac{\partial g_{l\gamma}\hat{V}_{\perp l}}{\partial x_\beta}\hat{v}_3\frac{\partial n}{\partial \hat{v}_3}}_{n_7} \\
&= -g_{3\beta}\frac{\partial n}{\partial \hat{v}_3}\left(\underbrace{\frac{e}{m}E^*_\beta}_{n_8} - \underbrace{\frac{\hat{d}g_{l\beta}\hat{V}_{\perp l}}{dt}}_{n_9}\right) - \underbrace{\frac{1}{m}g_{p\alpha}g_{q\mu}\frac{\partial g_{3\mu}}{\partial x_\alpha}\frac{\partial \pi_{pq}}{\partial \hat{v}_3}}_{n_{10}},
\end{aligned}
\tag{5.1}
$$

$$
\begin{aligned}
&(\delta_{\lambda\alpha} - g_{3\alpha}g_{3\lambda})\left(\underbrace{neE^*_\alpha}_{v_1} - \underbrace{\rho\frac{\hat{d}g_{l\alpha}\hat{V}_{\perp l}}{dt}}_{v_2} - \underbrace{\frac{\partial p}{\partial x_\alpha}}_{v_3} - \underbrace{\hat{v}_3 g_{3\beta}\frac{\partial g_{3\alpha}}{\partial x_\beta}\frac{\partial p}{\partial \hat{v}_3}}_{v_4}\right. \\
&\left. - \underbrace{g_{3\beta}\hat{v}_3^2\frac{\partial g_{3\alpha}}{\partial x_\beta}\rho}_{v_5} + \underbrace{g_{3\gamma}\frac{\partial g_{l\gamma}\hat{V}_{\perp l}}{\partial x_\alpha}\frac{\partial p}{\partial \hat{v}_3}}_{v_6} - \underbrace{\rho\hat{v}_3 g_{3\beta}\frac{\partial g_{l\alpha}\hat{V}_{\perp l}}{\partial x_\beta}}_{v_7} - \underbrace{\rho\hat{v}_3\frac{\hat{d}g_{3\alpha}}{dt}}_{v_8} - \underbrace{\frac{\hat{d}g_{3\alpha}}{dt}\frac{\partial p}{\partial \hat{v}_3}}_{v_9}\right) \\
&= \underbrace{g_{\nu\lambda}g_{\beta\alpha}\frac{\partial \pi_{\beta\nu}}{\partial x_\alpha}}_{v_{10}} + g_{\nu\lambda}\left(\underbrace{g_{p\delta}\frac{\hat{d}g_{3\delta}}{dt}}_{v_{11}} - \underbrace{g_{p\beta}g_{3\gamma}\frac{\partial g_{l\gamma}\hat{V}_{\perp l}}{\partial x_\beta}}_{v_{12}} + \underbrace{\hat{v}_3 g_{3\alpha}g_{q\mu}\frac{\partial g_{3\mu}}{\partial x_\alpha}}_{v_{13}}\right)\frac{\partial \pi_{p\nu}}{\partial \hat{v}_3} \\
&+ \left(\underbrace{\frac{\partial g_{p\alpha}g_{q\lambda}}{\partial x_\alpha}}_{v_{14}} + \underbrace{g_{p\alpha}g_{q\mu}g_{3\lambda}\frac{\partial g_{3\mu}}{\partial x_\alpha}}_{v_{15}} + \underbrace{g_{3\alpha}g_{q\mu}g_{p\lambda}\frac{\partial g_{3\mu}}{\partial x_\alpha}}_{v_{16}}\right)\pi_{pq} \\
&+ \underbrace{g_{\nu\lambda}\frac{\partial g_{3\alpha}}{\partial x_\alpha}\frac{\partial q_\nu}{\partial \hat{v}_3}}_{v_{17}} + \underbrace{g_{p\alpha}g_{q\mu}g_{\nu\lambda}\frac{\partial g_{3\mu}}{\partial x_\alpha}\frac{\partial \Theta_{pq\nu}}{\partial \hat{v}_3}}_{v_{18}},
\end{aligned}
\tag{5.2}
$$

$$
\begin{aligned}
&\underbrace{n\frac{\partial T}{\partial t}}_{t_1} + \underbrace{ng_{\beta\alpha}\hat{V}_{\perp\beta}\frac{\partial T}{\partial x_\alpha}}_{t_2} + \underbrace{ng_{3\alpha}\hat{v}_3\frac{\partial T}{\partial x_\alpha}}_{t_3} + \underbrace{p\frac{\partial g_{l\gamma}\hat{V}_{\perp l}}{\partial x_\beta}(\delta_{\gamma\beta} - g_{3\beta}g_{3\gamma})}_{t_4} \\
&+ \underbrace{\hat{v}_3\frac{\partial g_{3\alpha}}{\partial x_\alpha}p}_{t_5} + \underbrace{\frac{\partial g_{3\alpha}}{\partial x_\alpha}\frac{T^2}{m}\frac{\partial n}{\partial \hat{v}_3}}_{t_6} + \underbrace{\frac{\partial g_{3\alpha}}{\partial x_\alpha}\frac{3p}{m}\frac{\partial T}{\partial \hat{v}_3}}_{t_7} - \underbrace{\hat{v}_3 g_{3\beta}g_{3\gamma}\frac{\partial g_{l\gamma}\hat{V}_{\perp l}}{\partial x_\beta}n\frac{\partial T}{\partial \hat{v}_3}}_{t_8} \\
&= -g_{3\alpha}\left(\underbrace{\frac{e}{m}E_\alpha^*}_{t_9} - \underbrace{\frac{\hat{d}g_{l\alpha}\hat{V}_{\perp l}}{dt}}_{t_{10}}\right)n\frac{\partial T}{\partial \hat{v}_3} + \underbrace{\frac{T}{m}g_{p\alpha}g_{q\mu}\frac{\partial g_{3\mu}}{\partial x_\alpha}\frac{\partial \pi_{pq}}{\partial \hat{v}_3}}_{t_{11}} \\
&- g_{p\alpha}g_{q\mu}\left(\underbrace{\hat{v}_3\frac{\partial g_{3\mu}}{\partial x_\alpha}}_{t_{12}} + \underbrace{\frac{\partial g_{l\mu}\hat{V}_{\perp l}}{\partial x_\alpha}}_{t_{13}}\right)\pi_{pq} + \left(\underbrace{g_{p\beta}g_{3\gamma}\frac{\partial g_{l\gamma}\hat{V}_{\perp l}}{\partial x_\beta}}_{t_{14}} - \underbrace{g_{p\delta}\frac{\hat{d}g_{3\delta}}{dt}}_{t_{15}}\right)\frac{\partial q_p}{\partial \hat{v}_3} \\
&- \underbrace{\frac{\partial}{\partial x_\alpha}(g_{\beta\alpha}q_\beta)}_{t_{16}} - \underbrace{\hat{v}_3 g_{3\alpha}g_{q\mu}\frac{\partial g_{3\mu}}{\partial x_\alpha}\frac{\partial q_q}{\partial \hat{v}_3}}_{t_{17}} - \underbrace{g_{3\alpha}g_{q\mu}\frac{\partial g_{3\mu}}{\partial x_\alpha}q_q}_{t_{18}} \\
&- \underbrace{g_{p\alpha}g_{q\mu}\frac{\partial g_{3\mu}}{\partial x_\alpha}\frac{1}{m}\frac{\partial n\Xi_{pq}}{\partial \hat{v}_3}}_{t_{19}}.
\end{aligned}
\tag{5.3}
$$

To complete the set of equations, the following quasineutrality condition is also required

$$\sum_a \int d\hat{v}_3 n_a e_a = 0, \tag{5.4}$$

where the subscript 'a' is introduced to denote the particle species.

First, we note that for electrostatic modes the magnetic field is unperturbed, so that one has

$$\frac{\partial g_{\beta\alpha}}{\partial t} = 0.$$

Next, we introduce the basic orderings for electrostatic modes to reduce the basic set of equations. For electrostatic modes one has

$$\frac{\delta n}{n} \sim \frac{e\delta\varphi}{T} \quad \text{and} \quad \delta V_\perp \sim \frac{k_\perp \delta\varphi}{B}.$$

These orderings can be confirmed in the later analyses. Therefore, one has

$$\frac{\delta n}{n} \gg \frac{\delta V_\perp}{v_{th}} \sim k_\perp \rho \frac{\delta n}{n}, \tag{5.5}$$

where ρ is the Larmor radius and $k_\perp$ is perpendicular wave number. Using the formulas of transport coefficients in section 3.5, one can estimate their orderings as follows

$$\pi_{pq} \sim \frac{nT}{\omega_c} k_\perp \delta \hat{V}_\perp \sim p k_\perp \rho \frac{\delta \hat{V}_\perp}{v_{th}},$$
$$\nabla \cdot \mathbf{q} \sim \frac{nT}{Rm\omega_c} k_\perp \delta T \sim k_\perp \rho n \omega_t \, \delta T,$$
$$\Theta_{\alpha\beta\gamma} \sim \frac{nT}{m\omega_c} k_\perp \delta T \sim k_\perp \rho n v_T \, \delta T,$$
$$\Xi_{\alpha\beta} \sim \frac{T^2}{\omega_c} k_\perp \delta \hat{V}_\perp \sim k_\perp \rho m v_T T \, \delta V_\perp,$$

where R represents the curvature radius of the magnetic field line, the ω_t is the transit frequency, and v_T is the thermal velocity. Therefore, the terms related to these four transport coefficients in equations (5.1)–(5.3) can be neglected in the leading-order analyses and the basic set of equations is reduced to

$$\begin{aligned} &\frac{\partial n}{\partial t} + g_{\beta\alpha} \hat{V}_{\perp\beta} \frac{\partial n}{\partial x_\alpha} + g_{3\alpha} \hat{v}_3 \frac{\partial n}{\partial x_\alpha} + n \frac{\partial g_{l\gamma} \hat{V}_{\perp l}}{\partial x_\beta} (\delta_{\beta\gamma} - g_{3\beta} g_{3\gamma}) \\ &+ \underbrace{n \hat{v}_3 \frac{\partial g_{3\alpha}}{\partial x_\alpha}}_{n_5} + \underbrace{\frac{1}{m} \frac{\partial g_{3\alpha}}{\partial x_\alpha} \frac{\partial nT}{\partial \hat{v}_3}}_{n_6} - g_{3\beta} g_{3\gamma} \frac{\partial g_{l\gamma} \hat{V}_{\perp l}}{\partial x_\beta} \hat{v}_3 \frac{\partial n}{\partial \hat{v}_3} \\ &= -g_{3\alpha} \frac{\partial n}{\partial \hat{v}_3} \frac{e}{m} (E_\alpha + e_{\alpha\beta\gamma} g_{l\beta} \hat{V}_l B_\gamma), \end{aligned} \tag{5.6}$$

$$(\delta_{\lambda\alpha} - g_{3\alpha} g_{3\lambda}) \left(neE_\alpha + nee_{\alpha\beta\gamma} g_{l\beta} \hat{V}_l B_\gamma - \frac{\partial p}{\partial x_\alpha} \right) = 0, \tag{5.7}$$

$$\begin{aligned} &n \frac{\partial T}{\partial t} + n g_{\beta\alpha} \hat{V}_{\perp\beta} \frac{\partial T}{\partial x_\alpha} + n g_{3\alpha} \hat{v}_3 \frac{\partial T}{\partial x_\alpha} + p \frac{\partial g_{l\gamma} \hat{V}_{\perp l}}{\partial x_\beta} (\delta_{\gamma\beta} - g_{3\beta} g_{3\gamma}) \\ &+ \underbrace{\hat{v}_3 \frac{\partial g_{3\alpha}}{\partial x_\alpha} p}_{t_5} + \underbrace{\frac{\partial g_{3\alpha}}{\partial x_\alpha} \frac{T^2}{m} \frac{\partial n}{\partial \hat{v}_3}}_{t_6} + \underbrace{\frac{\partial g_{3\alpha}}{\partial x_\alpha} \frac{3p}{m} \frac{\partial T}{\partial \hat{v}_3}}_{t_7} = 0. \end{aligned} \tag{5.8}$$

Here, in reducing the momentum equation (5.7) terms 'v_4' and 'v_5' are neglected assuming $k_\perp \gg 1/R$ and terms 'v_6'–'v_9' are neglected noting that $\hat{V}_\perp \ll \hat{v}_3$. In reducing the energy transport equation (5.8), term 't_9' is neglected since $\partial T/\partial \hat{v}_3 = 0$ for equilibrium, as shown in chapter 3, and term 't_8' is much smaller than term 't_7' if T is perturbed.

5.2 The dispersion relation for electrostatic modes

In this section, the reduced basic set of equations (5.6)–(5.8) is used to investigate the linear stability of electrostatic modes.

First, we note that the perpendicular velocity can be obtained from the momentum equation (5.7):

$$\hat{\mathbf{V}}_\perp = \frac{\mathbf{E}\times\mathbf{B}}{B^2} + \frac{\mathbf{B}\times\nabla P}{neB^2} = \frac{\mathbf{B}\times\nabla\varphi}{B^2} + \frac{\mathbf{B}\times\nabla P}{neB^2}. \tag{5.9}$$

Using this expression, the continuity equation can be simplified term by term as follows. Note that the linearizations are introduced in the derivations without further explanations and the MHD equilibrium equalities in appendix A are used to express the field line curvature, etc. One has

$$\begin{aligned}\hat{\mathbf{V}}_\perp\cdot\nabla n &= \frac{\mathbf{E}\times\mathbf{B}}{B^2}\cdot\nabla n + \frac{\mathbf{B}\times\nabla P}{neB^2}\cdot\nabla n\\ &= -\frac{\mathbf{B}\times\nabla n}{B^2}\cdot\nabla\delta\varphi + \frac{\mathbf{B}\times\nabla T}{eB^2}\cdot\nabla\delta n\\ &\quad -\frac{\mathbf{B}\times\nabla n}{eB^2}\cdot\nabla\delta T,\end{aligned}$$

$$\begin{aligned}\nabla\cdot\hat{\mathbf{V}}_\perp &= \nabla\times\frac{\mathbf{B}}{B^2}\cdot\nabla\delta\varphi + \frac{1}{ne}\nabla\times\frac{\mathbf{B}}{B^2}\cdot\nabla\delta p\\ &\quad -\frac{\mathbf{B}\times\nabla T}{neB^2}\cdot\nabla\delta n + \frac{\mathbf{B}\times\nabla n}{neB^2}\cdot\nabla\delta T,\end{aligned}$$

and

$$\begin{aligned}(\mathbf{b}\cdot\nabla\hat{\mathbf{V}}_\perp)\cdot\mathbf{b} &= -(\mathbf{b}\cdot\nabla\mathbf{b})\cdot\hat{\mathbf{V}}_\perp\\ &= \frac{\mathbf{B}\times(\mathbf{b}\cdot\nabla\mathbf{b})}{B^2}\cdot\nabla\delta\varphi\\ &\quad +\frac{\mathbf{B}\times(\mathbf{b}\cdot\nabla\mathbf{b})}{neB^2}\cdot\nabla\delta p.\end{aligned}$$

Also, it is noted that the equilibrium density is given by the parallel Maxwellian distribution as shown in chapter 4

$$n = \left(\frac{m}{2\pi T}\right)^{1/2} n_0 \exp\left\{-\frac{m\hat{v}_3^2}{2T}\right\}. \tag{5.10}$$

Therefore, the continuity equation (5.6) becomes

$$
\begin{aligned}
&\frac{\partial \delta n}{\partial t} - \frac{\mathbf{B}\times\nabla n}{B^2}\cdot\nabla\delta\varphi + \underbrace{\frac{\mathbf{B}\times\nabla T}{eB^2}\cdot\nabla\delta n}_{x_1} - \underbrace{\frac{\mathbf{B}\times\nabla n}{eB^2}\cdot\nabla\delta T}_{x_2}\\
&+ g_{3\alpha}\hat{v}_3\frac{\partial \delta n}{\partial x_\alpha} + \underbrace{n\left(\frac{\mathbf{B}\times\nabla B^2}{B^4} + \frac{\mathbf{B}\times\nabla P}{B^4}\right)\cdot\nabla\delta\varphi}_{o_3}\\
&+ \underbrace{\frac{1}{e}\left(\frac{\mathbf{B}\times\nabla B^2}{B^4} + \frac{\mathbf{B}\times\nabla P}{B^4}\right)\cdot\nabla\delta p}_{o_4}\\
&- \underbrace{\frac{\mathbf{B}\times\nabla T}{eB^2}\cdot\nabla\delta n}_{x_1} + \underbrace{\frac{\mathbf{B}\times\nabla n}{eB^2}\cdot\nabla\delta T}_{x_2}\\
&- \underbrace{n\left(\frac{\mathbf{B}\times\nabla P}{B^4} + \frac{\mathbf{B}\times\nabla B^2}{2B^4}\right)\cdot\nabla\delta\varphi}_{o_3} - \underbrace{\frac{1}{e}\left(\frac{\mathbf{B}\times\nabla P}{B^4} + \frac{\mathbf{B}\times\nabla B^2}{2B^4}\right)\cdot\nabla\delta p}_{o_4}\\
&- \hat{v}_3\frac{\partial n}{\partial \hat{v}_3}\frac{\mathbf{B}\times\kappa}{B^2}\cdot\nabla\delta\varphi - \hat{v}_3\frac{\partial n}{\partial \hat{v}_3}\frac{1}{ne}\frac{\mathbf{B}\times\kappa}{B^2}\cdot\nabla\delta p\\
&+ \delta n\hat{v}_3\frac{\partial g_{3\alpha}}{\partial x_\alpha} + \frac{T}{m}\frac{\partial g_{3\alpha}}{\partial x_\alpha}\frac{\partial \delta n}{\partial \hat{v}_3} + \frac{1}{m}\frac{\partial g_{3\alpha}}{\partial x_\alpha}\frac{\partial n\delta T}{\partial \hat{v}_3}\\
&= -g_{3\alpha}\frac{\partial n}{\partial \hat{v}_3}\frac{e}{m}\delta E_\alpha.
\end{aligned}
$$

Noting the cancellations with the same indices 'xi' and the combinations with 'oi', the continuity equation is further reduced to

$$
\begin{aligned}
&\frac{\partial \delta n}{\partial t} - \frac{\mathbf{B}\times\nabla n}{B^2}\cdot\nabla\,\delta\varphi + g_{3\alpha}\hat{v}_3\frac{\partial \delta n}{\partial x_\alpha}\\
&+ n\frac{\mathbf{B}\times\nabla\ln B}{B^2}\cdot\nabla\,\delta\varphi + \frac{1}{e}\frac{\mathbf{B}\times\nabla\ln B}{B^2}\cdot\nabla\,\delta p\\
&- \hat{v}_3\frac{\partial n}{\partial \hat{v}_3}\frac{\mathbf{B}\times\kappa}{B^2}\cdot\nabla\,\delta\varphi - \hat{v}_3\frac{\partial n}{\partial \hat{v}_3}\frac{1}{ne}\frac{\mathbf{B}\times\kappa}{B^2}\cdot\nabla\,\delta p\\
&+ \delta n\hat{v}_3\frac{\partial g_{3\alpha}}{\partial x_\alpha} + \frac{T}{m}\frac{\partial g_{3\alpha}}{\partial x_\alpha}\frac{\partial \delta n}{\partial \hat{v}_3} + \frac{1}{m}\frac{\partial g_{3\alpha}}{\partial x_\alpha}\frac{\partial n\,\delta T}{\partial \hat{v}_3}\\
&= -g_{3\alpha}\hat{v}_3\frac{ne}{T}\frac{\partial\varphi}{\partial x_\alpha}.
\end{aligned}
\tag{5.11}
$$

As usual, one can introduce the so-called adiabatic transformation,

$$\delta n = -\frac{ne}{T}\delta\varphi + \delta n_h. \tag{5.12}$$

Equation (5.11) is then transformed to

$$\begin{aligned}
&\frac{\partial \delta n_h}{\partial t} + g_{3\alpha}\hat{v}_3 \frac{\partial \delta n_h}{\partial x_\alpha} \\
&+ T\left(\frac{1}{e}\frac{\mathbf{B} \times \nabla \ln B}{B^2} - \hat{v}_3 \frac{\partial n}{\partial \hat{v}_3}\frac{1}{ne}\frac{\mathbf{B} \times \kappa}{B^2}\right) \cdot \nabla \delta n_h \\
&+ n\left(\frac{1}{e}\frac{\mathbf{B} \times \nabla \ln B}{B^2} - \hat{v}_3 \frac{\partial n}{\partial \hat{v}_3}\frac{1}{ne}\frac{\mathbf{B} \times \kappa}{B^2}\right) \cdot \nabla \delta T \\
&+ \underbrace{\delta n_h \hat{v}_3 \frac{\partial g_{3\alpha}}{\partial x_\alpha}}_{n_5} + \underbrace{\frac{T}{m}\frac{\partial g_{3\alpha}}{\partial x_\alpha}\frac{\partial \delta n_h}{\partial \hat{v}_3}}_{n_6} + \frac{1}{m}\frac{\partial g_{3\alpha}}{\partial x_\alpha}\frac{\partial n\, \delta T}{\partial \hat{v}_3} \\
&= \frac{ne}{T}\frac{\partial \delta\varphi}{\partial t} + \frac{\mathbf{B} \times \nabla n}{B^2} \cdot \nabla \delta\varphi.
\end{aligned} \tag{5.13}$$

One needs the energy transport equation as well. Similarly, the energy transport equation (5.3) can be reduced to

$$\begin{aligned}
&\frac{\partial \delta T}{\partial t} - \frac{\mathbf{B} \times \nabla T}{B^2} \cdot \nabla \delta\varphi + \underbrace{\frac{T\mathbf{B} \times \nabla n}{neB^2} \cdot \nabla \delta T}_{o_1} - \underbrace{\frac{T\mathbf{B}\nabla T}{neB^2} \cdot \nabla \delta n}_{o_2} \\
&+ g_{3\alpha}\hat{v}_3 \frac{\partial \delta T}{\partial x_\alpha} + \underbrace{T\left(\frac{\mathbf{B} \times \nabla B^2}{B^4} + \frac{\mathbf{B} \times \nabla P}{B^4}\right) \cdot \nabla \delta\varphi}_{o_3} \\
&+ \underbrace{\frac{T}{ne}\left(\frac{\mathbf{B} \times \nabla B^2}{B^4} + \frac{\mathbf{B} \times \nabla P}{B^4}\right) \cdot \nabla \delta p}_{o_4} \\
&+ \underbrace{\frac{T\mathbf{B}\nabla n}{neB^2} \cdot \nabla \delta T}_{o_1} - \underbrace{\frac{T\mathbf{B}\nabla T}{neB^2} \cdot \nabla \delta n}_{o_2} \\
&- \underbrace{T\left(\frac{\mathbf{B} \times \nabla P}{B^4} + \frac{\mathbf{B} \times \nabla B^2}{2B^4}\right) \cdot \nabla \delta\varphi}_{o_3} - \underbrace{\frac{T}{ne}\left(\frac{\mathbf{B} \times \nabla P}{B^4} + \frac{\mathbf{B} \times \nabla B^2}{2B^4}\right) \cdot \nabla \delta p}_{o_4} \\
&+ \underbrace{\hat{v}_3 \frac{\partial g_{3\alpha}}{\partial x_\alpha}\delta p}_{t_5} + \underbrace{\frac{\partial g_{3\alpha}}{\partial x_\alpha}\frac{T^2}{m}\frac{\partial \delta n}{\partial \hat{v}_3}}_{t_6} + \underbrace{\frac{\partial g_{3\alpha}}{\partial x_\alpha}\frac{3p}{m}\frac{\partial \delta T}{\partial \hat{v}_3}}_{t_7} = 0.
\end{aligned}$$

Noting the cancellations with the same indices 'xi' and combinations with 'oi', the energy transport equation is further reduced to

$$\frac{\partial \delta T}{\partial t} + g_{3\alpha}\hat{v}_3\frac{\partial \delta T}{\partial x_\alpha} + 2\frac{T\mathbf{B}\nabla n}{neB^2}\cdot\nabla\,\delta T - 2\frac{T\mathbf{B}\nabla T}{neB^2}\cdot\nabla\,\delta n$$
$$+\frac{T}{e}\frac{\mathbf{B}\times\nabla\ln B}{B^2}\cdot\nabla\delta T + \frac{T^2}{ne}\frac{\mathbf{B}\times\nabla\ln B}{B^2}\cdot\nabla\,\delta n$$
$$+\underbrace{\hat{v}_3\frac{\partial g_{3\alpha}}{\partial x_\alpha}\delta p}_{t_5} + \underbrace{\frac{\partial g_{3\alpha}}{\partial x_\alpha}\frac{T^2}{m}\frac{\partial \delta n}{\partial \hat{v}_3}}_{t_6} + \underbrace{\frac{\partial g_{3\alpha}}{\partial x_\alpha}\frac{3p}{m}\frac{\partial \delta T}{\partial \hat{v}_3}}_{t_7} \tag{5.14}$$
$$=\frac{\mathbf{B}\times\nabla T}{B^2}\cdot\nabla\delta\varphi - T\frac{\mathbf{B}\times\nabla\ln B}{B^2}\cdot\nabla\,\delta\varphi.$$

One can further prove the term containing δT in equation (5.13) can be neglected. Without taking into account the effects from term 't_7', using either the terms containing $\delta\varphi$ or δn in equation (5.14) for ordering analyses, one obtains the following ordering estimate

$$\frac{\delta T}{T} \sim \frac{R}{a}k_\wedge\rho_a\frac{\delta n}{n},$$

where $k_\wedge$ is the wavenumber perpendicular to the magnetic field line, but lying on the magnetic flux surface. This ordering is the consequence of the different significance of the terms 'n_8' and 't_9' in their respective equations, (5.1) and (5.3). The term 't_9' is negligible due to $\partial T/\partial\hat{v}_3 = 0$ for equilibrium, while the term 'n_8' is finite since $\partial n/\partial\hat{v}_3 \neq 0$ for equilibrium. Therefore, one has

$$\frac{\delta T}{T} \ll \frac{\delta n}{n}$$

assuming $(R/a)k_\wedge\rho_a \ll 1$. As will be seen later on, the terms 'n_5' and 'n_6' in equation (5.13) and the term 't_7' in equation (5.14) can result in the changes of integration paths for respective equations, the terms 'n_5' and 'n_6' lead the integration path in equation (5.13) to follow the particle adiabatic invariant trajectory in the phase space, while term 't_7' in equation (5.14) causes the path to deviate the trajectory. This further reduces the δT effects. Therefore, equation (5.13) is further reduced to

$$\frac{\partial \delta n_h}{\partial t} + g_{3\alpha}\hat{v}_3\frac{\partial \delta n_h}{\partial x_\alpha} + T\left(\frac{1}{e}\frac{\mathbf{B}\times\nabla\ln B}{B^2} - \hat{v}_3\frac{\partial n}{\partial\hat{v}_3}\frac{1}{ne}\frac{\mathbf{B}\times\kappa}{B^2}\right)\cdot\nabla\,\delta n_h$$
$$+\underbrace{\delta n_h\hat{v}_3\frac{\partial g_{3\alpha}}{\partial x_\alpha}}_{n_5} + \underbrace{\frac{T}{m}\frac{\partial g_{3\alpha}}{\partial x_\alpha}\frac{\partial \delta n_h}{\partial\hat{v}_3}}_{n_6} \tag{5.15}$$
$$=\frac{ne}{T}\frac{\partial\delta\varphi}{\partial t} + \frac{\mathbf{B}\times\nabla n}{B^2}\cdot\nabla\,\delta\varphi.$$

Using equation (5.10) for the equilibrium density, one can find that

$$\frac{\mathbf{B}\times\nabla n}{B^2} = n\frac{\mathbf{B}\times\nabla\ln n_0}{B^2} + n\frac{\mathbf{B}\times\nabla\ln T}{B^2}\left(\frac{m\hat{v}_3}{2T} - \frac{1}{2}\right). \tag{5.16}$$

Equation (5.15), then, can be rewritten as

$$
\begin{aligned}
&(\hat{\mathbf{v}}_3 \cdot \nabla - i\omega + i\bar{\omega}_d)\delta n_h \\
&+ \underbrace{\delta n_h \hat{v}_3 \frac{\partial g_{3\alpha}}{\partial x_\alpha}}_{n_5} + \underbrace{\frac{T}{m}\frac{\partial g_{3\alpha}}{\partial x_\alpha}\frac{\partial \delta n_h}{\partial \hat{v}_3}}_{n_6} \\
&= -i\frac{e}{T}n(\omega - \bar{\omega}_*^T)\delta\varphi,
\end{aligned}
\tag{5.17}
$$

where

$$
i\bar{\omega}_d = \left(\frac{T}{e}\frac{\mathbf{B} \times \nabla \ln B}{B^2} + \frac{m\hat{v}_3^2}{e}\frac{\mathbf{B} \times \kappa}{B^2}\right) \cdot \nabla, \tag{5.18}
$$

$$
\bar{\omega}_*^T = \left[1 + \bar{\eta}\left(\frac{m\hat{v}_3^2}{2T} - \frac{1}{2}\right)\right]\bar{\omega}_*, \tag{5.19}
$$

$$
i\bar{\omega}_* = \frac{T\mathbf{B} \times \nabla n_0}{n_0 e B^2} \cdot \nabla, \tag{5.20}
$$

$$
\bar{\eta} = \frac{d \ln T}{d \ln n_0}. \tag{5.21}
$$

Note here that $\bar{\omega}_d$ in equation (5.18) resembles the magnetic drift in the guiding center theory [10–12], except $v_\perp^2$ in the guiding center theory is replaced by the thermal velocity T/m. This is because the average over the perpendicular velocity has been performed in the continuity equation (5.17). Letting

$$
\delta n_h = n\delta\tilde{n}_h,
$$

where n is the equilibrium density in equation (5.10), equation (5.17) is further reduced to

$$
(\hat{\mathbf{v}}_3 \cdot \nabla|_{\hat{v}_3,\iota} - i\omega + i\bar{\omega}_d)\delta\tilde{n}_h + \underbrace{\frac{T}{m}\frac{\partial g_{3\alpha}}{\partial x_\alpha}\frac{\partial \delta\tilde{n}_h}{\partial \hat{v}_3}}_{n_6} = -i\frac{e}{T}(\omega - \bar{\omega}_*^T)\delta\varphi. \tag{5.22}
$$

This is a partial differential equation of first order. One can find its characteristic:

$$
\frac{T}{m}\int^{\hat{x}_3} \nabla \cdot \mathbf{b}\, d\hat{x}_3 - \frac{\hat{v}_3^2}{2} = \iota. \tag{5.23}
$$

Here, $\partial g_{3\alpha}/\partial x_\alpha = \nabla \cdot \mathbf{b}$ has been used.

Noting that

$$\begin{aligned}\hat{v}_3 \frac{\partial \delta \tilde{n}_h}{\partial \hat{x}_3}\bigg|_\iota &= \hat{v}_3 \frac{\partial \delta \tilde{n}_h}{\partial \hat{x}_3}\bigg|_{\hat{v}_3} + \hat{v}_3 \frac{\partial \delta \tilde{n}_h}{\partial \hat{v}_3}\bigg|_{\hat{x}_3} \frac{\partial \hat{v}_3}{\partial \hat{x}_3}\bigg|_\iota \\ &= \hat{v}_3 \frac{\partial \delta \tilde{n}_h}{\partial \hat{x}_3}\bigg|_{\hat{v}_3} + \frac{T}{m}\nabla \cdot \mathbf{b} \frac{\partial \delta \tilde{n}_h}{\partial \hat{v}_3}\bigg|_{\hat{x}_3},\end{aligned}$$

equation (5.22) is reduced to a differential equation along this characteristic curve,

$$\hat{v}_3 \frac{\partial \delta \tilde{n}_h}{\partial \hat{x}_3}\bigg|_\iota - (i\omega - \bar{\omega}_d)\delta \tilde{n}_h = -i\frac{e}{T}\left(\omega - \bar{\omega}_*^T\right)\delta\varphi, \tag{5.24}$$

Here, $\hat{v}_3(\hat{x}_3, \iota)$ needs to be evaluated along the characteristic curve as defined in equation (5.23). It is also noted here that the particle species indices are suppressed.

It is interesting to analyze the characteristic in equation (5.23). Note that

$$\begin{aligned}\frac{\hat{v}_3^2}{2} &= \varepsilon - \mu B = \varepsilon - \mu B_0 - \mu \mathbf{b} \cdot \nabla B\, \delta \hat{x}_3 \\ &= \varepsilon - \mu B_0 + \mu B \nabla \cdot \mathbf{b}\, \delta \hat{x}_3,\end{aligned} \tag{5.25}$$

where ε and μ are the particle kinetic energy and magnetic moment, respectively. Since $\varepsilon - \mu B_0$ is an adiabatic invariant, one can see from equation (5.25) that the characteristic curve in equation (5.23) basically follows the particle adiabatic invariant trajectory in the phase space. This actually converts $\hat{\mathbf{v}}_3 \cdot \nabla|_{\hat{v}_3,t}$, holding $\hat{v}_3$ constant, in the first term of equation (5.22) to $\hat{\mathbf{v}}_3 \cdot \nabla|_{\varepsilon,\mu,t}$, holding (ε, μ) constant, in the first term of equation (5.24), leading it to become gyrokinetic-theory-like. Similar to the grad-B drift frequency in equation (5.18), here the perpendicular velocity in the last term in equation (5.25) is replaced by the thermal velocity in the characteristic curve in equation (5.23). It is reasonable since the perpendicular average has been performed in constructing the moment equation. This difference is compensated in the next order correction in the perpendicular magnetofluid theory, such as viscous tensor, etc.

Here, it is pointed out that by similar analyses, one can find that term 't_7' in equation (5.14), with a factor '3', causes actually the characteristic to deviate the particle adiabatic invariant trajectory in the phase space. Consequently, the density resonates with waves in equation (5.24), while the temperature in the energy equation (5.14) does not. This further assures the assumption that $\partial \ln T/\partial \hat{v}_3 \ll \partial \ln n/\partial \hat{v}_3$, i.e. the $\hat{v}_3$ dependence mainly concentrates in the density n as discussed on the introduction of Maxwellian distribution in equation (3.20).

Equation (5.24) is similar to the gyrokinetic equation. The differences lie in that equation (5.24) is given in the actual phase space, instead of the guiding center, and the average over the perpendicular velocity is implemented. The analyses for toroidal and slab branches of ion temperature gradient modes and other related analyses in reference [6] and references therein can be directly extended to the current perpendicular magnetofluid case. We do not further elaborate on them.

Instead, we employ the WKB ansatz, i.e. $\nabla = i\mathbf{k}$. to explain the formulation and physics. Using the adiabatic transform in equation (5.12) and the solution of equation (5.24), the quasineutrality condition in equation (5.4) is expressed as follows

$$\delta\varphi \sum_a \frac{e_a^2}{T_a} = \sum_a e_a \frac{1}{n_0} \int_{-\infty}^{+\infty} d\hat{v}_3 n_h$$
$$= -\delta\varphi \sum_a \frac{e_a^2}{T_a} \frac{1}{n_0} \int_{-\infty}^{+\infty} d\hat{v}_3 n \frac{\omega - \bar{\omega}_*^T}{k_\| \hat{v}_3 - \omega + \bar{\omega}_d}.$$

Therefore, one obtains the dispersion relation

$$\sum_a \frac{e_a^2}{T_a} + \sum_a \frac{e_a^2}{T_a} \frac{1}{n_0} \int_{-\infty}^{+\infty} d\hat{v}_3 n \frac{\omega - \bar{\omega}_*^T}{k_\| \hat{v}_3 - \omega + \bar{\omega}_d} = 0. \tag{5.26}$$

Here, one can see that the Landau damping, magnetic drift, and diamagnetic drift are all retained in the perpendicular magnetofluid theory. One can also include the trapped particle effects if the non-WKB treatment in the parallel direction is carried out.

5.3 Discussion: comparison with the gyrokinetic theory

We have demonstrated the usage of the collisionless perpendicular magnetofluid theory to study the electrostatic modes. The advantages of this formalism are obvious as compared with the Braginskii theory for the collisionless case [7, 8]. Although Braginskii's theory computes the thermal heat conductivity and gyroviscous tensor, the introduction of the parallel velocity leads the parallel kinetic effects to be left out. The kinetic-description-like dispersion relation in equation (5.26) cannot be obtained from Braginskii's formulation. The kinetic modifications are not only necessary for pressure, but also for the parallel velocity. The kinetic modifications appear not only in the particle-wave resonances in the dispersion relation but also through the integration path change in the continuity equation (5.24).

The electrostatic mode theory by the guiding center and gyrokinetic theories have been extensively reviewed in other books, for example, references [13, 14]. Comparing the dispersion relation in equation (5.26) with that by kinetic theory, one can see that the parallel kinetic effects, such as the trapped particle and wave-particle resonance effects, are fully retained. The magnetic and diamagnetic drift effects are also kept. But, they appear in a way with an average over the perpendicular velocity. For example, for magnetic drift velocity in equation (5.18) $v_\perp^2$ for grad B drift is replaced by the perpendicular temperature T. The current analyses are in the lowest order. The higher order analyses can take into account the perpendicular motion in further detail.

The kinetic theory is not necessarily better than the perpendicular magnetofluid description. The fluid thermodynamic description is sometimes unavoidable. The particle initial loading requires the fluid thermal parameters, such as the density, velocity, and temperature in the Maxwellian distribution. The boundary conditions for distribution function require the fluid thermodynamic description as well. One simply cannot specify the initial and boundary conditions in the six-dimensional space for all particles.

References

[1] Zheng L 2020 Perpendicular magnetofluid theory for magnetically confined plasmas in the collisionless limit *Phys. Plasmas* **27** 052503

[2] Rutherford P H and Frieman E A 1968 Drift instabilities in general magnetic field configurations *Phys. Fluids* **11** 569–85

[3] Taylor J B and Hastie R J 1968 Stability of general plasma equilibria: I. formal theory *Plasma Phys.* **10** 479

[4] Antonsen T M and Lane B 1980 Kinetic equations for low frequency instabilities in inhomogeneous plasmas *Phys. Fluids* **23** 1205–14

[5] Catto P J, Tang W M and Baldwin D E 1981 Generalized gyrokinetics *Plasma Phys.* **23** 639–50

[6] Zheng L J, Kotschenreuther M T and Van Dam J W 2007 Revisiting linear gyrokinetics to recover ideal magnetohydrodynamics and missing finite Larmor radius effects *Phys. Plasmas* **14** 072505

[7] Braginskii S I 1958 Transport phenomena in a complete ionized two-temperature plasma *ZhETF (J. Exp. Theor. Phys. USSR)* **33** 459
Braginskii S I 1958 *Soviet Phys. JETP* **6** 358–69

[8] Braginskii S I 1966 Transport processes *Reviews of Plasma Physics* vol 1 ed M Leontovich (New York: Consultants Bureau) pp 205–311

[9] Zheng L 2015 *Advanced Tokamak Stability Theory* (San Rafael, CA: Morgan & Claypool)

[10] Hazeltine R D 1972 *Recursive Derivation of Drift-kinetic Equation Plasma Phys.* **15** 77–80

[11] Hazeltine R D and Meiss J D 1992 *Plasma Confinement* (New York: Addison-Wesley)

[12] White R B 1989 *Theory of Tokamak Plasmas* (Amsterdam: North-Holland)

[13] Tang W M 1978 Microinstability theory in tokamaks *Nucl. Fusion* **18** 1089–160

[14] Miyamoto K 1989 *Plasma Physics for Nuclear Fusion* (Cambridge, MA: MIT Press)

Chapter 6

Electromagnetic modes in the perpendicular magnetofluid description

In this chapter, the perpendicular magnetofluid theory [1], as reviewed in chapter 3, will be used to study the electromagnetic modes in the collisionless limit. It is intended to demonstrate how to use the theory and to exploit the basic features of the model so that they can be compared with the gyrokinetic description [2–6], as well as the Braginskii one [7, 8]. The subtle toroidal structure of electromagnetic modes has been reviewed in the author's earlier books [9] and [10]. One can directly extend the current analyses to study the interchange, ballooning, Alfvén, kink modes, etc. since the non-ideal MHD effects usually become important only in the so-called singular layers. The readers are supposed to be familiar with the MHD analyses in the singular layers as reviewed in the author's earlier books on MHD [9] and [10], in which the references for MHD modes are detailed and will not be repeated here.

We will proceed with the studies of electromagnetic modes step by step. The equilibrium results in chapter 4 will be used in the analyses. After introducing the basic set of equations, the electromagnetic modes in the low frequency regime will be investigated. The perpendicular MHD equations [11] are obtained with the kinetic description of the parallel inertia effects. This results in enhanced apparent mass effects [12–14]. The intermediate and high frequency regimes will be considered afterwards to allow the wave-particle resonance and the parallel electric field effects to be described in the perpendicular magnetofluid formalism [15, 16]. The finite Larmor radius effects will be discussed next. In the end, the coupling of the electrostatic modes [15] to the electromagnetic modes will be addressed.

doi:10.1088/978-0-7503-3575-1ch6

6.1 Basic set of equations

In this section, the basic set of equations in the collisionless perpendicular magneto-fluid theory is described with respective terms labelled. They will be reduced for various analyses in the following sections.

The continuity, momentum, and energy transport equations have been given in equations (3.94), (3.95), and (3.97), respectively, in section 3.5. They are summarized as follows.

$$\begin{aligned}
&\underbrace{\frac{\partial n}{\partial t}}_{n_1} + \underbrace{g_{\beta\alpha}\hat{V}_{\perp\beta}\frac{\partial n}{\partial x_\alpha}}_{n_2} + \underbrace{g_{3\alpha}\hat{v}_3\frac{\partial n}{\partial x_\alpha}}_{n_3} + \underbrace{n\frac{\partial g_{l\gamma}\hat{V}_{\perp l}}{\partial x_\beta}(\delta_{\beta\gamma} - g_{3\beta}g_{3\gamma})}_{n_4} \\
&+ \underbrace{n\hat{v}_3\frac{\partial g_{3\alpha}}{\partial x_\alpha}}_{n_5} + \underbrace{\frac{1}{m}\frac{\partial g_{3\alpha}}{\partial x_\alpha}\frac{\partial nT}{\partial \hat{v}_3}}_{n_6} - \underbrace{g_{3\beta}g_{3\gamma}\frac{\partial g_{l\gamma}\hat{V}_{\perp l}}{\partial x_\beta}\hat{v}_3\frac{\partial n}{\partial \hat{v}_3}}_{n_7} \\
&= -g_{3\beta}\frac{\partial n}{\partial \hat{v}_3}\left(\underbrace{\frac{e}{m}E^*_\beta}_{n_8} - \underbrace{\frac{\hat{d}g_{l\beta}\hat{V}_{\perp l}}{dt}}_{n_9}\right) - \underbrace{\frac{1}{m}g_{p\alpha}g_{q\mu}\frac{\partial g_{3\mu}}{\partial x_\alpha}\frac{\partial \pi_{pq}}{\partial \hat{v}_3}}_{n_{10}},
\end{aligned} \tag{6.1}$$

$$\begin{aligned}
&(\delta_{\lambda\alpha} - g_{3\alpha}g_{3\lambda})\left(\underbrace{neE^*_\alpha}_{v_1} - \underbrace{\rho\frac{\hat{d}g_{l\alpha}\hat{V}_{\perp l}}{dt}}_{v_2} - \underbrace{\frac{\partial p}{\partial x_\alpha}}_{v_3} - \underbrace{\hat{v}_3 g_{3\beta}\frac{\partial g_{3\alpha}}{\partial x_\beta}\frac{\partial p}{\partial \hat{v}_3}}_{v_4}\right. \\
&\left. - \underbrace{g_{3\beta}\hat{v}_3^2\frac{\partial g_{3\alpha}}{\partial x_\beta}\rho}_{v_5} + \underbrace{g_{3\gamma}\frac{\partial g_{l\gamma}\hat{V}_{\perp l}}{\partial x_\alpha}\frac{\partial p}{\partial \hat{v}_3}}_{v_6} - \underbrace{\rho\hat{v}_3 g_{3\beta}\frac{\partial g_{l\alpha}\hat{V}_{\perp l}}{\partial x_\beta}}_{v_7} - \underbrace{\rho\hat{v}_3\frac{\hat{d}g_{3\alpha}}{dt}}_{v_8} - \underbrace{\frac{\hat{d}g_{3\alpha}}{dt}\frac{\partial p}{\partial \hat{v}_3}}_{v_9}\right) \\
&= \underbrace{g_{\nu\lambda}g_{\beta\alpha}\frac{\partial \pi_{\beta\nu}}{\partial x_\alpha}}_{v_{10}} + g_{\nu\lambda}\left(\underbrace{g_{p\delta}\frac{\hat{d}g_{3\delta}}{dt}}_{v_{11}} - \underbrace{g_{p\beta}g_{3\gamma}\frac{\partial g_{l\gamma}\hat{V}_{\perp l}}{\partial x_\beta}}_{v_{12}} + \underbrace{\hat{v}_3 g_{3\alpha}g_{p\mu}\frac{\partial g_{3\mu}}{\partial x_\alpha}}_{v_{13}}\right)\frac{\partial \pi_{p\nu}}{\partial \hat{v}_3} \\
&+ \left(\underbrace{\frac{\partial g_{p\alpha}g_{q\lambda}}{\partial x_\alpha}}_{v_{14}} + \underbrace{g_{p\alpha}g_{q\mu}g_{3\lambda}\frac{\partial g_{3\mu}}{\partial x_\alpha}}_{v_{15}} + \underbrace{g_{3\alpha}g_{q\mu}g_{p\lambda}\frac{\partial g_{3\mu}}{\partial x_\alpha}}_{v_{16}}\right)\pi_{pq} \\
&+ \underbrace{g_{\nu\lambda}\frac{\partial g_{3\alpha}}{\partial x_\alpha}\frac{\partial q_\nu}{\partial \hat{v}_3}}_{v_{17}} + \underbrace{g_{p\alpha}g_{q\mu}g_{\nu\lambda}\frac{\partial g_{3\mu}}{\partial x_\alpha}\frac{\partial \Theta_{pq\nu}}{\partial \hat{v}_3}}_{v_{18}},
\end{aligned} \tag{6.2}$$

and

$$
\begin{aligned}
&\underbrace{n\frac{\partial T}{\partial t}}_{t_1} + \underbrace{ng_{\beta\alpha}\hat{V}_{\perp\beta}\frac{\partial T}{\partial x_\alpha}}_{t_2} + \underbrace{ng_{3\alpha}\hat{v}_3\frac{\partial T}{\partial x_\alpha}}_{t_3} + \underbrace{p\frac{\partial g_{l\gamma}\hat{V}_{\perp l}}{\partial x_\beta}(\delta_{\gamma\beta} - g_{3\beta}g_{3\gamma})}_{t_4} \\
&+ \underbrace{\hat{v}_3\frac{\partial g_{3\alpha}}{\partial x_\alpha}p}_{t_5} + \underbrace{\frac{\partial g_{3\alpha}}{\partial x_\alpha}\frac{T^2}{m}\frac{\partial n}{\partial \hat{v}_3}}_{t_6} + \underbrace{\frac{\partial g_{3\alpha}}{\partial x_\alpha}\frac{3p}{m}\frac{\partial T}{\partial \hat{v}_3}}_{t_7} - \underbrace{\hat{v}_3 g_{3\beta}g_{3\gamma}\frac{\partial g_{l\gamma}\hat{V}_{\perp l}}{\partial x_\beta}n\frac{\partial T}{\partial \hat{v}_3}}_{t_8} \\
&= -g_{3\alpha}\left(\underbrace{\frac{e}{m}E^*_\alpha}_{t_9} - \underbrace{\frac{\hat{d}g_{l\alpha}\hat{V}_{\perp l}}{dt}}_{t_{10}}\right)n\frac{\partial T}{\partial \hat{v}_3} + \underbrace{\frac{T}{m}g_{p\alpha}g_{q\mu}\frac{\partial g_{3\mu}}{\partial x_\alpha}\frac{\partial \pi_{pq}}{\partial \hat{v}_3}}_{t_{11}} \\
&- g_{p\alpha}g_{q\mu}\left(\underbrace{\hat{v}_3\frac{\partial g_{3\mu}}{\partial x_\alpha}}_{t_{12}} + \underbrace{\frac{\partial g_{l\mu}\hat{V}_{\perp l}}{\partial x_\alpha}}_{t_{13}}\right)\pi_{pq} + \left(\underbrace{g_{p\beta}g_{3\gamma}\frac{\partial g_{l\gamma}\hat{V}_{\perp l}}{\partial x_\beta}}_{t_{14}} - \underbrace{g_{p\delta}\frac{\hat{d}g_{3\delta}}{dt}}_{t_{15}}\right)\frac{\partial q_p}{\partial \hat{v}_3} \\
&- \underbrace{\frac{\partial}{\partial x_\alpha}(g_{\beta\alpha}q_\beta)}_{t_{16}} - \underbrace{\hat{v}_3 g_{3\alpha}g_{q\mu}\frac{\partial g_{3\mu}}{\partial x_\alpha}\frac{\partial q_q}{\partial \hat{v}_3}}_{t_{17}} - \underbrace{g_{3\alpha}g_{q\mu}\frac{\partial g_{3\mu}}{\partial x_\alpha}q_q}_{t_{18}} \\
&- \underbrace{g_{p\alpha}g_{q\mu}\frac{\partial g_{3\mu}}{\partial x_\alpha}\frac{1}{m}\frac{\partial n\Xi_{pq}}{\partial \hat{v}_3}}_{t_{19}}.
\end{aligned}
\tag{6.3}
$$

To complete the set of equations, the following Maxwell equations are also required

$$\sum_a n_a e_a = 0, \tag{6.4}$$

$$\nabla \times \mathbf{E} = -\frac{\partial \mathbf{B}}{\partial t}, \tag{6.5}$$

$$\mu_0 \mathbf{J} = \nabla \times \mathbf{B}, \tag{6.6}$$

$$\nabla \cdot \mathbf{B} = 0. \tag{6.7}$$

Here, the Gauss equation has been replaced by the quasineutrality condition in equation (6.4).

6.2 Perpendicular MHD with the kinetic parallel inertia

In this section, we will apply the perpendicular magnetofluid theory to study the MHD modes in the low frequency regime: $\omega \ll \omega_{ti}$, where ω_{ti} is the ion transit frequency. Only the linear theory is considered. The perpendicular MHD equations [11] will be obtained with the kinetic modification for parallel inertia effects. It results in enhanced apparent mass effects [13, 14].

The equilibrium density and temperature have been solved in equation (4.7) in chapter 4:

$$n = n_0 \exp\left\{ -\frac{m\hat{v}_3^2/2 + e\varphi}{T} \right\} \quad \text{and} \quad \mathbf{b}\cdot\nabla(n_0, T) = \frac{\partial T}{\partial \hat{v}_3} = 0. \tag{6.8}$$

For simplicity, the equilibrium electric field is assumed to be negligible. These solutions will be used in the following analyses.

Neglecting the finite Larmor radius effects on the right-hand side, the momentum equation (6.2) is reduced to

$$\begin{aligned}(\delta_{\lambda\alpha} - g_{3\alpha}g_{3\lambda})\Bigg[&\underbrace{neE_\alpha + nee_{\alpha\beta\gamma}g_{l\beta}\hat{V}_l B_\gamma}_{v_1} - \underbrace{\rho\frac{\hat{d}g_{l\alpha}\hat{V}_{\perp l}}{dt}}_{v_2} - \underbrace{\frac{\partial p}{\partial x_\alpha}}_{v_3} - \underbrace{\hat{v}_3 g_{3\beta}\frac{\partial g_{3\alpha}}{\partial x_\beta}\frac{\partial p}{\partial \hat{v}_3}}_{v_4} \\ &- \underbrace{g_{3\beta}\hat{v}_3^2\frac{\partial g_{3\alpha}}{\partial x_\beta}\rho}_{v_5} + \underbrace{g_{3\gamma}\frac{\partial g_{l\gamma}\hat{V}_{\perp l}}{\partial x_\alpha}\frac{\partial p}{\partial \hat{v}_3}}_{v_6} - \underbrace{\rho\hat{v}_3 g_{3\beta}\frac{\partial g_{l\alpha}\hat{V}_{\perp l}}{\partial x_\beta}}_{v_7} - \underbrace{\rho\hat{v}_3\frac{\hat{d}g_{3\alpha}}{dt}}_{v_8} - \underbrace{\frac{\hat{d}g_{3\alpha}}{dt}\frac{\partial p}{\partial \hat{v}_3}}_{v_9}\Bigg] = 0.\end{aligned} \tag{6.9}$$

To study the MHD modes, one needs to derive the momentum equation for the single fluid. This is achieved by integrating over $\hat{v}_3$ and the summation over all particle species. It is defined that

$$\langle\cdots\rangle_s = \sum_a \frac{1}{n_0}\int_{-\infty}^{+\infty} d\hat{v}_3(\cdots). \tag{6.10}$$

To simplify equation (6.9), one needs to estimate the orderings. As is known, the term 'v_3' is the pressure driving term and, therefore, can be used as the reference term. Since our major interest is to study the modes coupled to the shear Alfvén waves, the operator $\nabla\cdot(\mathbf{B}/B^2)\times\cdots$ will be applied on the momentum equation (6.9) to derive the so-called vorticity equation.

We first analyze terms 'v_6' and 'v_7'. If $g_{3\alpha}$ is a perturbed quantity, noting that

$$\frac{\partial n}{\partial \hat{v}_3} = -\frac{mv_3}{T}n \quad \text{and} \quad \frac{\partial T}{\partial \hat{v}_3} = 0, \tag{6.11}$$

one has terms 'v_6' and 'v_7' cancel each other. Here, equation (6.8) has been used. If $g_{3\alpha}$ is an equilibrium quantity, the terms 'v_6' and 'v_7' are of order $\hat{V}_\perp/\hat{v}_3$ smaller than the term 'v_3'. Therefore, the terms 'v_6' and 'v_7' can be neglected. Similarly, one can prove that the terms 'v_8' and 'v_9' can be neglected noting the definition of d/dt. Using equation (6.11), one can show that p and ρ in terms 'v_4' and 'v_5' must be the perturbed quantities. Therefore, equation (6.9) is reduced to

$$
\begin{aligned}
&(\delta_{\lambda\alpha} - g_{3\alpha}g_{3\lambda})\Bigg[\underbrace{neE_\alpha + nee_{\alpha\beta\gamma}g_{l\beta}\hat{V}_l B_\gamma}_{v_1} - \underbrace{\rho\frac{\hat{d}g_{l\alpha}\hat{V}_{\perp l}}{dt}}_{v_2} \\
&- \underbrace{\frac{\partial p}{\partial x_\alpha}}_{v_3} - \underbrace{\hat{v}_3 g_{3\beta}\frac{\partial g_{3\alpha}}{\partial x_\beta}\frac{\partial \delta p}{\partial \hat{v}_3}}_{v_4} - \underbrace{g_{3\beta}\hat{v}_3^2\frac{\partial g_{3\alpha}}{\partial x_\beta}\delta\rho}_{v_5}\Bigg] = 0.
\end{aligned}
\tag{6.12}
$$

Equation (6.12) depends on the particle species and the parallel velocity $\hat{v}_3$. In principle, one can obtain the single fluid equation through summing them together. The difficulty lies in the treatment of the convective term $\rho\mathbf{V}_\perp \cdot \nabla\mathbf{V}_\perp$. It is nonlinear. In the conventional two fluid theory, the large mass ratio between ion and electron is exploited in constructing a single fluid equation. This strategy is not applicable simply in the current situation since there are multiple ion components with different parallel speeds. One needs to deal with it case by case. First, we can neglect the electron inertia in the single fluid equation as well. Second, when the FLR effects are neglected, the diamagnetic drift velocity can be ignored. The dominant perpendicular perturbed velocity is the $\mathbf{E} \times \mathbf{B}$ drift velocity in this case. One then has

$$
\langle \rho\mathbf{V}_\perp \cdot \nabla\mathbf{V}_\perp\rangle_s = \rho_s \mathbf{V}_{\mathbf{E}\times\mathbf{B}} \cdot \nabla\mathbf{V}_{\mathbf{E}\times\mathbf{B}}.
$$

Note that $g_{3\beta}\frac{\partial g_{3\alpha}}{\partial x_\beta} = \mathbf{b} \cdot \nabla\mathbf{b} = \kappa$. Therefore, after the integration over $\hat{v}_3$ and the summation over particle species, equation (6.12) is transformed to the single fluid momentum equation

$$
\rho_s \frac{d\mathbf{V}_\perp}{dt} = \mathbf{J} \times \mathbf{B} - \nabla_\perp p_s - \kappa\left(\langle mn\hat{v}_3^2\rangle_s - p_s\right), \tag{6.13}
$$

where

$$
\rho_s = \langle mn\rangle_s, \qquad p_s = \langle p\rangle_s, \qquad \mathbf{J} = \langle ne\mathbf{V}_\perp\rangle_s + \mathbf{J}_\parallel,
$$

To determine the pressure-related terms in equation (6.13), the continuity and energy transport equations are needed. Neglecting the finite Larmor radius effects in the term 'n_{10}', the continuity equation (6.1) is reduced to

$$
\begin{aligned}
&\underbrace{\frac{\partial n}{\partial t}}_{n_1} + \underbrace{g_{\beta\alpha}\hat{V}_{\perp\beta}\frac{\partial n}{\partial x_\alpha}}_{n_2} + \underbrace{g_{3\alpha}\hat{v}_3\frac{\partial n}{\partial x_\alpha}}_{n_3} + \underbrace{n\frac{\partial g_{l\gamma}\hat{V}_{\perp l}}{\partial x_\beta}(\delta_{\beta\gamma} - g_{3\beta}g_{3\gamma})}_{n_4} \\
&+ \underbrace{n\hat{v}_3\frac{\partial g_{3\alpha}}{\partial x_\alpha}}_{n_5} + \underbrace{\frac{T}{m}\frac{\partial g_{3\alpha}}{\partial x_\alpha}\frac{\partial n}{\partial \hat{v}_3}}_{n_6} - \underbrace{g_{3\beta}g_{3\gamma}\frac{\partial g_{l\gamma}\hat{V}_{\perp l}}{\partial x_\beta}\hat{v}_3\frac{\partial n}{\partial \hat{v}_3}}_{n_7} \\
&= -g_{3\beta}\frac{\partial n}{\partial \hat{v}_3}\left(\underbrace{\frac{e}{m}E_\beta^*}_{n_8} - \underbrace{\frac{\hat{d}g_{l\beta}\hat{V}_{\perp l}}{dt}}_{n_9}\right).
\end{aligned}
\tag{6.14}
$$

Note that

$$g_{3\alpha}\frac{\hat{d}g_{l\alpha}\hat{V}_{\perp l}}{dt}\frac{\partial n}{\partial \hat{v}_3} = \mathbf{b}\cdot\frac{\hat{d}\delta\mathbf{V}_\perp}{dt}\frac{\partial n}{\partial\hat{v}_3} = -\delta\mathbf{V}_\perp\cdot\frac{\hat{d}\mathbf{b}}{dt}\frac{\partial n}{\partial\hat{v}_3}$$

$$\sim \delta\mathbf{V}_\perp\frac{n}{v_T}\mathbf{V}_\perp\cdot\nabla\mathbf{b} \ll \text{term “}n_3\text{”}.$$

It is also noted that $\mathbf{b}\cdot\mathbf{E}^* = 0$. Therefore, the terms '$n_8$' and '$n_9$' can be neglected. Equation (6.14) then becomes

$$\underbrace{\frac{\partial n}{\partial t}}_{n_1} + \underbrace{g_{\beta\alpha}\hat{V}_{\perp\beta}\frac{\partial n}{\partial x_\alpha}}_{n_2} + \underbrace{g_{3\alpha}\hat{v}_3\frac{\partial n}{\partial x_\alpha}}_{n_3} + \underbrace{n\frac{\partial g_{l\gamma}\hat{V}_{\perp l}}{\partial x_\beta}(\delta_{\beta\gamma} - g_{3\beta}g_{3\gamma})}_{n_4}$$
$$+ \underbrace{n\hat{v}_3\frac{\partial g_{3\alpha}}{\partial x_\alpha}}_{n_5} + \underbrace{\frac{T}{m}\frac{\partial g_{3\alpha}}{\partial x_\alpha}\frac{\partial n}{\partial\hat{v}_3}}_{n_6} - \underbrace{g_{3\beta}g_{3\gamma}\frac{\partial g_{l\gamma}\hat{V}_{\perp l}}{\partial x_\beta}\hat{v}_3\frac{\partial n}{\partial \hat{v}_3}}_{n_7} = 0. \tag{6.15}$$

We also note that the linearization of term 'n_3' gives rise to

$$g_{3\alpha}\hat{v}_3\frac{\partial n}{\partial x_\alpha} = \hat{v}_3(\delta\mathbf{b}\cdot\nabla n + \mathbf{b}\cdot\nabla\delta n). \tag{6.16}$$

Here, it has been noted that $g_{3\alpha}\partial/\partial x_\alpha = \hat{e}_3 e_\alpha \partial/\partial x_\alpha = \mathbf{b}\cdot\nabla$ and $\mathbf{b}\cdot\nabla n = 0$ in equilibrium. Note further that the momentum equation (6.12) in lowest order gives

$$\mathbf{E} + \mathbf{V}_\perp\times\mathbf{B} = 0, \tag{6.17}$$

as in the ideal MHD theory orderings. Similar to the ideal MHD, the time dependence is assumed to $\exp\{-i\omega t\}$ and the field line displacement $\boldsymbol{\xi}_\perp = \delta\mathbf{V}_\perp/(-i\omega)$ is introduced. Using Faraday's law in equation (6.5), one obtains

$$\delta\mathbf{B} = -\frac{1}{i\omega}\nabla\times(\mathbf{V}_\perp\times\mathbf{B}) = \nabla\times(\boldsymbol{\xi}_\perp\times\mathbf{B}). \tag{6.18}$$

Using this equation, one can further derive

$$\delta\mathbf{B}\cdot\nabla n = \nabla\cdot[(\boldsymbol{\xi}_\perp\times\mathbf{B})\times\nabla n] = \nabla\cdot(\boldsymbol{\xi}_\perp\cdot\nabla n\mathbf{B}) = \mathbf{B}\cdot\nabla(\boldsymbol{\xi}_\perp\cdot\nabla n).$$

Note also that the term 'n_4' is reduced as follows

$$\begin{aligned} -n\frac{\partial g_{l\gamma}\hat{V}_{\perp l}}{\partial x_\beta}g_{3\beta}g_{3\gamma} &= -n(\mathbf{b}\cdot\nabla\delta\mathbf{V}_\perp)\cdot\mathbf{b} \\ &= n(\mathbf{b}\cdot\nabla\mathbf{b})\cdot\delta\mathbf{V}_\perp \\ &= -i\omega n\boldsymbol{\kappa}\cdot\boldsymbol{\xi}, \end{aligned}$$

and

$$
\begin{aligned}
-\hat{v}_3 g_{3\beta} g_{3\gamma} \frac{\partial g_{l\gamma} \hat{V}_{\perp l}}{\partial x_\beta} \frac{\partial n}{\partial \hat{v}_3} &= \frac{m\hat{v}_3^2}{T} n(\mathbf{b} \cdot \nabla \delta \mathbf{V}_\perp) \cdot \mathbf{b} \\
&= -\frac{m\hat{v}_3^2}{T} n(\mathbf{b} \cdot \nabla \mathbf{b}) \cdot \delta \mathbf{V}_\perp \\
&= i\omega \frac{m\hat{v}_3^2}{T} n\boldsymbol{\kappa} \cdot \boldsymbol{\xi}.
\end{aligned}
$$

Using these simplifications, the continuity equation (6.15) is reduced to

$$
\begin{aligned}
&-i\omega(\delta n + \boldsymbol{\xi}_\perp \cdot \nabla n) + \hat{v}_3 \mathbf{b} \cdot \nabla \delta n + \hat{v}_3 \mathbf{b} \cdot \nabla(\boldsymbol{\xi}_\perp \cdot \nabla n) \\
&+ \delta n \hat{v}_3 \frac{\partial g_{3\alpha}}{\partial x_\alpha} + \frac{T}{m} \frac{\partial g_{3\alpha}}{\partial x_\alpha} \frac{\partial \delta n}{\partial \hat{v}_3} + \frac{1}{m} \frac{\partial g_{3\alpha}}{\partial x_\alpha} \frac{\partial n \delta T}{\partial \hat{v}_3} \\
&= in\omega \nabla \cdot \boldsymbol{\xi}_\perp + i\omega n \boldsymbol{\kappa} \cdot \boldsymbol{\xi} - i\omega \frac{m\hat{v}_3^2}{T} n\boldsymbol{\kappa} \cdot \boldsymbol{\xi}.
\end{aligned}
\tag{6.19}
$$

The energy transport equation can be reduced similarly. The reduction of the energy transport equation is also discussed in the electrostatic case in chapter 5. Neglecting the FLR effects on the right and noting that $\partial T/\partial \hat{v}_3 = 0$ in equilibrium, equation (6.3) is reduced to

$$
\begin{aligned}
&\underbrace{n\frac{\partial T}{\partial t}}_{t_1} + \underbrace{n g_{\beta\alpha} \hat{V}_{\perp\beta} \frac{\partial T}{\partial x_\alpha}}_{t_2} + \underbrace{n g_{3\alpha} \hat{v}_3 \frac{\partial T}{\partial x_\alpha}}_{t_3} + \underbrace{p \frac{\partial g_{l\gamma} \hat{V}_{\perp l}}{\partial x_\beta} (\delta_{\gamma\beta} - g_{3\beta} g_{3\gamma})}_{t_4} \\
&+ \underbrace{\hat{v}_3 \frac{\partial g_{3\alpha}}{\partial x_\alpha} p}_{t_5} + \underbrace{\frac{\partial g_{3\alpha}}{\partial x_\alpha} \frac{T^2}{m} \frac{\partial n}{\partial \hat{v}_3}}_{t_6} + \underbrace{\frac{\partial g_{3\alpha}}{\partial x_\alpha} \frac{3p}{m} \frac{\partial T}{\partial \hat{v}_3}}_{t_7} = 0.
\end{aligned}
\tag{6.20}
$$

Using the reductions leading to the continuity equation (6.19), equation (6.20) can be further reduced to

$$
\begin{aligned}
&-i\omega n(\delta T + \boldsymbol{\xi}_\perp \cdot \nabla T) + \underbrace{n\hat{v}_3 \mathbf{b} \cdot \nabla \delta T}_{t_3} + n\hat{v}_3 \mathbf{b} \cdot \nabla(\boldsymbol{\xi}_\perp \cdot \nabla T) \\
&+ \delta n T \hat{v}_3 \frac{\partial g_{3\alpha}}{\partial x_\alpha} + \frac{T^2}{m} \frac{\partial g_{3\alpha}}{\partial x_\alpha} \frac{\partial \delta n}{\partial \hat{v}_3} + \underbrace{\frac{3p}{m} \frac{\partial g_{3\alpha}}{\partial x_\alpha} \frac{\partial \delta T}{\partial \hat{v}_3}}_{t_7} \\
&= inT\omega \nabla \cdot \boldsymbol{\xi}_\perp + i\omega n T \boldsymbol{\kappa} \cdot \boldsymbol{\xi} - i\omega \frac{m\hat{v}_3^2}{T} nT\boldsymbol{\kappa} \cdot \boldsymbol{\xi}.
\end{aligned}
\tag{6.21}
$$

The analyses of the energy transport equation (6.21) are similar to the continuity equation (6.19) to be described later on. The convective part can be extracted: $\delta T = -\boldsymbol{\xi} \cdot \nabla T + \delta T_g$. To follow the characteristic to be derived later on in equation (6.25) for the density, the period of δT_g in the term 't_3' has to be shorter by a factor of 1/3 due to the factor 3 in the term 't_7' in equation (6.21). Note that the source terms to δT_g on the right-hand side of equation (6.21), which are similar to the right-hand side of the continuity equation (6.24) to be analyzed later on, are periodic in 2π. They are decoupled to δT_g along the characteristic curve. Therefore, one can neglect δT_g as compared with the contribution of δn_g defined later in equation (6.23). In passing, it is noted that $\delta T = -\boldsymbol{\xi} \cdot \nabla T$ means $\mathbf{b} \cdot \nabla T = 0$ in the nonlinear expression. Consequently, one has

$$\delta p = -\boldsymbol{\xi} \cdot \nabla p + n\delta\tilde{n}_g T. \tag{6.22}$$

These analyses also show that the term containing δT in the continuity equation (6.19) can be neglected.

Let us return to analyze the continuity equation (6.19). Letting

$$\delta\tilde{n} = -\boldsymbol{\xi}_\perp \cdot \nabla n + n\delta\tilde{n}_g, \tag{6.23}$$

where the equilibrium density n is given equation (6.8), one obtains

$$\begin{aligned} &- i\omega\delta\tilde{n}_g + \hat{v}_3\mathbf{b} \cdot \nabla |_{\hat{v}_3,t}\, \delta\tilde{n}_g + \frac{T}{m}\frac{\partial g_{3\alpha}}{\partial x_\alpha}\frac{\partial\delta\tilde{n}_g}{\partial\hat{v}_3} \\ &= i\omega\nabla \cdot \boldsymbol{\xi}_\perp + i\omega\boldsymbol{\kappa} \cdot \boldsymbol{\xi} - i\omega\frac{m\hat{v}_3^2}{T}\boldsymbol{\kappa} \cdot \boldsymbol{\xi} + \hat{v}_3\frac{\partial g_{3\alpha}}{\partial x_\alpha}\boldsymbol{\xi}_\perp \cdot \nabla \ln T. \end{aligned} \tag{6.24}$$

This is a partial differential equation of first order. As with the electrostatic stability theory in chapter 5, one can find its characteristic:

$$\frac{T}{m}\int^{\hat{x}_3} \nabla \cdot \mathbf{b}d\hat{x}_3 - \frac{\hat{v}_3^2}{2} = \iota. \tag{6.25}$$

Along the characteristic curve, equation (6.24) is reduced to

$$\begin{aligned} \hat{v}_3\mathbf{b} \cdot \nabla |_\iota\, \delta\tilde{n}_g - i\omega\delta\tilde{n}_g &= i\omega\nabla \cdot \boldsymbol{\xi}_\perp + i\omega\boldsymbol{\kappa} \cdot \boldsymbol{\xi} - i\omega\frac{m\hat{v}_3^2}{T}\boldsymbol{\kappa} \cdot \boldsymbol{\xi} \\ &+ \hat{v}_3\frac{\partial g_{3\alpha}}{\partial x_\alpha}\boldsymbol{\xi}_\perp \cdot \nabla \ln T. \end{aligned} \tag{6.26}$$

Here, $\hat{v}_3$ needs to be evaluated along the characteristic curve as defined in equation (6.25). As discussed for equation (5.24) in the electrostatic mode case in chapter 5, one can see that the characteristic curve basically follows the particle adiabatic invariant trajectory in the phase space. This transformation actually converts $\hat{\mathbf{v}}_3 \cdot \nabla|_{\hat{v}_3,t}$, holding $\hat{v}_3$ constant, in the second term of equation (6.24) to $\hat{\mathbf{v}}_3 \cdot \nabla|_{\varepsilon,\mu,t}$, holding (ε, μ) constant, in the first term of equation (6.26) leading it to become gyrokinetic-theory-like. The only major difference lies in the average over the perpendicular velocity.

Equation (6.26) can be used to construct the equations for even and odd parts of $\delta\tilde{n}_g$ with respect to the parallel velocity:

$$\begin{aligned}&|\hat{v}_3|\mathbf{b}\cdot\nabla|_t(\delta\tilde{n}_g^+ - \delta\tilde{n}_g^-) - i\omega(\delta\tilde{n}_g^+ + \delta\tilde{n}_g^-)\\&= 2i\omega\nabla\cdot\boldsymbol{\xi}_\perp + 2i\omega\boldsymbol{\kappa}\cdot\boldsymbol{\xi} - 2i\omega\frac{m\hat{v}_3^2}{T}\boldsymbol{\kappa}\cdot\boldsymbol{\xi},\end{aligned}\tag{6.27}$$

$$\begin{aligned}&|\hat{v}_3|\mathbf{b}\cdot\nabla|_t(\delta\tilde{n}_g^+ + \delta\tilde{n}_g^-) - i\omega(\delta\tilde{n}_g^+ - \delta\tilde{n}_g^-)\\&= 2|\hat{v}_3|\frac{\partial g_{3\alpha}}{\partial x_\alpha}\boldsymbol{\xi}_\perp\cdot\nabla\ln T,\end{aligned}\tag{6.28}$$

where the superscript $\pm$ refers to the sign of parallel velocity. To obtain the equation for the even part $\delta\tilde{n}_g^+ + \delta\tilde{n}_g^-$, we apply the operator $|\hat{v}_3|\mathbf{b}\cdot\nabla|_t$ on equation (6.28), yielding

$$\begin{aligned}&|\hat{v}_3|\mathbf{b}\cdot\nabla|_t\left[|\hat{v}_3|\mathbf{b}\cdot\nabla|_t(\delta\tilde{n}_g^+ + \delta\tilde{n}_g^-)\right] - i\omega\,|\hat{v}_3|\mathbf{b}\cdot\nabla|_t(\delta\tilde{n}_g^+ - \delta\tilde{n}_g^-)\\&= 2\,|\hat{v}_3|\mathbf{b}\cdot\nabla|_t\left(|\hat{v}_3|\frac{\partial g_{3\alpha}}{\partial x_\alpha}\boldsymbol{\xi}_\perp\cdot\nabla\ln T\right).\end{aligned}$$

Inserting equation (6.27), one finally obtains

$$\begin{aligned}&|\hat{v}_3|\mathbf{b}\cdot\nabla|_t\left[|\hat{v}_3|\mathbf{b}\cdot\nabla|_t(\delta\tilde{n}_g^+ + \delta\tilde{n}_g^-)\right] + \omega^2(\delta\tilde{n}_g^+ + \delta\tilde{n}_g^-)\\&= -2\,\omega^2\nabla\cdot\boldsymbol{\xi}_\perp - 2\omega^2\boldsymbol{\kappa}\cdot\boldsymbol{\xi} + 2\omega^2\frac{m\hat{v}_3^2}{T}\boldsymbol{\kappa}\cdot\boldsymbol{\xi}\\&+ 2\,|\hat{v}_3|\mathbf{b}\cdot\nabla|_t\left(|\hat{v}_3|\frac{\partial g_{3\alpha}}{\partial x_\alpha}\boldsymbol{\xi}_\perp\cdot\nabla\ln T\right).\end{aligned}\tag{6.29}$$

Noting that

$$-\boldsymbol{\kappa}\left(\langle m\delta n\hat{v}_3^2\rangle_s - \delta p_s\right) = -\boldsymbol{\kappa}\left(\left\langle m\delta n_g\hat{v}_3^2\right\rangle_s - \delta p_{gs}\right),$$

where the convective contribution (the first term on the right-hand side of equation (6.23)) is cancelled, the momentum equation (6.13) becomes

$$\begin{aligned}-\rho_s\omega^2\boldsymbol{\xi}_\perp &= \nabla\times(\nabla\times\boldsymbol{\xi}_\perp\times\mathbf{B})\times\mathbf{B} + \mathbf{J}\times\nabla\times(\boldsymbol{\xi}_\perp\times\mathbf{B})\\&+ \nabla_\perp(\boldsymbol{\xi}_\perp\cdot\nabla p_s) - \nabla_\perp\delta p_{gs}\\&- \boldsymbol{\kappa}\left(\frac{m}{2}\left\langle n\left(\delta\tilde{n}_g^+ + \delta\tilde{n}_g^-\right)\hat{v}_3^2\right\rangle_s - \delta p_{gs}\right),\end{aligned}\tag{6.30}$$

where equation (6.23) has been used,

$$\delta p_{gs} = \sum_a\frac{T_a}{n_0}\int_0^{+\infty}d\hat{v}_3 n\left(\delta\tilde{n}_g^+ + \delta\tilde{n}_g^-\right)_a,$$

and $\delta\tilde{n}_g^+ + \delta\tilde{n}_g^-$ is governed by equation (6.29). Equations (6.30) and (6.29) constitute the complete set of equations for the MHD modes in the low frequency regime in the collisionless perpendicular magnetofluid description.

It is interesting to compare the current perpendicular magnetofluid theory with the ideal MHD theory. The singular layer physics reviewed in the author's earlier books [9, 10] is required to follow the following discussion. We first discuss the momentum equation. Equation (6.30) is similar to the linear ideal MHD equation, expect the last two terms. The plasma compressibility contribution $-\Gamma p_s \nabla \cdot \boldsymbol{\xi}$ in the ideal MHD description is replaced by the new contribution from the pressure-related parallel moments—the last two terms in equation (6.30), where Γ is the ratio of specific heats. This removes the unphysical parallel displacement $\boldsymbol{\xi}_\parallel$ from the ideal MHD description. The parallel compressibility is now described by the continuity equation (6.29). The basic set of equations (6.30) and (6.29) resembles that in the gyrokinetic description in the low frequency regime as reviewed in reference [9]

The continuity equation (6.29) bears also considerable similarity with the ideal MHD equation of parallel motion:

$$\Gamma P \mathbf{B} \cdot \nabla \left(\frac{1}{B^2} \mathbf{B} \cdot \nabla \nabla \cdot \boldsymbol{\xi} \right) + \rho_s \omega^2 \nabla \cdot \boldsymbol{\xi} = \rho_s \omega^2 \nabla \cdot \boldsymbol{\xi}_\perp \tag{6.31}$$

Both are the second order differential equations, except equation (6.29) is given in the particle phase space and equation (6.31) is in configuration space. The individual particle responses with different parallel velocities are kept in the current perpendicular magnetofluid description. From the MHD analyses in references [9] and [10], one can see that for the suppression of compressional Alvén modes, one has $\nabla \cdot \boldsymbol{\xi}_\perp = -2\boldsymbol{\kappa} \cdot \boldsymbol{\xi}$. This indicates the right-hand sides of equations (6.29) and (6.31) have a similar structure, except for the last term in equation (6.29).

Further analyses of the basic set of equations (6.29) and (6.30) are similar to those reviewed in the author's earlier books [9, 10] for ideal MHD theory. We do not elaborate them further, except for pointing out the enhancement of the apparent mass effect in the current perpendicular magnetofluid description. Since the non-ideal MHD effects are only important in the singular layer, this is the major modification on the ideal MHD theory in the low frequency regime.

From the ideal MHD equation of parallel motion, equation (6.31), one obtains

$$p \nabla \cdot \boldsymbol{\xi} \sim \frac{\omega^2 \nabla \cdot \boldsymbol{\xi}_\perp}{\omega_{si}^2 - \omega^2},$$

where $\omega_{si} = v_T / Rq$ represents the frequency of parallel acoustic wave and v_T is the ion thermal speed. Instead, using the continuity equation (6.29), one obtains

$$\delta \tilde{n}_g^+ + \delta \tilde{n}_g^- \sim \frac{\omega^2 \tilde{n} \nabla \cdot \boldsymbol{\xi}_\perp}{(\hat{v}_3^2 / v_T^2) \omega_{si}^2 - \omega^2}.$$

For the low frequency regime: $\omega \ll \omega_{ti}$, one has

$$T \frac{\delta \tilde{n}_g^+ + \delta \tilde{n}_g^-}{p \nabla \cdot \boldsymbol{\xi}_\perp} \sim \frac{v_T^2}{\hat{v}_3^2},$$

Note that the particle number with small parallel speed is proportional to $\hat{v}_3/v_T$. Therefore, one has

$$\frac{\delta p_s}{p\nabla \cdot \boldsymbol{\xi}_\perp} \sim \frac{v_T}{\hat{v}_3}. \tag{6.32}$$

This indicates that in the perpendicular magnetofluid description, the coupling of the parallel fluid motion is enhanced by a factor $v_T/\hat{v}_3$ from the particles with small parallel velocities, as also found in the kinetic description [13, 14]. Since $\hat{v}_3$ has a lower limit in the toroidal confinement, the enhancement factor is about $\sqrt{R/a}$.

6.3 Wave-particle resonance and parallel electric field effects

In this section, the low frequency theory in section 6.2 is extended to the frequency regime with the mode frequency being comparable with the ion transit frequency. The same equilibrium conditions are considered. The wave-particle resonance and parallel electric field effects become involved in this comparable frequency regime. To focus on these effects, the FLR effects are excluded.

Neglecting the finite Larmor radius effects on the right-hand side, the momentum equation (6.2) is reduced to

$$\begin{aligned}(\delta_{\lambda\alpha} - g_{3\alpha}g_{3\lambda})\Bigg[&\underbrace{neE_\alpha + nee_{\alpha\beta\gamma}g_{l\beta}\hat{V}_l B_\gamma}_{v_1} - \underbrace{\rho\frac{\hat{d}g_{l\alpha}\hat{V}_{\perp l}}{dt}}_{v_2} \\ &- \underbrace{\frac{\partial p}{\partial x_\alpha}}_{v_3} - \underbrace{\hat{v}_3 g_{3\beta}\frac{\partial g_{3\alpha}}{\partial x_\beta}\frac{\partial \delta p}{\partial \hat{v}_3}}_{v_4} - \underbrace{g_{3\beta}\hat{v}_3^2\frac{\partial g_{3\alpha}}{\partial x_\beta}\delta\rho}_{v_5}\Bigg] = 0.\end{aligned} \tag{6.33}$$

Here, as discussed in the reduction of the momentum equation (6.12) in section 6.2, the terms 'v_6'–'v_9' in equation (6.33) have been neglected and p and ρ in terms 'v_4' and 'v_5' are given as the perturbed quantities

The scalar and vector potentials are introduced to represent the electromagnetic field:

$$\delta\mathbf{E} = -\frac{\partial\delta\mathbf{A}}{\partial t} - \nabla\delta\varphi, \qquad \delta\mathbf{B} = \nabla\times\delta\mathbf{A}. \tag{6.34}$$

As proved in reference [17] and also been seen in chapter 7, if the field lines do not reconnect, $\delta\mathbf{A}_\parallel$ can be absorbed into $\nabla_\parallel\delta\varphi$ so that one can choose the representation $\delta\mathbf{A}_\parallel = 0$. Introducing the field displacement $\boldsymbol{\xi}_\perp$, one express $\delta\mathbf{A} = \boldsymbol{\xi}_\perp \times \mathbf{B}$.

One can solve the perpendicular velocity from equation (6.33). Because the FLR effects are neglected, only the $\mathbf{E}\times\mathbf{B}$ velocity from the term 'v_1' is left.

$$\delta\mathbf{V}_\perp = -\omega\boldsymbol{\xi}_\perp + \frac{\mathbf{B}\times\nabla\delta\varphi}{B^2}, \tag{6.35}$$

where equation (6.34) has been used. The perpendicular velocity is present in all three moment equations. In the momentum equation (6.33), it appears explicitly only in the inertia term. We can estimate the orderings of the velocity induced by the parallel electric field $\delta\varphi$ as follows.

$$\frac{\mathbf{B}\times\nabla\delta\varphi}{B^2} \sim \frac{k_\perp}{B}\delta\varphi \sim k_\perp\rho_a\frac{v_T}{R}\frac{\omega^2}{\omega^2+\omega_t^2}\xi_\wedge \ll \omega\xi_\wedge. \tag{6.36}$$

Here, the quasineutrality condition to be derived later on is used to estimate the $\delta\varphi$ ordering. This ordering estimate indicates that the velocity induced by the parallel electric field $\delta\varphi$ can be neglected in the inertia term in equation (6.33). Nevertheless, it is pointed out that the parallel electric field effects from $\delta\varphi$ cannot be neglected in the continuity and energy equations, as will be addressed later on.

Because the velocity induced by the parallel electric field $\delta\varphi$ can be neglected in the inertia term, the further reduction of the momentum equation is the same as equation (6.13) in section 6.2, Integrating over $\hat{v}_3$ and summing over particle species, one obtains

$$\rho_s\frac{d\mathbf{V}_\perp}{dt} = \mathbf{J}\times\mathbf{B} - \nabla_\perp p_s - \kappa\left(\langle mn\hat{v}_3^2\rangle_s - p_s\right), \tag{6.37}$$

where

$$\rho_s = \langle mn\rangle_s, \qquad p_s = \langle p\rangle_s, \qquad \mathbf{J} = \langle ne\mathbf{V}_\perp\rangle_s + \mathbf{J}_\parallel,$$

$$\langle\cdots\rangle_s = \sum_a\left(\frac{m}{2\pi T}\right)^{1/2}\int_{-\infty}^{+\infty} d\hat{v}_3(\cdots).$$

To reduce the momentum equation (6.37), one needs to determine the pressure-related moments in it. The continuity and energy transport equations are needed. Similar to the discussion of equation (6.21) in section 6.2, the contribution from the perturbed temperature can be neglected, except for the convective term

$$\delta T = -\boldsymbol{\xi}\cdot\nabla T. \tag{6.38}$$

Therefore, only the continuity equation needs to be analyzed. Neglecting the finite Larmor radius effects and using the similar ordering analyses in section 6.2, the continuity equation (6.1) is reduced to

$$\begin{aligned}
&\underbrace{\frac{\partial n}{\partial t}}_{n_1} + \underbrace{g_{\beta\alpha}\hat{V}_{\perp\beta}\frac{\partial n}{\partial x_\alpha}}_{n_2} + \underbrace{g_{3\alpha}\hat{v}_3\frac{\partial n}{\partial x_\alpha}}_{n_3} + \underbrace{n\frac{\partial g_{l\gamma}\hat{V}_{\perp l}}{\partial x_\beta}(\delta_{\beta\gamma} - g_{3\beta}g_{3\gamma})}_{n_4}\\
&+ \underbrace{n\hat{v}_3\frac{\partial g_{3\alpha}}{\partial x_\alpha}}_{n_5} + \underbrace{\frac{T}{m}\frac{\partial g_{3\alpha}}{\partial x_\alpha}\frac{\partial n}{\partial\hat{v}_3}}_{n_6} - \underbrace{g_{3\beta}g_{3\gamma}\frac{\partial g_{l\gamma}\hat{V}_{\perp l}}{\partial x_\beta}\hat{v}_3\frac{\partial n}{\partial\hat{v}_3}}_{n_7}\\
&= \underbrace{-\frac{ne}{T}\hat{v}_3\mathbf{b}\cdot\nabla\delta\varphi}_{n_8}.
\end{aligned} \tag{6.39}$$

Here, the parallel electric effect from $\delta\varphi$ in the term 'n_8' has been retained as compared to equation (6.15). Using similar simplification leading to equation (6.19) in section 6.2, the continuity equation (6.39) is reduced to

$$\begin{aligned}
&- i\omega(\delta n + \boldsymbol{\xi}_\perp \cdot \nabla n) + \hat{v}_3 \mathbf{b} \cdot \nabla \delta n + \hat{v}_3 \mathbf{b} \cdot \nabla(\boldsymbol{\xi}_\perp \cdot \nabla n) \\
&+ \delta n \hat{v}_3 \frac{\partial g_{3\alpha}}{\partial x_\alpha} + \frac{T}{m} \frac{\partial g_{3\alpha}}{\partial x_\alpha} \frac{\partial \delta n}{\partial \hat{v}_3} \\
&= in\omega \nabla \cdot \boldsymbol{\xi}_\perp + i\omega n \boldsymbol{\kappa} \cdot \boldsymbol{\xi} - i\omega \frac{m\hat{v}_3^2}{T} n \boldsymbol{\kappa} \cdot \boldsymbol{\xi} - \frac{ne}{T} \hat{v}_3 \mathbf{b} \cdot \nabla \delta\varphi
\end{aligned} \tag{6.40}$$

To extract the adiabatic part, we let

$$\delta \tilde{n} = -\frac{ne}{T} \delta\varphi + \delta n_h. \tag{6.41}$$

The continuity equation (6.40) is reduced to

$$\begin{aligned}
&- i\omega(\delta n_h + \boldsymbol{\xi}_\perp \cdot \nabla n) + \hat{v}_3 \mathbf{b} \cdot \nabla \delta n_h \\
&+ \hat{v}_3 \mathbf{b} \cdot \nabla(\boldsymbol{\xi}_\perp \cdot \nabla n) + \frac{T}{m} \frac{\partial g_{3\alpha}}{\partial x_\alpha} \frac{\partial \delta n_h}{\partial \hat{v}_3} + \frac{T}{m} \frac{\partial g_{3\alpha}}{\partial x_\alpha} \frac{\partial \delta n}{\partial \hat{v}_3} \\
&= in\omega \nabla \cdot \boldsymbol{\xi}_\perp + i\omega n \boldsymbol{\kappa} \cdot \boldsymbol{\xi} - i\omega \frac{m\hat{v}_3^2}{T} n \boldsymbol{\kappa} \cdot \boldsymbol{\xi} - i\omega \frac{ne}{T} \delta\varphi.
\end{aligned}$$

To extract the convective part, we further let that

$$\delta \tilde{n}_h = -\boldsymbol{\xi}_\perp \cdot \nabla n + n \delta \tilde{n}_g \tag{6.42}$$

and obtain

$$\begin{aligned}
&- i\omega \delta \tilde{n}_g + \hat{v}_3 \mathbf{b} \cdot \nabla |_{\hat{v}_3, \iota} \delta \tilde{n}_g + \frac{T}{m} \frac{\partial g_{3\alpha}}{\partial x_\alpha} \frac{\partial \delta \tilde{n}_g}{\partial \hat{v}_3} \\
&= i\omega \nabla \cdot \boldsymbol{\xi}_\perp + i\omega \boldsymbol{\kappa} \cdot \boldsymbol{\xi} - i\omega \frac{m\hat{v}_3^2}{T} \tilde{n} \boldsymbol{\kappa} \cdot \boldsymbol{\xi} \\
&+ \hat{v}_3 \frac{\partial g_{3\alpha}}{\partial x_\alpha} \boldsymbol{\xi}_\perp \cdot \nabla \ln T - i\omega \frac{e}{T} \delta\varphi.
\end{aligned} \tag{6.43}$$

This is a partial differential equation of first order. As with the electrostatic stability theory in chapter 5, one can find its characteristic:

$$\frac{T}{m} \int^{\hat{x}_3} \nabla \cdot \mathbf{b} d\hat{x}_3 - \frac{\hat{v}_3^2}{2} = \iota. \tag{6.44}$$

Along the characteristic curve, equation (6.43) is reduced to

$$\begin{aligned}
\hat{v}_3 \mathbf{b} \cdot \nabla |_\iota \delta \tilde{n}_g - i\omega \delta \tilde{n}_g = & i\omega \nabla \cdot \boldsymbol{\xi}_\perp + i\omega \boldsymbol{\kappa} \cdot \boldsymbol{\xi} - i\omega \frac{m\hat{v}_3^2}{T} \boldsymbol{\kappa} \cdot \boldsymbol{\xi} \\
& + \hat{v}_3 \frac{\partial g_{3\alpha}}{\partial x_\alpha} \boldsymbol{\xi}_\perp \cdot \nabla \ln T - i\omega \frac{e}{T} \delta\varphi
\end{aligned} \tag{6.45}$$

Here, $\hat{v}_3$ needs to be evaluated along the characteristic curve as defined in equation (6.44). As discussed for equation (5.24) in the electrostatic mode case in chapter 5, one can see that the characteristic curve basically follows the particle adiabatic invariant trajectory in the phase space. This transformation actually converts $\hat{\mathbf{v}}_3 \cdot \nabla|_{\hat{v}_3,t}$, holding $\hat{v}_3$ constant, in the second term of equation (6.43) to $\hat{\mathbf{v}}_3 \cdot \nabla|_{\varepsilon,\mu,t}$, roughly holding (ε, μ) constant, in the first term of equation (6.45) leading it to become gyrokinetic-theory-like. The only major difference lies in the average over the perpendicular velocity.

Similar to the derivation of equation (6.29) in section 6.2, equation (6.45) can be used to construct the equation for the even part of the perturbed density

$$
\begin{aligned}
&|\hat{v}_3|\mathbf{b} \cdot \nabla|_t \left[|\hat{v}_3|\mathbf{b} \cdot \nabla|_t(\delta\tilde{n}_g^+ + \delta\tilde{n}_g^-)\right] + \omega^2(\delta\tilde{n}_g^+ + \delta\tilde{n}_g^-) \\
&= -2\omega^2\nabla \cdot \boldsymbol{\xi}_\perp - 2\omega^2\boldsymbol{\kappa} \cdot \boldsymbol{\xi} + 2\omega^2\frac{m\hat{v}_3^2}{T}\boldsymbol{\kappa} \cdot \boldsymbol{\xi} + 2\omega^2\frac{e}{T}\delta\varphi \\
&+ 2\,|\hat{v}_3|\mathbf{b} \cdot \nabla|_t \left(|\hat{v}_3|\frac{\partial g_{3\alpha}}{\partial x_\alpha}\boldsymbol{\xi}_\perp \cdot \nabla \ln T\right).
\end{aligned}
\tag{6.46}
$$

To complete the basic set of equations, the quasineutrality condition is required to determine the parallel electric potential $\delta\varphi$:

$$
\sum_a e_a\frac{1}{n_0}\int_0^{+\infty} d\hat{v}_3(\delta n_a^+ + \delta n_a^-) = 0.
$$

Using the transforms in equations (6.41) and (6.42), the quasineutrality condition becomes

$$
\delta\varphi \sum_a \frac{e_a^2}{T_a} = \sum_a e_a\frac{1}{n_0}\int_0^{+\infty} d\hat{v}_3 n\left(\delta\tilde{n}_{ga}^+ + \delta\tilde{n}_{ga}^-\right). \tag{6.47}
$$

where the quasineutrality in equilibrium has been imposed.

The momentum equation (6.37) can be further reduced similarly to the reduction in section 6.2. Noting that

$$
-\boldsymbol{\kappa}\left(\langle m\delta n\hat{v}_3^2\rangle_s - \delta p_s\right) = -\boldsymbol{\kappa}\left(\left\langle m\delta n_g\hat{v}_3^2\right\rangle_s - \delta p_{gs}\right),
$$

where the convective contributions in equations (6.38) and (6.42) are cancelled and the first term on the right of equation (6.41) in the adiabatic transformation does not contribute due to the quasineutrality condition. the momentum equation (6.37) then becomes

$$
\begin{aligned}
-\rho_s\omega^2\boldsymbol{\xi}_\perp &= \nabla \times (\nabla \times \boldsymbol{\xi}_\perp \times \mathbf{B}) \times \mathbf{B} + \mathbf{J} \times \nabla \times (\boldsymbol{\xi}_\perp \times \mathbf{B}) \\
&+ \nabla_\perp(\boldsymbol{\xi}_\perp \cdot \nabla p_s) - \nabla_\perp\delta p_{gs} \\
&- \boldsymbol{\kappa}\left(\frac{m}{2}\left\langle n\left(\delta\tilde{n}_g^+ + \delta\tilde{n}_g^-\right)\hat{v}_3^2\right\rangle_s - \delta p_{gs}\right),
\end{aligned}
\tag{6.48}
$$

where

$$\delta p_{gs} = \sum_a \frac{T_a}{n_0} \int_0^{+\infty} d\hat{v}_3 n\left(\delta\tilde{n}_g^+ + \delta\tilde{n}_g^-\right)_a$$

and $\delta\tilde{n}_g^+ + \delta\tilde{n}_g^-$ satisfies the continuity equation (6.46). The momentum equation (6.48), the quasineutrality condition, equation (6.47), and the continuity equation (6.46) constitute the complete set of equations for the MHD modes in the comparable frequency regime in the collisionless perpendicular magnetofluid description. This basic set of equations resembles that in the gyrokinetic description in the comparable frequency regime [15].

Let us discuss the parallel electric effects. From equation (6.46), one has the estimate

$$n_{ag}^+ + n_{ag}^- \sim \frac{\omega^2}{\hat{v}_3^2/R^2q^2 + \omega^2} n\nabla \cdot \boldsymbol{\xi}_\perp. \tag{6.49}$$

Inserting it into equation (6.47), one gets the ordering estimate

$$\frac{e}{T}\delta\varphi \sim \frac{\omega^2}{\hat{v}_3^2/R^2q^2 + \omega^2} \nabla \cdot \boldsymbol{\xi}_\perp. \tag{6.50}$$

Applying this ordering estimate on the right-hand side of equation (6.46), one can see that the fourth term on the right, i.e. the parallel electric field effect, is of order $\mathcal{O}(\omega^2/(\hat{v}_3^2/R^2q^2 + \omega^2))$. Here, the estimate has not yet counted in the so-called small parallel speed effects as discussed at the end of section 6.2 for the enhancement of the apparent mass effect. Generally speaking, when $\hat{v}_3^2/R^2q^2$ is much larger than ω^2, the parallel electric field effects become negligible. Noting that $\hat{v}_3$ depends on the individual particle speed, there is an enhancement of small particle speed. Counting in the reduction in particle number with small parallel speed, the parallel electric field effects $\delta\varphi$ follow the reduction of order $\mathcal{O}(\omega^2/(\hat{v}_T\hat{v}_3/R^2q^2)$ in the low frequency case. Again, it is noted that $\hat{v}_3$ has a lower limit in the toroidal confinement $\hat{v}_3 \sim v_T\sqrt{a/R}$.

6.4 Finite Larmor radius effects

In this section, we consider the finite Larmor radius effects (FLR) in the collisionless perpendicular magnetofluid description. To focus on this, the low frequency regime, as discussed in section 6.2, is studied. We first reduce the gyroviscous tensor and then apply it to study the MHD modes. Since the FLR effects are only important in the singular layer, the MHD singular layer theory is applied in the analyses. Since the MHD theory has been extensively reviewed in the author's earlier books [9, 10], the singular layer treatment will not be further elaborated. The current theory modifies the analyses based on the Braginskii gyroviscous tensor [7, 8].

6.4.1 Gyroviscosity for electromagnetic modes

In this subsection, the gyroviscous tensor is reduced for electromagnetic modes. The analyses are carried out in the singular layer. The operator $\nabla \cdot \frac{\mathbf{B}}{B^2} \times \cdots$ is applied on the gyroviscous tensor in order to see its role in the vorticity equation. The results will be used to study the FLR effects on the MHD modes in the next subsections.

From equation (3.98) in chapter 3, the gyroviscous tensor π can be expressed as follows:

$$\pi_{\alpha\beta} = \eta_3 W_{3\alpha\beta},$$

where α, $\beta = 1$, 2 and

$$\eta_3 = nT/2\omega_c,$$
$$W_{3\alpha\beta} = \frac{1}{2}\left(\delta^{\perp}_{\alpha\mu} e_{\beta\gamma\nu} + \delta^{\perp}_{\beta\nu} e_{\alpha\gamma\mu}\right) b_\gamma W_{\mu\nu}.$$

Here, the rate-of-strain tensor is given as follows:

$$W_{\mu\nu} = W^{\perp}_{\mu\nu} + W^{\parallel}_{\mu\nu}, \tag{6.51}$$

$$W^{\perp}_{\mu\nu} = g_{\mu\alpha} g_{\nu\beta} \frac{\partial g_{l\beta} \hat{V}_{\perp l}}{\partial x_\alpha} + g_{\nu\alpha} g_{\mu\beta} \frac{\partial g_{l\beta} \hat{V}_{\perp l}}{\partial x_\alpha} - \delta^{\perp}_{\mu\nu} \frac{\partial g_{l\alpha} \hat{V}_{\perp l}}{\partial x_\alpha}, \tag{6.52}$$

$$W^{\parallel}_{\mu\nu} = \left(g_{\mu\alpha} g_{\nu\beta} \frac{\partial g_{3\beta}}{\partial x_\alpha} + g_{\nu\alpha} g_{\mu\beta} \frac{\partial g_{3\beta}}{\partial x_\alpha} - \delta^{\perp}_{\mu\nu} \frac{\partial g_{3\alpha}}{\partial x_\alpha} \right) \left(\hat{v}_3 + \frac{T}{m} \frac{\partial \ln n}{\partial \hat{v}_3} \right). \tag{6.53}$$

Note that both $W_{\mu\nu}$ and $\pi_{\alpha\beta}$ are symmetric.

The linear treatment is presented. The wavelength is assumed to be much smaller than the equilibrium scale length. As will be seen in the next subsection, the gyroviscous tensor enters the vorticity equation in the manner $\nabla \cdot \frac{\mathbf{B}}{B^2} \times \nabla \cdot \Pi$, where $\Pi_{\delta\sigma} = g_{\alpha\delta} g_{\beta\sigma} \pi_{\alpha\beta}$. We define the diamagnetic frequency due to the pressure gradient as follows

$$i\omega_*^p = \frac{\mathbf{B} \times \nabla P}{neB^2} \cdot \nabla. \tag{6.54}$$

As $W_{\mu\nu}$ in equation (6.51) contains two parts: $W^{\perp}_{\mu\nu}$ and $W^{\parallel}_{\mu\nu}$, we will treat them one by one.

The $W^{\perp}_{\mu\nu}$ contributions

We first calculate the contributions from $W^{\perp}_{\mu\nu}$ in equation (6.52). The calculations are tedious, We try to simplify the notations in this subsubsection and only in this subsubsection. Since the equilibrium velocity is assumed to be small, $\hat{\mathbf{V}}_\perp$ in equation (6.52) should be the perturbed velocity $\delta\hat{\mathbf{V}}_\perp$. Without causing ambiguity, the tag δ is suppressed for $\delta\hat{\mathbf{V}}_\perp$ in the derivations in this subsubsection.

We make a general reduction of $\nabla \cdot \Pi^\perp$ for each term in equation (6.52). The last term can be proved to be negligible as follows.

$$\frac{1}{2}\left(\delta^{\perp}_{\alpha\mu}e_{\beta\gamma\nu} + \delta^{\perp}_{\beta\nu}e_{\alpha\gamma\mu}\right)b_{\gamma}\delta^{\perp}_{\mu\nu}\partial_{\lambda}V_{\perp\lambda}$$
$$= \frac{1}{2}(e_{\beta\gamma\alpha} + e_{\alpha\gamma\beta})b_{\gamma}\partial_{\lambda}V_{\perp\lambda} = 0.$$

The first term in equation (6.52) can be reduced as follows.

$$\begin{aligned}
&g_{\alpha\delta}g_{\beta\sigma}\frac{1}{2}\delta^{\perp}_{\alpha\mu}e_{\beta\gamma\nu}b_{\gamma}\left(g_{\mu i}g_{\nu j}\frac{\partial g_{lj}\hat{V}_{\perp l}}{\partial x_i}\right)\\
&= (g_{\beta\sigma}e_{\beta\gamma\nu}b_{\gamma}g_{\nu j})\left(g_{\alpha\delta}g_{\mu i}\frac{1}{2}\delta^{\perp}_{\alpha\mu}\right)\frac{\partial V_{\perp j}}{\partial x_i}\\
&= \frac{1}{2}(\hat{\mathbf{e}}_{\beta}\cdot\mathbf{e}_{\sigma})e_{\beta\gamma\nu}b_{\gamma}(\hat{\mathbf{e}}_{\nu}\cdot\mathbf{e}_{j})\left(g_{\alpha\delta}g_{\mu i}\delta^{\perp}_{\alpha\mu}\right)\frac{\partial V_{\perp j}}{\partial x_i}\\
&= \frac{1}{2}\mathbf{e}_{\sigma}\times\mathbf{b}\cdot\mathbf{e}_{j}(\delta^{\perp}_{\delta i} - g_{3\delta}g_{3i})\frac{\partial V_{\perp j}}{\partial x_i}\\
&= \frac{1}{2}\delta^{\perp}_{\delta i}e_{\sigma\gamma j}b_{\gamma}\frac{\partial V_{\perp j}}{\partial x_i} - \frac{1}{2}g_{3\delta}g_{3i}e_{\sigma\gamma j}b_{\gamma}\frac{\partial V_{\perp j}}{\partial x_i}.
\end{aligned} \tag{6.55}$$

Here, the last term can be further reduced as follows

$$-\frac{1}{2}g_{3\delta}g_{3i}e_{\sigma\gamma j}b_{\gamma}\frac{\partial V_{\perp j}}{\partial x_i} = -\frac{1}{2}\mathbf{b}\mathbf{b}\times(\mathbf{b}\cdot\nabla)\mathbf{V}_{\perp}. \tag{6.56}$$

When applying $\nabla\cdot$, it becomes of order $\mathcal{O}(k_{\parallel}^{2}V_{\wedge})$ and therefore is negligible. The second term in equation (6.52) can be reduced as follows.

$$\begin{aligned}
&g_{\alpha\delta}g_{\beta\sigma}\frac{1}{2}\delta^{\perp}_{\alpha\mu}e_{\beta\gamma\nu}b_{\gamma}\left(g_{\nu i}g_{\mu j}\frac{\partial g_{lj}\hat{V}_{\perp l}}{\partial x_i}\right)\\
&= (g_{\beta\sigma}e_{\beta\gamma\nu}b_{\gamma}g_{\nu i})\left(g_{\alpha\delta}g_{\mu j}\frac{1}{2}\delta^{\perp}_{\alpha\mu}\right)\frac{\partial V_{\perp j}}{\partial x_i}\\
&= \frac{1}{2}(\hat{\mathbf{e}}_{\beta}\cdot\mathbf{e}_{\sigma})e_{\beta\gamma\nu}b_{\gamma}(\hat{\mathbf{e}}_{\nu}\cdot\mathbf{e}_{i})\left(g_{\alpha\delta}g_{\mu j}\delta^{\perp}_{\alpha\mu}\right)\frac{\partial V_{\perp j}}{\partial x_i}\\
&= \frac{1}{2}\mathbf{e}_{\sigma}\times\mathbf{b}\cdot\mathbf{e}_{i}\left(\delta^{\perp}_{\delta j} - g_{3\delta}g_{3j}\right)\\
&= \frac{1}{2}\delta^{\perp}_{\delta j}e_{\sigma\gamma i}b_{\gamma}\frac{\partial V_{\perp j}}{\partial x_i} - \frac{1}{2}g_{3\delta}g_{3j}e_{\sigma\gamma i}b_{\gamma}\frac{\partial V_{\perp j}}{\partial x_i}.
\end{aligned} \tag{6.57}$$

Here, the last term becomes

$$\begin{aligned}\frac{1}{2}g_{3\delta}g_{3j}e_{\sigma\gamma i}b_\gamma\frac{\partial V_{\perp j}}{\partial x_i} &= -\frac{1}{2}g_{3\delta}e_{\sigma\gamma i}b_\gamma\frac{\partial g_{3j}}{\partial x_i}V_{\perp j}\\ &= -\frac{1}{2}\mathbf{b}[(\mathbf{b}\times\nabla)\mathbf{b}]\cdot\mathbf{V}_\perp.\end{aligned} \tag{6.58}$$

When applying $\nabla\cdot$, it becomes of order $\mathcal{O}(k_\parallel V_\wedge/R)$ and therefore is negligible. Finally, one has

$$\Pi^\perp_{\alpha\beta} = \eta_3 W^x_{3\alpha\beta}, \tag{6.59}$$

where α, $\beta = 1, 2$ and

$$\begin{aligned}W^x_{3\alpha\beta} &= \frac{1}{2}\left(\delta^\perp_{\alpha\mu}e_{\beta\gamma\nu} + \delta^\perp_{\beta\nu}e_{\alpha\gamma\mu}\right)b_\gamma W^x_{\mu\nu}.\\ W^x_{\mu\nu} &= \partial_\nu V_{\perp\mu} + \partial_\mu V_{\perp\nu}.\end{aligned} \tag{6.60}$$

In the momentum equation, one needs to know $\nabla\cdot\Pi^\perp$. Using equation (6.59), one has

$$\begin{aligned}\nabla\cdot\Pi^\perp &= (\partial_\alpha\eta_3)W^x_{3\alpha\beta} + \eta_3\partial_\alpha W^x_{3\alpha\beta}\\ &= (\partial_\alpha\eta_3)W^x_{3\alpha\beta} + \frac{1}{2}\eta_3\partial_\alpha\left[\left(\delta^\perp_{\alpha\mu}e_{\beta\gamma\nu} + \delta^\perp_{\beta\nu}e_{\alpha\gamma\mu}\right)b_\gamma W^x_{\mu\nu}\right].\end{aligned} \tag{6.61}$$

Since we only focus on the FLR effects in the singular layer in tokamak geometry, the mode features in the singular layer [12, 18], as also reviewed in references [9, 10], are used here.

$$\mathbf{V}_\perp = V^S\frac{\nabla S}{|\nabla\psi|^2} + V^u\frac{\mathbf{B}\times\nabla S}{B^2}. \tag{6.62}$$

Here, S labels the magnetic surface; $x = S - S_0$ with S_0 being the reference rational surface; $\mathbf{B}\approx\nabla S\times\nabla u$ with $u = \psi_S'(S_0)\theta - \chi_S'(S_0)\zeta$; ψ and χ are, respectively, the poloidal and toroidal fluxes, and θ and ζ are, respectively, the generalized poloidal and toroidal angles. For the compressional Alfvén modes that are suppressed, one also has

$$\nabla\cdot\mathbf{V}_\perp = -2\boldsymbol{\kappa}\cdot\mathbf{V}_\perp, \tag{6.63}$$

where $\boldsymbol{\kappa}$ is the field line curvature. For localized modes of the first kind [12, 18], one has $\lambda_\perp \ll \lambda_\wedge$. Here, $\lambda_\perp$ is the wavelength perpendicular to the magnetic surface and $\lambda_\wedge$ denotes the wavelength in the binormal direction, i.e. in the direction perpendicular to the magnetic field line but on the magnetic surface. This leads to $V_\perp = (V^S/|\nabla S|) \ll (V^u|\nabla S|/B) = V_\wedge$. Equation (6.63) in leading order yields

$$\frac{\partial V^S}{\partial x} = -ik_u V^u, \tag{6.64}$$

where k_u is the wavenumber in the binormal direction.

We first analyze the ordering of the gyroviscous effects. From a formal analysis, one would find that

$$\nabla \cdot \Pi^{\perp} \sim k_{\perp}^2 \rho_i^2 neBV_{\wedge} \qquad \text{so that} \quad \nabla \cdot \frac{\mathbf{B}}{B^2} \times \nabla \cdot \Pi^{\perp} \sim k_{\perp}^3 \rho_i^2 neV_{\wedge}.$$

As will be seen, $\nabla \cdot \Pi^{\perp}$ in leading order is a gradient vector so that the application of the operator $\nabla \cdot \frac{\mathbf{B}}{B^2} \times \cdots$ leads the perpendicular wave number $k_{\perp}$ to become $1/R$. Here, R and a are introduced to denote, respectively, the major and minor radii. Consequently, one has

$$\nabla \cdot \frac{\mathbf{B}}{B^2} \times \nabla \cdot \Pi^{\perp} \sim k_{\perp}^2 \rho_i^2 \frac{ne}{R} V_{\wedge}. \tag{6.65}$$

This is the leading order. We nevertheless will keep as well the terms of next order

$$\nabla \cdot \frac{\mathbf{B}}{B^2} \times \nabla \cdot \Pi^{\perp} \sim k_{\perp} k_{\wedge} \rho_i^2 \frac{ne}{a} V_{\wedge} \tag{6.66}$$

since they are of the same order as the inertia effects due to the diamagnetic drift. As will be seen in the next subsection, the so-called gyroviscous cancellation [18, 19] is in this order in the toroidal geometry.

With the consideration mentioned above, we can now simplify equation (6.61) term by term. We first calculate $(\partial_\alpha \eta_3) W_{3\alpha\beta}^x$ part. Since the $(\partial_\perp \eta_3) \sim \eta_3/a$, the terms containing $\partial V^u/\partial x$ need to be kept when operating with $\nabla \cdot \frac{\mathbf{B}}{B^2} \times \cdots$.

$$\begin{aligned}
(\partial_\alpha \eta_3) W_{3\alpha\beta}^x &= \frac{1}{2}(\partial_\alpha \eta_3)\Big[\big(\delta_{\alpha\mu}^{\perp} e_{\beta\gamma\nu} + \delta_{\beta\nu}^{\perp} e_{\alpha\gamma\mu}\big) b_\gamma W_{\mu\nu}^x\Big] \\
&= \frac{1}{2}(\partial_\alpha \eta_3)\Big[e_{\beta\gamma\nu} b_\gamma W_{\alpha\nu}^x + e_{\alpha\gamma\mu} b_\gamma W_{\mu\beta}^x\Big] \\
&= \frac{1}{2}(\partial_\alpha \eta_3) e_{\beta\gamma\nu} b_\gamma (\partial_\nu V_{\perp\alpha} + \partial_\alpha V_{\perp\nu}) \\
&\quad + \frac{1}{2}(\partial_\alpha \eta_3) e_{\alpha\gamma\mu} b_\gamma (\partial_\beta V_{\perp\mu} + \partial_\mu V_{\perp\beta}) \\
&= \frac{1}{2}(\mathbf{b} \times \nabla \mathbf{V}_\perp) \cdot \nabla \eta_3 + \frac{1}{2}\mathbf{b} \times (\nabla \eta_3 \cdot \nabla)\mathbf{V}_\perp \\
&\quad - \frac{1}{2}(\nabla \mathbf{V}_\perp) \cdot (\mathbf{b} \times \nabla \eta_3) - \frac{1}{2}(\mathbf{b} \times \nabla \eta_3) \cdot \nabla \mathbf{V}_\perp \\
&\rightarrow \frac{1}{2}(\mathbf{b} \times \nabla \mathbf{V}_\perp) \cdot \nabla \eta_3 + \frac{1}{2}\nabla \eta_3 \cdot \nabla(\mathbf{b} \times \mathbf{V}_\perp) \\
&\quad - \frac{1}{2}(\nabla \mathbf{V}_\perp) \cdot (\mathbf{b} \times \nabla \eta_3) - \frac{1}{2}(\mathbf{b} \times \nabla \eta_3) \cdot \nabla \mathbf{V}_\perp \\
&= \nabla \eta_3 \cdot \nabla(\mathbf{b} \times \mathbf{V}_\perp) - (\mathbf{b} \times \nabla \eta_3) \cdot \nabla \mathbf{V}_\perp \\
&\quad - \frac{1}{2}\mathbf{b} \cdot \nabla \eta_3 \nabla \times \mathbf{V}_\perp + \frac{1}{2}\mathbf{b} \nabla \eta_3 \cdot \nabla \times \mathbf{V}_\perp.
\end{aligned} \tag{6.67}$$

Here, in the reduction

$$\frac{1}{2}\mathbf{b} \times (\nabla\eta_3 \cdot \nabla)\mathbf{V}_\perp \to \frac{1}{2}\nabla\eta_3 \cdot \nabla(\mathbf{b} \times \mathbf{V}_\perp),$$

it has been noted that

$$\nabla \cdot \frac{\mathbf{B}}{B^2} \times [(\nabla\eta_3 \cdot \nabla\mathbf{b}) \times \mathbf{V}_\perp] = 0.$$

Also, it has been noted that

$$\mathbf{A} \times (\nabla \times \mathbf{B}) = (\nabla\mathbf{B}) \cdot \mathbf{A} - \mathbf{A} \cdot \nabla\mathbf{B}$$

and therefore

$$\begin{aligned}\frac{1}{2}(\mathbf{b} \times \nabla\mathbf{V}_\perp) \cdot \nabla\eta_3 &= \frac{1}{2}\mathbf{b} \times (\nabla\eta_3 \cdot \nabla)\mathbf{V}_\perp + \frac{1}{2}\mathbf{b} \times [\nabla\eta_3 \times \nabla \times \mathbf{V}_\perp]\\ &= \frac{1}{2}\mathbf{b} \times (\nabla\eta_3 \cdot \nabla)\mathbf{V}_\perp - \frac{1}{2}\mathbf{b} \cdot \nabla\eta_3\nabla \times \mathbf{V}_\perp\\ &\quad + \frac{1}{2}\nabla\eta_3\mathbf{b} \cdot \nabla \times \mathbf{V}_\perp, - \frac{1}{2}(\nabla\mathbf{V}_\perp) \cdot (\mathbf{b} \times \nabla\eta_3)\\ &= -\frac{1}{2}(\mathbf{b} \times \nabla\eta_3) \cdot \nabla\mathbf{V}_\perp - \frac{1}{2}(\mathbf{b} \times \nabla\eta_3) \times \nabla \times \mathbf{V}_\perp\\ &= -\frac{1}{2}(\mathbf{b} \times \nabla\eta_3) \cdot \nabla\mathbf{V}_\perp + \frac{1}{2}\mathbf{b}\nabla\eta_3 \cdot \nabla \times \mathbf{V}_\perp - \frac{1}{2}\nabla\eta_3\mathbf{b} \cdot \nabla \times \mathbf{V}_\perp.\end{aligned}$$

The second part in equation (6.61) is reduced as follows.

$$\begin{aligned}\partial_\alpha(\delta^\perp_{\alpha\mu}e_{\beta\gamma\nu}b_\gamma W^x_{\mu\nu}) &= e_{\beta\gamma\nu}(\partial_\mu V_{\perp\nu} + \partial_\nu V_{\perp\mu})\partial_\mu b_\gamma + e_{\beta\gamma\nu}b_\gamma\partial_\mu(\partial_\mu V_{\perp\nu} + \partial_\nu V_{\perp\mu})\\ &= e_{\beta\gamma\nu}\partial_\mu V_{\perp\nu}\partial_\mu b_\gamma - \nabla \times [(\mathbf{V}_\perp \cdot \nabla)\mathbf{b}] + \mathbf{b} \times \nabla^2\mathbf{V}_\perp\\ &= \frac{1}{2}\nabla^2(\mathbf{b} \times \mathbf{V}_\perp) + \frac{1}{2}\mathbf{b} \times \nabla^2\mathbf{V}_\perp - \nabla \times [(\mathbf{V}_\perp \cdot \nabla)\mathbf{b}].\end{aligned} \tag{6.68}$$

The third part in equation (6.61) is reduced as follows.

$$\begin{aligned}\partial_\alpha(\delta^\perp_{\beta\nu}e_{\alpha\gamma\mu}b_\gamma W^x_{\mu\nu}) &= e_{\alpha\gamma\mu}(\partial_\mu V_{\perp\beta} + \partial_\beta V_{\perp\mu})\partial_\alpha b_\gamma + e_{\alpha\gamma\mu}b_\gamma\partial_\alpha(\partial_\mu V_{\perp\beta} + \partial_\beta V_{\perp\mu})\\ &= (\nabla \times \mathbf{b} \cdot \nabla)\mathbf{V}_\perp + \nabla(\nabla \times \mathbf{b} \cdot \mathbf{V}_\perp) + b_\gamma\partial_\beta(e_{\alpha\gamma\mu}\partial_\alpha V_{\perp\mu})\\ &= \mathbf{b} \times \nabla^2\mathbf{V}_\perp + (\nabla \times \mathbf{b} \cdot \nabla)\mathbf{V}_\perp + \nabla(\nabla \times \mathbf{b} \cdot \mathbf{V}_\perp) - (\mathbf{b} \cdot \nabla)\nabla \times \mathbf{V}_\perp,\end{aligned} \tag{6.69}$$

where it has been noted that

$$\begin{aligned}b_\gamma\partial_\beta(e_{\alpha\gamma\mu}\partial_\alpha V_{\perp\mu}) &= -\nabla(\nabla \times \mathbf{V}_\perp) \cdot \mathbf{b}\\ &= -\mathbf{b} \cdot \nabla(\nabla \times \mathbf{V}_\perp) + \mathbf{b} \times [\nabla \times (\nabla \times \mathbf{V}_\perp)]\\ &= -\mathbf{b} \cdot \nabla(\nabla \times \mathbf{V}_\perp) + \mathbf{b} \times \nabla^2\mathbf{V}_\perp - \mathbf{b} \times \nabla\nabla \cdot \mathbf{V}_\perp.\end{aligned}$$

Inserting equations (6.67)–(6.69) into equation (6.61), one obtains

$$\begin{aligned}\nabla\cdot\Pi^{\perp} = {} & \nabla\eta_3\cdot\nabla(\mathbf{b}\times\mathbf{V}_{\perp}) + (\nabla\eta_3\times\mathbf{b}\cdot\nabla)\mathbf{V}_{\perp} \\ & - \frac{1}{2}(\mathbf{b}\cdot\nabla\eta_3)\nabla\times\mathbf{V}_{\perp} + \frac{1}{2}\mathbf{b}\nabla\eta_3\cdot\nabla\times\mathbf{V}_{\perp} \\ & + \frac{\eta_3}{2}\left\{\frac{1}{2}\nabla^2(\mathbf{b}\times\mathbf{V}_{\perp}) + \frac{3}{2}\mathbf{b}\times\nabla^2\mathbf{V}_{\perp} - \nabla\times[(\mathbf{V}_{\perp}\cdot\nabla)\mathbf{b}]\right. \\ & \left. + (\nabla\times\mathbf{b}\cdot\nabla)\mathbf{V}_{\perp} + \nabla(\nabla\times\mathbf{b}\cdot\mathbf{V}_{\perp}) - (\mathbf{b}\cdot\nabla)\nabla\times\mathbf{V}_{\perp}\right\}.\end{aligned} \tag{6.70}$$

Note here that, although there are second derivative terms in equation (6.70), they do not give rise to a third order derivative when applying the operator $\nabla\cdot\frac{\mathbf{B}}{B^2}\times\nabla\cdots$. This is because their highest-order-derivative terms form a gradient vector as follows.

$$\begin{aligned}&\frac{\eta_3}{4}\nabla^2(\mathbf{b}\times\mathbf{V}_{\perp}) + \frac{3\eta_4}{4}\mathbf{b}\times\nabla^2\mathbf{V}_{\perp} = \frac{4\eta_3}{4}\mathbf{b}\times\nabla^2\mathbf{V}_{\perp} + \cdots \\ &= -\eta_3\mathbf{b}\times(\nabla\times\nabla\times\mathbf{V}_{\perp}) + \cdots = \nabla(\eta_3\mathbf{b}\cdot\nabla\times\mathbf{V}_{\perp}) + \cdots.\end{aligned}$$

Here, the terms containing only the first order derivatives of $\mathbf{V}_{\perp}$ are not detailed.

We now simplify equation (6.70) term by term with the operator $\nabla\cdot\frac{\mathbf{B}}{B^2}\times\nabla\cdots$ applied. The singular layer decomposition in equation (6.62) is considered. The first term on the right-hand side of equation (6.70) is reduced as follows.

$$\begin{aligned}\nabla\cdot\frac{\mathbf{B}}{B^2}\times[\nabla\eta_3\cdot\nabla(\mathbf{b}\times\mathbf{V}_{\perp})] &= -\frac{1}{B}\nabla\cdot(\nabla\eta_3\cdot\nabla\mathbf{V}_{\perp}) \\ &= -\frac{1}{B}(\nabla\eta_3\cdot\nabla\nabla\cdot\mathbf{V}_{\perp}) \\ &= -\frac{1}{B}(\nabla\eta_3\cdot\nabla 2\kappa\cdot\mathbf{V}_{\perp}) \to 0,\end{aligned}$$

where equation (6.63) has been used. This term is negligible since it contains only the first order derivative on V^u. The second term on the right-hand side of equation (6.70) is reduced as follows.

$$\begin{aligned}&\nabla\cdot\frac{\mathbf{B}}{B^2}\times[(\nabla\eta_3\times\mathbf{b}\cdot\nabla)\mathbf{V}_{\perp}] \\ &= (\nabla\eta_3\times\mathbf{b}\cdot\nabla)\nabla\cdot\frac{\mathbf{B}}{B^2}\times\mathbf{V}_{\perp} \\ &= -\frac{|\nabla S|^2}{B^2}(\nabla\eta_3\times\mathbf{b}\cdot\nabla)\frac{\partial V^u}{\partial x} \\ &= \frac{|\nabla S|^2}{B^2}\frac{m}{2eB}iBk_u p_S'\frac{\partial V^u}{\partial x} - \frac{|\nabla S|^2}{B^2}\frac{mnT}{2e}\left(\nabla\frac{1}{B}\times\mathbf{b}\cdot\nabla S\right)\frac{\partial^2 V^u}{\partial x^2} \\ &= i\frac{1}{2}\omega_*^p mn\frac{|\nabla S|^2}{B^2}\frac{\partial V^u}{\partial x} + \frac{mnT}{4eB^2 p_S'}|\nabla S|^2\mathbf{B}\cdot\nabla\sigma\frac{\partial^2 V^u}{\partial x^2}.\end{aligned}$$

Here, it has been noted that

$$
\begin{aligned}
&-2\boldsymbol{\kappa}\cdot\frac{\mathbf{B}\times\nabla p}{B^2}=\mathbf{B}\times\nabla p\cdot\nabla\frac{1}{B^2}=\nabla\cdot\mathbf{J}_\perp\\
&=-\nabla\cdot\mathbf{J}_\parallel=-\nabla\cdot\left(\frac{\mathbf{J}\cdot\mathbf{B}}{B^2}\mathbf{B}\right)=-\mathbf{B}\cdot\nabla\sigma.
\end{aligned}
\tag{6.71}
$$

The third term on the right-hand side of equation (6.70) yields

$$
-\nabla\cdot\frac{\mathbf{B}}{B^2}\times\left[\frac{1}{2}(\mathbf{b}\cdot\nabla\eta_3)\nabla\times\mathbf{V}_\perp\right]=-\nabla\cdot\left[\frac{1}{2}(\mathbf{b}\cdot\nabla\eta_3)\nabla\left(\frac{\mathbf{B}}{B^2}\cdot\mathbf{V}_\perp\right)\right]=0.
$$

The fourth term on the right-hand side of equation (6.70) yields

$$
\nabla\cdot\frac{\mathbf{B}}{B^2}\times\left(\frac{1}{2}\mathbf{b}\nabla\eta_3\cdot\nabla\times\mathbf{V}_\perp\right)=0.
$$

The fifth term on the right-hand side of equation (6.70) gives rise to

$$
\begin{aligned}
&\frac{\eta_3}{2}\frac{1}{2}\nabla^2(\mathbf{b}\times\mathbf{V}_\perp)=\frac{\eta_3}{2}(\nabla_i\mathbf{b}\times\nabla_i\mathbf{V}_\perp)+\frac{\eta_3}{2}\frac{1}{2}(\mathbf{b}\times\nabla^2\mathbf{V}_\perp)\\
&=\frac{\eta_3}{2}(\nabla_i\mathbf{b}\times\nabla_i\mathbf{V}_\perp)-\frac{\eta_3}{2}\frac{1}{2}\mathbf{b}\times(\nabla\times\nabla\times\mathbf{V}_\perp)\\
&+\frac{\eta_3}{2}\frac{1}{2}[\mathbf{b}\times\nabla(\nabla\cdot\mathbf{V}_\perp)].
\end{aligned}
\tag{6.72}
$$

The sixth term on the right-hand side of equation (6.70) gives rise to

$$
\frac{\eta_3}{2}\frac{3}{2}\mathbf{b}\times\nabla^2\mathbf{V}_\perp=\frac{\eta_3}{2}\frac{3}{2}\mathbf{b}\times\nabla(\nabla\cdot\mathbf{V}_\perp)-\frac{\eta_3}{2}\frac{3}{2}\mathbf{b}\times(\nabla\times\nabla\times\mathbf{V}_\perp).
\tag{6.73}
$$

The fifth and sixth terms in equations (6.72) and (6.73) can be combined to yield

$$
\begin{aligned}
\nabla\cdot\frac{\mathbf{B}}{B^2}\times\left[\frac{\eta_3}{2}(\nabla_i\mathbf{b}\times\nabla_i\mathbf{V}_\perp)\right]&=-\nabla\cdot\left[\frac{\eta_3}{2}(\nabla_i\mathbf{b})\cdot\frac{\mathbf{B}}{B^2}\times\nabla_i\mathbf{V}_\perp)\right]\\
&+\nabla\cdot\left[\frac{\eta_3}{2}\nabla_i\mathbf{b}\times(\nabla_i\mathbf{V}_\perp)\cdot\frac{\mathbf{B}}{B^2}\right]\\
&=\nabla\cdot\left[\frac{\eta_3}{2}\nabla_i\mathbf{b}\times\left(\nabla_i\mathbf{V}_\perp\cdot\frac{\mathbf{B}}{B^2}\right)\right]=0,
\end{aligned}
$$

$$
\begin{aligned}
\nabla\cdot\frac{\mathbf{B}}{B^2}\times\left[\frac{\eta_3}{2}\frac{3+1}{2}\mathbf{b}\times\nabla(\nabla\cdot\mathbf{V}_\perp)\right]&=-\frac{1}{B}\eta_3\nabla^2(\nabla\cdot\mathbf{V}_\perp)=\frac{2}{B}\eta_3\nabla^2(\boldsymbol{\kappa}\cdot\mathbf{V}_\perp)\\
&=\frac{mnT}{2eB^2p_S'}|\nabla S|^2\mathbf{B}\cdot\nabla\sigma\frac{\partial^2V^u}{\partial x^2},
\end{aligned}
$$

and

$$
\begin{aligned}
&\nabla \cdot \frac{\mathbf{B}}{B^2} \times \left[-\frac{\eta_3}{2}\frac{3+1}{2}\mathbf{b} \times (\nabla \times \nabla \times \mathbf{V}_\perp) \right] \\
&= \nabla \cdot \frac{\eta_3}{B}(\nabla \times \nabla \times \mathbf{V}_\perp) \\
&= -\frac{mp_S'}{2eB^2}\nabla^2 V^S - \frac{mp}{2e}|\nabla S|^2 \nabla \frac{1}{B^2} \cdot \frac{\mathbf{B} \times \nabla S}{B^2}\frac{\partial^2 V^u}{\partial x^2} \\
&= \frac{mp_S'}{2eB^2}|\nabla S|^2 ik_u \frac{\partial V^u}{\partial x} - \frac{mp}{2eB^2 p_S'}|\nabla S|^2 \mathbf{B} \cdot \nabla \sigma \frac{\partial^2 V^u}{\partial x^2} \\
&= i\frac{1}{2}\omega_*^p \frac{mn}{B^2}|\nabla S|^2 \frac{\partial V^u}{\partial x} + \frac{mp}{2eB^2 p_S'}|\nabla S|^2 \mathbf{B} \cdot \nabla \sigma \frac{\partial^2 V^u}{\partial x^2},
\end{aligned}
$$

where equation (6.64) has been used. The seventh term on the right-hand side of equation (6.70) yields

$$
\begin{aligned}
-\nabla \cdot \frac{\mathbf{B}}{B^2} \times \left\{ \frac{\eta_3}{2}\nabla \times [(\mathbf{V}_\perp \cdot \nabla)\mathbf{b}] \right\} &= -\nabla \cdot \left\{ \frac{\eta_3}{2}\nabla \left[\frac{\mathbf{B}}{B^2} \cdot (\mathbf{V}_\perp \cdot \nabla)\mathbf{b} \right] \right\} \\
&= 0.
\end{aligned}
$$

The eighth term on the right-hand side of equation (6.70) yields

$$
\begin{aligned}
\nabla \cdot \frac{\mathbf{B}}{B^2} \times \left[\frac{\eta_3}{2}(\nabla \times \mathbf{b} \cdot \nabla)\mathbf{V}_\perp \right] &= \frac{\eta_3}{2}(\nabla \times \mathbf{b} \cdot \nabla)\nabla \cdot \frac{\mathbf{B}}{B^2} \times \mathbf{V}_\perp \\
&= -\frac{\eta_3}{2}\frac{|\nabla S|^2}{B^2}\nabla \times \mathbf{b} \cdot \nabla S \frac{\partial^2 V^u}{\partial x^2} \\
&= \frac{mnT}{8eB^2 p_S'}|\nabla S|^2 \mathbf{B} \cdot \nabla \sigma \frac{\partial^2 V^u}{\partial x^2}.
\end{aligned}
$$

Here, it has been noted that

$$
\nabla \times \mathbf{b} \cdot \nabla S = \nabla \times \frac{\mathbf{B}}{B} \cdot \nabla S = \nabla \frac{1}{B} \times \mathbf{B} \cdot \nabla S = -\frac{B}{2p_S'}\mathbf{B} \cdot \nabla \sigma,
$$

where $\mathbf{J} \cdot \nabla S = 0$ and equation (6.71) have been used. The ninth term on the right-hand side of equation (6.70) yields

$$
\nabla \cdot \frac{\mathbf{B}}{B^2} \times \left[\frac{\eta_3}{2}\nabla(\nabla \times \mathbf{b} \cdot \mathbf{V}_\perp) \right] = -\frac{\mathbf{B}}{B^2} \cdot \nabla \times \left[\frac{\eta_3}{2}\nabla(\nabla \times \mathbf{b} \cdot \mathbf{V}_\perp) \right] = 0.
$$

The last term on the right-hand side of equation (6.70) yields

$$
\begin{aligned}
& -\nabla\cdot\frac{\mathbf{B}}{B^2}\times\left[\frac{\eta_3}{2}(\mathbf{b}\cdot\nabla)\nabla\times\mathbf{V}_\perp\right] \\
&=\frac{\mathbf{B}}{B^2}\cdot\nabla\times\left[\frac{\eta_3}{2}(\mathbf{b}\cdot\nabla)\nabla\times\mathbf{V}_\perp\right] \\
&=-\frac{\eta_3}{2}(\mathbf{b}\cdot\nabla)\nabla^2\frac{\mathbf{B}}{B^2}\cdot\mathbf{V}_\perp+\frac{\eta_3}{2}\left(\mathbf{b}\cdot\nabla\frac{\mathbf{B}}{B^2}\right)\cdot\nabla^2\mathbf{V}_\perp \\
&=\frac{\eta_3}{2}|\nabla S|^2\kappa\cdot\frac{\mathbf{B}\times\nabla S}{B^2}\frac{\partial^2 V^u}{\partial x^2} \\
&=\frac{mnT}{8eB^2p_S'}|\nabla S|^2\mathbf{B}\cdot\nabla\sigma\frac{\partial^2 V^u}{\partial x^2}.
\end{aligned}
$$

Collecting the results above, one finally obtains

$$
\begin{aligned}
&\nabla\cdot\left(\frac{\mathbf{B}}{B^2}\times\nabla\cdot\delta\Pi^\perp\right) \\
&=i\omega_*^{\,p}nm\frac{|\nabla S|^2}{B^2}\frac{\partial\delta V^u}{\partial x}+\frac{3mnT}{2eB^2p_S'}|\nabla S|^2(\mathbf{B}\cdot\nabla\sigma)\frac{\partial^2\delta V^u}{\partial x^2}.
\end{aligned}
\tag{6.74}
$$

Here, the δ tag on $\mathbf{V}_\perp$ has been added back. The orderings of the first and second terms of equation (6.74) are given, respectively, by equations (6.66) and (6.65). This result will be used to study the FLR effects in the next subsubsection.

The $W^{\|}_{\mu\nu}$ contributions

We now calculate the effects of $W^{\|}_{\alpha\beta}$ in equation (6.53) in the vorticity equation in the term: $\nabla\cdot\left[\frac{\mathbf{B}}{B^2}\times(\nabla\cdot\Pi^{\|})\right]$. By definition, one has

$$\Pi^{\|}_{\alpha\beta}=\eta_3 W^{v}_{3\alpha\beta},$$

where α, $\beta = 1$, 2 and

$$
\begin{aligned}
W^{v}_{3\alpha\beta}&=\frac{1}{2}\left(\delta^{\perp}_{\alpha\mu}e_{\beta\gamma\nu}+\delta^{\perp}_{\beta\nu}e_{\alpha\gamma\mu}\right)b_\gamma W^{v}_{\mu\nu}. \\
W^{v}_{\mu\nu}&=\left(\frac{\partial g_{3\nu}}{\partial x_\mu}+\frac{\partial g_{3\mu}}{\partial x_\nu}\right)\left(\hat{v}_3+\frac{T}{m}\frac{\partial\ln n}{\partial\hat{v}_3}\right).
\end{aligned}
\tag{6.75}
$$

Note that for the equilibrium density in equation (4.7) in chapter 4, the factor $\left(\hat{v}_3 n + \frac{T}{m}\frac{\partial n}{\partial \hat{v}_3}\right)$ vanishes for equilibrium density. Therefore, one has

$$\delta\Pi^{\|}_{\alpha\beta} = \frac{T}{2\omega_c}\frac{1}{2}\left(\delta^{\perp}_{\alpha\mu}e_{\beta\gamma\nu} + \delta^{\perp}_{\beta\nu}e_{\alpha\gamma\mu}\right)b_\gamma\left(\frac{\partial g_{3\nu}}{\partial x_\mu} + \frac{\partial g_{3\mu}}{\partial x_\nu}\right)\left(\delta n\hat{v}_3 + \frac{T}{m}\frac{\partial \delta n}{\partial \hat{v}_3}\right). \tag{6.76}$$

From the continuity equation (6.1), one can find that

$$\delta n \sim \frac{1}{\omega_{ti}} n\nabla \cdot \delta \mathbf{V}_\perp \sim \frac{1}{\omega_{ti}} n\delta V_\wedge / R,$$

where equation (6.63) has been used. Using this ordering estimate, one obtains

$$\nabla \cdot \frac{\mathbf{B}}{B^2} \times \nabla \cdot \delta\Pi^{\|} \sim k_\perp^2 \rho_i^2 \frac{ne}{R}\delta V_\wedge. \tag{6.77}$$

This is the same order as the $\Pi^\perp$ contribution in equation (6.65). Therefore, we need to keep the $\Pi^{\|}$ effects. By a straightforward reduction, one has

$$\nabla \cdot \frac{\mathbf{B}}{B^2} \times \nabla \cdot \delta\Pi^{\|} = -\frac{\mathbf{B}}{B^2} \cdot \nabla \times \nabla \cdot \delta\Pi^{\|}, \tag{6.78}$$

where $\delta\Pi^{\|}$ is given in equation (6.76). The results depend on the solution of the perturbed continuity equation. This shows that the parallel fluid velocity in the Braginskii gyroviscous tensor needs a kinetic description.

6.4.2 Finite Larmor radius effects and the gyroviscous cancellation

In this subsection, the finite Larmor radius (FLR) effects are investigated using the gyroviscous tensor obtained in subsection 6.4.1. The conventional gyroviscous cancellation is demonstrated and the toroidal FLR effects are emphasized. To focus on the FLR effects, the low frequency regime: $\omega \ll \omega_{ti}$ as studied in section 6.2 is considered. FLR effects arise from two aspects: the diamagnetic drift velocity and the gyroviscosity. They have to be considered in parallel for ordering consistency. The modifications on the continuity, momentum, and energy transport equations need to be considered. Since the FLR effects are only important in the singular layer, the vorticity equation is employed for describing the theory. The equilibrium solution given in equation (4.7) in chapter 4 will be employed.

We first consider the momentum equation. From equation (6.2), the momentum equation with FLR effects included is given as follows.

$$
\begin{aligned}
&(\delta_{\lambda\alpha} - g_{3\alpha}g_{3\lambda})\left(\underbrace{neE_\alpha + nee_{\alpha\beta\gamma}g_{l\beta}\hat{V}_l B_\gamma}_{v_1} - \underbrace{\rho\frac{\hat{d}g_{l\alpha}\hat{V}_{\perp l}}{dt}}_{v_2} - \underbrace{\frac{\partial p}{\partial x_\alpha}}_{v_3} - \underbrace{\hat{v}_3 g_{3\beta}\frac{\partial g_{3\alpha}}{\partial x_\beta}\frac{\partial p}{\partial \hat{v}_3}}_{v_4}\right.\\
&\left. - \underbrace{g_{3\beta}\hat{v}_3^2\frac{\partial g_{3\alpha}}{\partial x_\beta}\rho}_{v_5} + \underbrace{g_{3\gamma}\frac{\partial g_{l\gamma}\hat{V}_{\perp l}}{\partial x_\alpha}\frac{\partial p}{\partial \hat{v}_3}}_{v_6} - \underbrace{\rho\hat{v}_3 g_{3\beta}\frac{\partial g_{l\alpha}\hat{V}_{\perp l}}{\partial x_\beta}}_{v_7} - \underbrace{\rho\hat{v}_3\frac{\hat{d}g_{3\alpha}}{dt}}_{v_8} - \underbrace{\frac{\hat{d}g_{3\alpha}}{dt}\frac{\partial p}{\partial \hat{v}_3}}_{v_9}\right)\\
&= \underbrace{g_{\nu\lambda}g_{\beta\alpha}\frac{\partial \pi_{\beta\nu}}{\partial x_\alpha}}_{v_{10}} + g_{\nu\lambda}\left(\underbrace{g_{p\delta}\frac{\hat{d}g_{3\delta}}{dt}}_{v_{11}} - \underbrace{g_{p\beta}g_{3\gamma}\frac{\partial g_{l\gamma}\hat{V}_{\perp l}}{\partial x_\beta}}_{v_{12}} + \underbrace{\hat{v}_3 g_{3\alpha}g_{p\mu}\frac{\partial g_{3\mu}}{\partial x_\alpha}}_{v_{13}}\right)\frac{\partial \pi_{p\nu}}{\partial \hat{v}_3} \qquad (6.79)\\
&+\left(\underbrace{\frac{\partial g_{p\alpha}g_{q\lambda}}{\partial x_\alpha}}_{v_{14}} + \underbrace{g_{p\alpha}g_{q\mu}g_{3\lambda}\frac{\partial g_{3\mu}}{\partial x_\alpha}}_{v_{15}} + \underbrace{g_{3\alpha}g_{q\mu}g_{p\lambda}\frac{\partial g_{3\mu}}{\partial x_\alpha}}_{v_{16}}\right)\pi_{pq}\\
&+\underbrace{g_{\nu\lambda}\frac{\partial g_{3\alpha}}{\partial x_\alpha}\frac{\partial q_\nu}{\partial \hat{v}_3}}_{v_{17}} + \underbrace{g_{p\alpha}g_{q\mu}g_{\nu\lambda}\frac{\partial g_{3\mu}}{\partial x_\alpha}\frac{\partial \Theta_{pq\nu}}{\partial \hat{v}_3}}_{v_{18}}.
\end{aligned}
$$

To simplify equation (6.79), we need to estimate the orderings. First, similar to the analyses in section 6.2, one can find that the terms 'v_6'–'v_9' are of order $V_\perp/\hat{v}_3$ smaller than the term 'v_3' and, therefore, can be neglected; Also, p and ρ in terms 'v_4' and 'v_5' must be the perturbed quantities. Next, we analyze the terms related to the FLR effects on the right hand side. Since the integration over $\hat{v}_3$ will be performed to derive the single fluid equation, the terms 'v_{17}' and 'v_{18}' vanish and the term 'v_{13}' and 'v_{16}' cancel after the integration. The terms 'v_{11}' and 'v_{12}' are negligible since they are of order $V_\perp/\hat{v}_3$ smaller than the term 'v_{10}'. The term 'v_{15}' is proportional to $g_{3\lambda}$, i.e. to **b**, and, therefore, does not contribute to the vorticity equation. Finally, the terms 'v_{10}' and 'v_{14}' can be combined to give rise to $\partial\Pi_{\alpha\lambda}/\partial x_\alpha$ as analyzed in subsection 6.4.1. Therefore, equation (6.79) can be reduced to

$$
\begin{aligned}
&(\delta_{\lambda\alpha} - g_{3\alpha}g_{3\lambda})\\
&\left(\underbrace{neE_\alpha + nee_{\alpha\beta\gamma}g_{l\beta}\hat{V}_l B_\gamma}_{v_1} - \underbrace{\rho\frac{\hat{d}g_{l\alpha}\hat{V}_{\perp l}}{dt}}_{v_2} - \underbrace{\frac{\partial p}{\partial x_\alpha}}_{v_3}\right. \qquad (6.80)\\
&\left. - \underbrace{\hat{v}_3 g_{3\beta}\frac{\partial g_{3\alpha}}{\partial x_\beta}\frac{\partial p}{\partial \hat{v}_3}}_{v_4} - \underbrace{g_{3\beta}\hat{v}_3^2\frac{\partial g_{3\alpha}}{\partial x_\beta}\rho}_{v_5}\right) - \underbrace{\frac{\partial \Pi_{\beta\nu}}{\partial x_\alpha}}_{v_{10}} = 0.
\end{aligned}
$$

Using equation (6.80), one can express the perpendicular velocity as follows

$$\mathbf{V}_\perp = \frac{\mathbf{E}\times\mathbf{B}}{B^2} + \frac{\mathbf{B}\times\nabla p}{neB^2} + \frac{\rho\mathbf{B}\times \hat{d}g_{la}\hat{V}_{\perp l}/dt}{neB^2} + \cdots \tag{6.81}$$

Here, the contributions from terms 'v_4' and 'v_5' can be neglected, since the radial wavelength in the ∇p term is much smaller than the equilibrium scale length and the contributions from $\Pi_{\beta\nu}$ can be dropped in view of that $k_\perp\rho_i \ll 1$. Since the parallel electric field is excluded, similar to the discussion in section 6.2, we represent the perturbed electric field as follows $\delta\mathbf{E} = -\partial\delta\mathbf{A}/\partial t$ with $\delta\mathbf{A} = \boldsymbol{\xi}_\perp \times \mathbf{B}$. In the linearization, one obtains the following orderings

$$\frac{\mathbf{E}\times\mathbf{B}}{B^2} \sim \omega\xi_\theta,$$

$$\frac{\mathbf{B}\times\nabla p}{neB^2} \sim \frac{\mathbf{B}\times\nabla\boldsymbol{\xi}\cdot\nabla p}{neB^2} \sim k_\perp\rho_a\frac{v_T}{a}\xi_r \sim \omega_*^p\xi_\theta,$$

$$\frac{\rho\mathbf{B}\times \hat{d}g_{la}\hat{V}_{\perp l}/dt}{neB^2} \sim \frac{\omega}{\omega_c}\omega\xi_\theta.$$

Therefore, the third term in equation (6.81) can be neglected as well. Finally, the perpendicular velocity can be expressed as

$$\delta\mathbf{V}_\perp = -i\omega\boldsymbol{\xi}_\perp + \frac{\mathbf{B}\times\nabla\delta p}{neB^2}. \tag{6.82}$$

Here, only perturbed quantities under the gradient are kept, neglecting the small terms such as

$$-\frac{\mathbf{B}\times\nabla p}{neB^2}\frac{\delta n}{n} \quad \text{and} \quad \frac{1}{ne}\delta\left(\frac{\mathbf{B}^2}{B}\right)\times\nabla p.$$

Since the FLR effects are only important in the singular layer, we next express the diamagnetic drift velocity in the layer. We first introduce the expansion

$$\boldsymbol{\xi}_\perp = \xi^\psi\frac{\nabla\psi}{|\nabla\psi|^2} + \xi^u\frac{\mathbf{B}\times\nabla\psi}{B^2}. \tag{6.83}$$

Since it is in the singular layer, from references [12, 18], one has in the lowest order

$$\frac{\partial\xi^\psi}{\partial x} + ik_u\xi^u = 0, \qquad \text{with } \lambda_\perp \gg \lambda_\wedge. \tag{6.84}$$

As will be proved later on in this subsection, one has

$$\delta p = -\boldsymbol{\xi}\cdot\nabla p + \delta p_g = -\xi^\psi p' + \delta p_g$$

with $\delta p_g \sim (\omega/\omega_{ti})^2(p\xi_\wedge/R)$. The δp_g effects on the perturbed velocity are negligible in the current low frequency regime. Therefore, one has

$$\begin{aligned}\frac{\mathbf{B}\times\nabla\delta p}{neB^2} &= -ip'\frac{\mathbf{B}\times\mathbf{k}_u}{neB^2}\xi^\psi + p'\frac{\mathbf{B}\times\nabla\psi}{neB^2}\frac{\partial\xi^\psi}{\partial x}\\ &= i\omega_*^p\left(\xi^\psi\frac{\nabla\psi}{|\nabla\psi|^2} + \xi^u\frac{\mathbf{B}\times\nabla\psi}{B^2}\right),\end{aligned}$$

where equation (6.84) has been used and $\mathbf{k}_u = k_u\mathbf{B}\times\nabla\psi/B^2$ has been noted. Finally, one obtains

$$\delta\mathbf{V}_\perp = -i(\omega - \omega_*^p)\boldsymbol{\xi}_\perp. \tag{6.85}$$

As in section 6.2, to study the MHD modes, one needs to derive the momentum equation for the single fluid. This is achieved by integrating over $\hat{v}_3$ and the summation over all particle species:

$$\langle\cdots\rangle_s = \sum_a \frac{1}{n_0}\int_{-\infty}^{+\infty} d\hat{v}_3(\cdots).$$

This leads to

$$\left\langle\rho\frac{d\delta\mathbf{V}_\perp}{dt}\right\rangle_s = \mathbf{J}\times\mathbf{B} - \nabla_\perp p_s - \boldsymbol{\kappa}\left(\langle mn\hat{v}_3^2\rangle_s - p_s\right) + [\nabla\cdot\Pi]_s, \tag{6.86}$$

where

$$\rho_s = \langle mn\rangle_s, \qquad p_s = \langle p\rangle_s, \qquad \mathbf{J} = \langle ne\mathbf{V}_\perp\rangle_s + \mathbf{J}_\parallel,$$

$$\frac{d}{dt} = \frac{\partial}{\partial t} + \mathbf{V}_\perp\cdot\nabla.$$

We can derive the vorticity equation by applying operator $\nabla\cdot(\mathbf{B}/B^2)\times\cdots$ on equation (6.86),

$$\begin{aligned}&\sum_a\frac{1}{n_0}\int_{-\infty}^{+\infty} d\hat{v}_3 i\rho_a(\omega - \omega_*^p)\nabla\cdot\frac{\mathbf{B}}{B^2}\times\delta\mathbf{V}\\ &= \mathbf{B}\cdot\nabla\frac{\mathbf{B}\cdot\delta\mathbf{J}}{B^2} + \delta\mathbf{B}\cdot\nabla\sigma - \mathbf{J}\cdot\nabla\frac{\mathbf{B}\cdot\delta\mathbf{B}}{B^2}\\ &\quad + \nabla\times\frac{\mathbf{B}}{B^2}\cdot\nabla_\perp\delta p_s + \frac{\mathbf{B}\times\boldsymbol{\kappa}}{B^2}\cdot\nabla\left(\langle mn\hat{v}_3^2\rangle_s - p_s\right)\\ &\quad - \sum_a\frac{1}{n_0}\int_{-\infty}^{+\infty} d\hat{v}_3\nabla\cdot\left[\frac{\mathbf{B}}{B^2}\times(\nabla\cdot\delta\Pi)\right]\end{aligned} \tag{6.87}$$

Let us prove the so-called gyroviscous cancellation here. In section 6.4.1 we have obtained in equations (6.74) and (6.78) that

$$\begin{aligned}&\nabla\cdot\left[\frac{\mathbf{B}}{B^2}\times(\nabla\cdot\delta\Pi)\right]\\ &= i\omega_*^p\rho\frac{|\nabla\psi|^2}{B^2}\frac{\partial v^u}{\partial x} + \frac{3mp}{2eB^2p_s'}|\nabla\psi|^2(\mathbf{B}\cdot\nabla\sigma)\frac{\partial^2 v^u}{\partial x^2}\\ &\quad + \frac{\mathbf{B}}{B^2}\cdot\nabla\times\nabla\cdot\delta\Pi^\parallel.\end{aligned} \tag{6.88}$$

Noting that $\delta V^u \gg \delta V^\psi$, one has

$$\nabla \cdot \frac{\mathbf{B}}{B^2} \times \delta \mathbf{V}_\perp = -\frac{|\nabla\psi|^2}{B^2}\frac{\partial v^u}{\partial x}. \tag{6.89}$$

Therefore, the first term on the right-hand side of equation (6.88) cancels with the inertia term proportional to ω_*^p on the left-hand side of equation (6.87). This is the so-called gyroviscous cancellation [19, 20], which leads the frequency shift in the inertia term $(\omega - \omega_{*s}^p)^2$ to become $\omega(\omega - \omega_{*s}^p)$. Nevertheless, one can see there is only partial cancellation in the toroidal geometry. The last two terms in equation (6.88) remain and they are larger. Counting the cancellation, equation (6.87) becomes

$$\begin{aligned}
&-\rho_s\omega(\omega - \omega_{*s}^p)\frac{|\nabla\psi^2|}{B^2}\frac{\partial \xi^u}{\partial x} + i(\omega - \omega_{*s}^p)p_s\frac{3m|\nabla\psi|^2(\mathbf{B}\cdot\nabla\sigma)}{2eB^2p'}\frac{\partial^2\xi^u}{\partial x^2}\\
&+\frac{\mathbf{B}}{B^2}\cdot\nabla\times\nabla\cdot\delta\Pi^{\|}\\
&=\mathbf{B}\cdot\nabla\frac{\mathbf{B}\cdot\delta\mathbf{J}}{B^2} + \delta\mathbf{B}\cdot\nabla\sigma - \mathbf{J}\cdot\nabla\frac{\mathbf{B}\cdot\delta\mathbf{B}}{B^2} + \nabla\times\frac{\mathbf{B}}{B^2}\cdot\nabla\delta p_s\\
&+\frac{\mathbf{B}\times\boldsymbol{\kappa}}{B^2}\cdot\nabla\left(\langle m\delta n\hat{v}_3^2\rangle_s - \delta p_s\right).
\end{aligned} \tag{6.90}$$

Here, it is also noted that $\mathbf{B}\cdot\nabla\sigma/p'$ is independent of p and T is independent of $\hat{v}_3$ in lowest order.

To reduce the vorticity equation (6.90), one needs to determine the pressure-related moments in it. Similar to the discussion of equation (6.21) in section 6.2, the contribution from the perturbed temperature can be neglected, except for the convective term

$$\delta T = -\boldsymbol{\xi}\cdot\nabla T. \tag{6.91}$$

Therefore, only the continuity equation (6.1) needs to be analyzed,

$$\begin{aligned}
&\underbrace{\frac{\partial n}{\partial t}}_{n_1} + \underbrace{g_{\beta\alpha}\hat{V}_{\perp\beta}\frac{\partial n}{\partial x_\alpha}}_{n_2} + \underbrace{g_{3\alpha}\hat{v}_3\frac{\partial n}{\partial x_\alpha}}_{n_3} + \underbrace{n\frac{\partial g_{l\gamma}\hat{V}_{\perp l}}{\partial x_\beta}(\delta_{\beta\gamma} - g_{3\beta}g_{3\gamma})}_{n_4}\\
&+\underbrace{n\hat{v}_3\frac{\partial g_{3\alpha}}{\partial x_\alpha}}_{n_5} + \underbrace{\frac{1}{m}\frac{\partial g_{3\alpha}}{\partial x_\alpha}\frac{\partial nT}{\partial \hat{v}_3}}_{n_6} - \underbrace{g_{3\beta}g_{3\gamma}\frac{\partial g_{l\gamma}\hat{V}_{\perp l}}{\partial x_\beta}\hat{v}_3\frac{\partial n}{\partial \hat{v}_3}}_{n_7}\\
&= -g_{3\beta}\frac{\partial n}{\partial \hat{v}_3}\left(\underbrace{\frac{e}{m}E_\beta^*}_{n_8} - \underbrace{\frac{\hat{d}g_{l\beta}\hat{V}_{\perp l}}{dt}}_{n_9}\right) - \underbrace{\frac{1}{m}g_{p\alpha}g_{q\mu}\frac{\partial g_{3\mu}}{\partial x_\alpha}\frac{\partial \pi_{pq}}{\partial \hat{v}_3}}_{n_{10}},
\end{aligned} \tag{6.92}$$

Let us analyze the orderings of equation (6.92). The term 'n_8' can be neglected since $\mathbf{E}_\| = 0$. The term 'n_{10}' is neglected for $k_\perp\rho_i \ll 1$. The ordering of the term 'n_9' for equilibrium diamagnetic velocity in equation (6.81) is analyzed as follows:

$$\text{term } `n9' \sim \mathbf{b} \cdot \frac{\mathbf{B} \times \nabla p}{neB^2} \cdot \nabla \mathbf{V}_\perp \frac{n}{v_T} \sim \mathbf{V}_\perp \cdot \frac{\mathbf{B} \times \nabla n}{neB^2} \cdot \nabla \mathbf{b} \frac{n}{v_T}$$
$$\sim \omega_* n \frac{\lambda_\wedge}{R} \frac{V_\perp}{v_T}.$$

It can be neglected as compared to the term 'n_1'. Therefore, the continuity equation (6.92) is reduced to

$$\begin{aligned}
&\underbrace{\frac{\partial n}{\partial t}}_{n_1} + \underbrace{g_{\beta\alpha} \hat{V}_{\perp\beta} \frac{\partial n}{\partial x_\alpha}}_{n_2} + \underbrace{g_{3\alpha} \hat{v}_3 \frac{\partial n}{\partial x_\alpha}}_{n_3} + \underbrace{n \frac{\partial g_{l\gamma} \hat{V}_{\perp l}}{\partial x_\beta} (\delta_{\beta\gamma} - g_{3\beta} g_{3\gamma})}_{n_4} \\
&+ \underbrace{n \hat{v}_3 \frac{\partial g_{3\alpha}}{\partial x_\alpha}}_{n_5} + \underbrace{\frac{1}{m} \frac{\partial g_{3\alpha}}{\partial x_\alpha} \frac{\partial nT}{\partial \hat{v}_3}}_{n_6} - \underbrace{g_{3\beta} g_{3\gamma} \frac{\partial g_{l\gamma} \hat{V}_{\perp l}}{\partial x_\beta} \hat{v}_3 \frac{\partial n}{\partial \hat{v}_3}}_{n_7} = 0.
\end{aligned} \tag{6.93}$$

One can linearize equation (6.93) term by term. The derivations are similar to the electrostatic mode case in chapter 5 and the low frequency case in section 6.2. Using equation (6.81), one obtains

$$\begin{aligned}
\hat{\mathbf{V}}_\perp \cdot \nabla n &= \frac{\mathbf{E} \times \mathbf{B}}{B^2} \cdot \nabla n + \frac{\mathbf{B} \times \nabla P}{neB^2} \cdot \nabla n \\
&= - i\omega \boldsymbol{\xi} \cdot \nabla n + \frac{\mathbf{B} \times \nabla T}{eB^2} \cdot \nabla \delta n - \frac{\mathbf{B} \times \nabla n}{eB^2} \cdot \nabla \delta T,
\end{aligned}$$

$$\begin{aligned}
\nabla \cdot \hat{\mathbf{V}}_\perp &= - i\omega \nabla \cdot \boldsymbol{\xi} + \frac{1}{ne} \nabla \times \frac{\mathbf{B}}{B^2} \cdot \nabla \delta p \\
&\quad - \frac{\mathbf{B} \times \nabla T}{neB^2} \cdot \nabla \delta n + \frac{\mathbf{B} \times \nabla n}{neB^2} \cdot \nabla \delta T,
\end{aligned}$$

and

$$\begin{aligned}
(\mathbf{b} \cdot \nabla \hat{\mathbf{V}}_\perp) \cdot \mathbf{b} &= - (\mathbf{b} \cdot \nabla \mathbf{b}) \cdot \hat{\mathbf{V}}_\perp \\
&= i\omega (\mathbf{b} \cdot \nabla \mathbf{b}) \cdot \boldsymbol{\xi} + \frac{\mathbf{B} \times (\mathbf{b} \cdot \nabla \mathbf{b})}{neB^2} \cdot \nabla \delta p.
\end{aligned}$$

Noting also that

$$\begin{aligned}
g_{3\alpha} \hat{v}_3 \frac{\partial n}{\partial x_\alpha} &= \hat{v}_3 (\delta \mathbf{b} \cdot \nabla n + \mathbf{b} \cdot \nabla \delta n) \\
&= \hat{v}_3 [\mathbf{b} \cdot \nabla (\boldsymbol{\xi} \cdot \nabla n) + \mathbf{b} \cdot \nabla \delta n],
\end{aligned}$$

equation (6.93) is reduced to

$$
\begin{aligned}
&- i\omega(\delta n + \boldsymbol{\xi}_\perp \cdot \nabla n) + \hat{v}_3 \mathbf{b} \cdot \nabla \delta n + \hat{v}_3 \mathbf{b} \cdot \nabla(\boldsymbol{\xi}_\perp \cdot \nabla n) \\
&+ \underbrace{\frac{\mathbf{B} \times \nabla T}{eB^2} \cdot \nabla \delta n}_{x_1} - \underbrace{\frac{\mathbf{B} \times \nabla n}{eB^2} \cdot \nabla \delta T}_{x_2} \\
&+ \underbrace{\frac{1}{e}\left(\frac{\mathbf{B} \times \nabla B^2}{B^4} + \frac{\mathbf{B} \times \nabla P}{B^4}\right) \cdot \nabla \delta p}_{o_4} \\
&- \underbrace{\frac{\mathbf{B} \times \nabla T}{eB^2} \cdot \nabla \delta n}_{x_1} + \underbrace{\frac{\mathbf{B} \times \nabla n}{eB^2} \cdot \nabla \delta T}_{x_2} \\
&- \underbrace{\frac{1}{e}\left(\frac{\mathbf{B} \times \nabla P}{B^4} + \frac{\mathbf{B} \times \nabla B^2}{2B^4}\right) \cdot \nabla \delta p}_{o_4} - \hat{v}_3 \frac{\partial n}{\partial \hat{v}_3} \frac{1}{ne} \frac{\mathbf{B} \times \boldsymbol{\kappa}}{B^2} \cdot \nabla \delta p \\
&+ \delta n \hat{v}_3 \frac{\partial g_{3\alpha}}{\partial x_\alpha} + \frac{T}{m} \frac{\partial g_{3\alpha}}{\partial x_\alpha} \frac{\partial \delta n}{\partial \hat{v}_3} + \frac{1}{m} \frac{\partial g_{3\alpha}}{\partial x_\alpha} \frac{\partial n \delta T}{\partial \hat{v}_3} \\
&= in\omega \nabla \cdot \boldsymbol{\xi}_\perp + i\omega n \boldsymbol{\kappa} \cdot \boldsymbol{\xi} - i\omega \frac{m\hat{v}_3^2}{T} n \boldsymbol{\kappa} \cdot \boldsymbol{\xi}.
\end{aligned}
$$

Noting the cancellations with the same indices 'xi' and the combinations with 'oi', the continuity equation is further reduced to

$$
\begin{aligned}
&- i\omega(\delta n + \boldsymbol{\xi}_\perp \cdot \nabla n) + \hat{v}_3 \mathbf{b} \cdot \nabla \delta n + \hat{v}_3 \mathbf{b} \cdot \nabla(\boldsymbol{\xi}_\perp \cdot \nabla n) \\
&+ T\left(\frac{1}{e} \frac{\mathbf{B} \times \nabla \ln B}{B^2} - \hat{v}_3 \frac{\partial n}{\partial \hat{v}_3} \frac{1}{ne} \frac{\mathbf{B} \times \boldsymbol{\kappa}}{B^2}\right) \cdot \nabla \delta n \\
&+ n\left(\frac{1}{e} \frac{\mathbf{B} \times \nabla \ln B}{B^2} - \hat{v}_3 \frac{\partial n}{\partial \hat{v}_3} \frac{1}{ne} \frac{\mathbf{B} \times \boldsymbol{\kappa}}{B^2}\right) \cdot \nabla \delta T \\
&+ \delta n_h \hat{v}_3 \frac{\partial g_{3\alpha}}{\partial x_\alpha} + \frac{T}{m} \frac{\partial g_{3\alpha}}{\partial x_\alpha} \frac{\partial \delta n_h}{\partial \hat{v}_3} + \frac{1}{m} \frac{\partial g_{3\alpha}}{\partial x_\alpha} \frac{\partial n \delta T}{\partial \hat{v}_3} \\
&= in\omega \nabla \cdot \boldsymbol{\xi}_\perp + i\omega n \boldsymbol{\kappa} \cdot \boldsymbol{\xi} - i\omega \frac{m\hat{v}_3^2}{T} n \boldsymbol{\kappa} \cdot \boldsymbol{\xi}.
\end{aligned}
\tag{6.94}
$$

To further reduce the continuity equation, we extract the convective part of perturbed density,

$$
\delta n = -\boldsymbol{\xi} \cdot \nabla n + n \delta \tilde{n}_g. \tag{6.95}
$$

Note that in the singular layer, using equation (6.84), one has

$$
\begin{aligned}
T \frac{1}{e} \frac{\mathbf{B} \times \nabla \ln B}{B^2} \cdot \nabla(-\boldsymbol{\xi} \cdot \nabla n) &= T\left(\frac{1}{e} \frac{\mathbf{B} \times \nabla \ln B}{B^2} \cdot \nabla n\right) ik_u \xi^u \\
&= ik_u n' T \boldsymbol{\kappa} \cdot \boldsymbol{\xi}.
\end{aligned}
$$

The similar reduction can be applied to the temperature convective term in equation (6.94). Finally, equation (6.94) becomes

$$
\begin{aligned}
&(\hat{\mathbf{v}}_3 \cdot \nabla |_{\hat{v}_3,t} - i\omega + i\bar{\omega}_d)\delta\tilde{n}_g + \frac{T}{m}\frac{\partial g_{3\alpha}}{\partial x_\alpha}\frac{\partial \delta\tilde{n}_g}{\partial \hat{v}_3} \\
&= i\omega\nabla \cdot \boldsymbol{\xi}_\perp + i\tilde{n}\omega_*^{\ p}\boldsymbol{\kappa} \cdot \boldsymbol{\xi} + i(\omega - \omega_*^{\ p})\boldsymbol{\kappa} \cdot \boldsymbol{\xi} - i(\omega - \omega_*^{\ p})\frac{m\hat{v}_3^2}{T}T\boldsymbol{\kappa} \cdot \boldsymbol{\xi} \\
&+ \hat{v}_3\frac{\partial g_{3\alpha}}{\partial x_\alpha}\boldsymbol{\xi}_\perp \cdot \nabla \ln T,
\end{aligned}
\tag{6.96}
$$

where $\bar{\omega}_d$ has been defined in equation (5.18) in section 5.2

$$
i\bar{\omega}_d = \left(\frac{T}{e}\frac{\mathbf{B} \times \nabla \ln B}{B^2} + \frac{m\hat{v}_3^2}{e}\frac{\mathbf{B} \times \boldsymbol{\kappa}}{B^2}\right) \cdot \nabla,
$$

Equation (6.96) is a partial differential equation of first order. As with the electrostatic stability theory in section 5.2, one can find its characteristic:

$$
\frac{T}{m}\int^{\hat{x}_3} \nabla \cdot \mathbf{b} d\hat{x}_3 - \frac{\hat{v}_3^2}{2} = \iota. \tag{6.97}
$$

The continuity equation (6.96) is then reduced to

$$
\begin{aligned}
&(\hat{v}_3\mathbf{b} \cdot \nabla_\iota - i\omega + i\bar{\omega}_d)\delta\tilde{n}_g \\
&= i\omega\nabla \cdot \boldsymbol{\xi}_\perp + i\tilde{n}\omega_*^{\ p}\boldsymbol{\kappa} \cdot \boldsymbol{\xi} + i(\omega - \omega_*^{\ p})\boldsymbol{\kappa} \cdot \boldsymbol{\xi} \\
&- i(\omega - \omega_*^{\ p})\frac{m\hat{v}_3^2}{T}T\boldsymbol{\kappa} \cdot \boldsymbol{\xi} + \hat{v}_3\frac{\partial g_{3\alpha}}{\partial x_\alpha}\boldsymbol{\xi}_\perp \cdot \nabla \ln T,
\end{aligned}
\tag{6.98}
$$

where $\mathbf{b} \cdot \nabla_\iota$ is the derivative along the characteristic as explained in section 5.2. This transformation actually converts $\hat{\mathbf{v}}_3 \cdot \nabla|_{\hat{v}_3,t}$, holding $\hat{v}_3$ constant, in the first term of equation (6.96) to $\hat{\mathbf{v}}_3 \cdot \nabla|_{\varepsilon,\mu,t}$, roughly holding (ε, μ) constant, in the first term of equation (6.98) leading it to become gyrokinetic-theory-like. The only major difference lies in the average over the perpendicular velocity.

The vorticity equation (6.90) can be further reduced similar to the reduction in section 6.2. Using equations (6.91) and (6.95), one can prove that

$$
-\boldsymbol{\kappa}\left(\langle m\delta n\hat{v}_3^2\rangle_s - \delta p_s\right) = -\boldsymbol{\kappa}\left(\left\langle m\delta n_g\hat{v}_3^2\right\rangle_s - \delta p_{gs}\right).
$$

The vorticity equation (6.90) therefore, becomes

$$\begin{aligned}
&-\rho_s\omega(\omega-\omega_{*s}^{p})\frac{|\nabla\psi^2|}{B^2}\frac{\partial\xi^u}{\partial x}+i(\omega-\omega_{*s}^{p})p_s\frac{3m|\nabla\psi|^2(\mathbf{B}\cdot\nabla\sigma)}{2eB^2p'}\frac{\partial^2\xi^u}{\partial x^2}\\
&+\frac{\mathbf{B}}{B^2}\cdot\nabla\times\nabla\cdot\delta\Pi^{\parallel}\\
&=\mathbf{B}\cdot\nabla\frac{\mathbf{B}\cdot\delta\mathbf{J}}{B^2}+\delta\mathbf{B}\cdot\nabla\sigma-\mathbf{J}\cdot\nabla\frac{\mathbf{B}\cdot\delta\mathbf{B}}{B^2}\\
&-\nabla\times\frac{\mathbf{B}}{B^2}\cdot\nabla(\boldsymbol{\xi}_\perp\cdot\nabla p_s)+\nabla\times\frac{\mathbf{B}}{B^2}\cdot\nabla\delta p_{gs}\\
&+\frac{\mathbf{B}\times\boldsymbol{\kappa}}{B^2}\cdot\nabla\left[\frac{m}{2}\left\langle n\left(\delta\tilde{n}_g^{+}+\delta\tilde{n}_g^{-}\right)\hat{v}_3^2\right\rangle_s-\delta p_{gs}\right],
\end{aligned}\tag{6.99}$$

where

$$\delta p_{gs}=\sum_a\frac{T_a}{n_0}\int_0^{+\infty}d\hat{v}_3 n\left(\delta\tilde{n}_g^{+}+\delta\tilde{n}_g^{-}\right)_a$$

and the parallel viscous tensor $\delta\Pi^{\parallel}$ defined in equation (6.76) can be further reduced as follows,

$$\begin{aligned}
\left\langle\delta\Pi^{\parallel}_{\alpha\beta}\right\rangle_s&=\frac{T}{2\omega_c}\frac{1}{2}\left(\delta^{\perp}_{\alpha\mu}e_{\beta\gamma\nu}+\delta^{\perp}_{\beta\nu}e_{\alpha\gamma\mu}\right)b_\gamma\left(\frac{\partial g_{3\nu}}{\partial x_\mu}+\frac{\partial g_{3\mu}}{\partial x_\nu}\right)\left\langle\left(\hat{v}_3\delta n+\frac{T}{m}\frac{\partial\delta n}{\partial\hat{v}_3}\right)\right\rangle_s\\
&=\frac{T}{2\omega_c}\frac{1}{2}\left(\delta^{\perp}_{\alpha\mu}e_{\beta\gamma\nu}+\delta^{\perp}_{\beta\nu}e_{\alpha\gamma\mu}\right)b_\gamma\left(\frac{\partial g_{3\nu}}{\partial x_\mu}+\frac{\partial g_{3\mu}}{\partial x_\nu}\right)\left\langle\left(\hat{v}_3\delta n_g+\frac{T}{m}\frac{\partial\delta n_g}{\partial\hat{v}_3}\right)\right\rangle_s.
\end{aligned}$$

Here, $\delta n_g=n\delta\tilde{n}_g$ and $\delta\tilde{n}_g$ is governed by the continuity equation (6.98).

Equations (6.98) and (6.99) constitute the complete set of equations to study the finite Larmor radius effects on the MHD modes in the low frequency regime. The results modify the Braginskii gyroviscous description [7, 8] by removing the fluid description for the parallel motion. Because the parallel fluid velocity is absent in the current formalism, the fourth order derivative term in the Braginskii's gyroviscous description disappears [21].

It is interesting to discuss the finite Larmor radius effects in the current perpendicular magnetofluid description to compare with the gyrokinetic results. From equation (6.99), one can see that the FLR effects are not only present in the form of frequency shift: $\omega^2\to\omega(\omega-\omega_*^{P})$ in the inertia term, The additional gyroviscous term (the second on the left) is actually larger and can lead to a third order derivative term on ξ^ψ in the singular layer equation noting equation (6.84), besides the parallel viscosity effects. This is different from the conventional gyrokinetic theory, in which the only FLR effects are described solo by the Bessel function J_0 and J_1 and in the lowest order only gives rise to the frequency shift $\omega^2\to\omega(\omega-\omega_*^{P})$ [16]. But, the current results are consistent with the revised gyrokinetic theory in references [6, 9]. This further confirms that the FLR effects are stronger than the conventional-gyrokinetics picture solo based on the Bessel functions.

6.5 Coupling of the electrostatic drift waves

In this section, the coupling of the electrostatic drift waves to the MHD modes will be addressed. Because the electrostatic drift waves concern about the frequency regime: $\omega \sim \omega_*$ [22], as has also been shown in chapter 5. The finite Larmor radius effects on the MHD modes, as described in section 6.4.2, need to be included. We will then extend the theory in section 6.4.2 to include the electrostatic drift wave effects described in chapter 5. The equilibrium solution given in equation (4.7) in chapter 4 will be employed.

Based on the similar ordering analyses in equations (6.80) and (6.93) in section 6.4.2, the following reduced momentum and continuity equations can be used as the starting set of equations.

$$
\begin{aligned}
&(\delta_{\lambda\alpha} - g_{3\alpha}g_{3\lambda})\Bigg(\underbrace{neE_\alpha + nee_{\alpha\beta\gamma}g_{l\beta}\hat{V}_l B_\gamma}_{v_1} - \underbrace{\rho\frac{\hat{d}g_{l\alpha}\hat{V}_{\perp l}}{dt}}_{v_2} - \underbrace{\frac{\partial p}{\partial x_\alpha}}_{v_3} \\
&\quad \underbrace{- \hat{v}_3 g_{3\beta}\frac{\partial g_{3\alpha}}{\partial x_\beta}\frac{\partial p}{\partial \hat{v}_3}}_{v_4} - \underbrace{g_{3\beta}\hat{v}_3^2\frac{\partial g_{3\alpha}}{\partial x_\beta}\rho}_{v_5}\Bigg) - \underbrace{\frac{\partial \Pi_{\beta\nu}}{\partial x_\alpha}}_{v_{10}} = 0.
\end{aligned}
\tag{6.100}
$$

$$
\begin{aligned}
&\underbrace{\frac{\partial n}{\partial t}}_{n_1} + \underbrace{g_{\beta\alpha}\hat{V}_{\perp\beta}\frac{\partial n}{\partial x_\alpha}}_{n_2} + \underbrace{g_{3\alpha}\hat{v}_3\frac{\partial n}{\partial x_\alpha}}_{n_3} + \underbrace{n\frac{\partial g_{l\gamma}\hat{V}_{\perp l}}{\partial x_\beta}\left(\delta_{\beta\gamma} - g_{3\beta}g_{3\gamma}\right)}_{n_4} \\
&\underbrace{+ n\hat{v}_3\frac{\partial g_{3\alpha}}{\partial x_\alpha}}_{n_5} + \underbrace{\frac{1}{m}\frac{\partial g_{3\alpha}}{\partial x_\alpha}\frac{\partial nT}{\partial \hat{v}_3}}_{n_6} - \underbrace{g_{3\beta}g_{3\gamma}\frac{\partial g_{l\gamma}\hat{V}_{\perp l}}{\partial x_\beta}\hat{v}_3\frac{\partial n}{\partial \hat{v}_3}}_{n_7} = -g_{3\alpha}\frac{\partial n}{\partial \hat{v}_3}\frac{e}{m}E_\alpha.
\end{aligned}
\tag{6.101}
$$

Here, the parallel electric field effects on the right-hand side of equation (6.101) have been kept. The isothermal condition $\mathbf{b} \cdot \nabla T = 0$ can also be used as the energy transport equation as equation (6.91) in section 6.4.2.

$$
\delta T = -\boldsymbol{\xi} \cdot \nabla T. \tag{6.102}
$$

The electric and magnetic fields are represented as follows

$$
\begin{aligned}
&\delta\mathbf{E} = -\frac{\partial \delta\mathbf{A}}{\partial t} - \nabla\delta\varphi, \qquad \delta\mathbf{B} = \nabla \times \delta\mathbf{A}, \\
&\delta\mathbf{A} = \boldsymbol{\xi} \times \mathbf{B}.
\end{aligned}
$$

We first calculate the perpendicular velocity from the momentum equation (6.101).

$$\mathbf{V}_\perp = \frac{\mathbf{E}\times\mathbf{B}}{B^2} + \frac{\mathbf{B}\times\nabla p}{neB^2} + \frac{\rho\mathbf{B}\times\hat{d}g_{la}\hat{V}_{\perp l}/dt}{neB^2} + \cdots. \tag{6.103}$$

In the linearization, one obtains the following orderings

$$\begin{aligned}
\frac{(\partial\delta\mathbf{A}/\partial t)\times\mathbf{B}}{B^2} &\sim \omega\xi_\theta,\\
\frac{\mathbf{B}\times\nabla\delta\varphi}{B^2} &\sim \frac{k_\perp}{B}\varphi \sim k_\perp\rho_a\frac{v_T}{R}\frac{\omega^2}{\omega^2+\omega_t^2}\xi_\wedge,\\
\frac{\mathbf{B}\times\nabla p}{neB^2} &\sim \frac{\mathbf{B}\times\nabla\boldsymbol{\xi}\cdot\nabla p}{neB^2} \sim k_\perp\rho_a\frac{v_T}{a}\xi_r \sim \omega_*^p\xi_\theta,\\
\frac{\rho\mathbf{B}\times\hat{d}g_{la}\hat{V}_{\perp l}/dt}{neB^2} &\sim \frac{\omega}{\omega_c}\omega\xi_\theta.
\end{aligned}$$

Therefore, the third term on the right-hand side of equation (6.103) can be neglected as well. The perturbed perpendicular velocity can be expressed as

$$\delta\mathbf{V}_\perp = -i\omega\boldsymbol{\xi}_\perp + \frac{\mathbf{B}\times\nabla\delta\varphi}{B^2} + \frac{\mathbf{B}\times\nabla\delta p}{neB^2}. \tag{6.104}$$

As will be seen later on, one has

$$\delta p = -\boldsymbol{\xi}\cdot\nabla p - ne\delta\varphi + \delta n_g T.$$

Inserting it to equation (6.104), one can find that the $\delta\varphi$ contributions cancel each other. Therefore,

$$\delta\mathbf{V}_\perp = -i\omega\boldsymbol{\xi}_\perp - \frac{\mathbf{B}\times\nabla\boldsymbol{\xi}\cdot\nabla p}{neB^2} + \frac{\mathbf{B}\times\nabla\delta n_g T}{neB^2}. \tag{6.105}$$

The last term is of order $\mathcal{O}((\omega/\omega_{ti})^2(p\xi_\wedge/R))$. Therefore, different from the low frequency case in equation (6.85) in section 6.4.2, its contribution to the diamagnetic velocity cannot be ignored. Therefore, one has

$$\delta\mathbf{V}_\perp = -i(\omega - \omega_*^p)\boldsymbol{\xi}_\perp + \frac{\mathbf{B}\times\nabla\left(\delta n_g T\right)}{neB^2}. \tag{6.106}$$

As in section 6.4.2, to study the MHD modes, one needs to derive the momentum equation for the single fluid. This is achieved by integrating over $\hat{v}_3$ and the summation over all particle species:

$$\langle\cdots\rangle_s = \sum_a \frac{1}{n_0}\int_{-\infty}^{+\infty} d\hat{v}_3(\cdots).$$

This leads equation (6.100) to become

$$\left\langle \rho \frac{d\mathbf{V}_\perp}{dt} \right\rangle_s = \mathbf{J} \times \mathbf{B} - \nabla_\perp p_s - \kappa\left(\langle mn\hat{v}_3^2 \rangle_s - p_s\right) + [\nabla \cdot \Pi]_s, \tag{6.107}$$

where

$$\rho_s = \langle mn \rangle_s, \qquad p_s = \langle p \rangle_s, \qquad \mathbf{J} = \langle ne\mathbf{V}_\perp \rangle_s + \mathbf{J}_\parallel,$$
$$\frac{d_s}{dt} = \frac{\partial}{\partial t} + \mathbf{V}_\perp \cdot \nabla.$$

We can derive the vorticity equation by applying operator $\nabla \cdot (\mathbf{B}/B^2) \times \cdots$ on equation (6.107).

$$\begin{aligned}
&\sum_a \frac{1}{n_0} \int_{-\infty}^{+\infty} d\hat{v}_3 i\rho_a(\omega - \omega_*^p)\nabla \cdot \frac{\mathbf{B}}{B^2} \times \delta\mathbf{V}_\perp \\
&= \mathbf{B} \cdot \nabla \frac{\mathbf{B} \cdot \delta\mathbf{J}}{B^2} + \delta\mathbf{B} \cdot \nabla\sigma - \mathbf{J} \cdot \nabla \frac{\mathbf{B} \cdot \delta\mathbf{B}}{B^2} \\
&+ \nabla \times \frac{\mathbf{B}}{B^2} \cdot \nabla_\perp \delta p_s + \frac{\mathbf{B} \times \kappa}{B^2} \cdot \nabla\left(\langle mn\hat{v}_3^2 \rangle_s - p_s\right) \\
&- \sum_a \frac{1}{n_0} \int_{-\infty}^{+\infty} d\hat{v}_3 \nabla \cdot \left[\frac{\mathbf{B}}{B^2} \times (\nabla \cdot \delta\Pi)\right],
\end{aligned} \tag{6.108}$$

where ω_*^p is defined in equation (6.54). As shown in details in section 6.4.2, using equations (6.74) and (6.78) in section 6.4.1 to prove the so-called gyroviscous cancellation, equation (6.108) is reduced to

$$\begin{aligned}
&-\rho_s\omega(\omega - \omega_{*s}^p)\frac{|\nabla\psi^2|}{B^2}\frac{\partial\xi^u}{\partial x} + i(\omega - \omega_{*s}^p)p_s \frac{3m|\nabla\psi|^2(\mathbf{B} \cdot \nabla\sigma)}{2eB^2p'}\frac{\partial^2\xi^u}{\partial x^2} \\
&+ \frac{\mathbf{B}}{B^2} \cdot \nabla \times \nabla \cdot \delta\Pi^{\parallel} + i\left\langle \rho(\omega - \omega_*^p)|\nabla\psi|^2 \frac{mT}{e}\frac{d^2\delta n_g}{dx^2} \right\rangle_s \\
&= \mathbf{B} \cdot \nabla \frac{\mathbf{B} \cdot \delta\mathbf{J}}{B^2} + \delta\mathbf{B} \cdot \nabla\sigma - \mathbf{J} \cdot \nabla \frac{\mathbf{B} \cdot \delta\mathbf{B}}{B^2} + \nabla \times \frac{\mathbf{B}}{B^2} \cdot \nabla\delta p_s \\
&+ \frac{\mathbf{B} \times \kappa}{B^2} \cdot \nabla\left(\langle m\delta n\hat{v}_3^2 \rangle_s - \delta p_s\right).
\end{aligned} \tag{6.109}$$

Here, different from equation (6.87), the additional diamagnetic effects in the last term equation (6.106) have been taken into account.

Next, we reduce the continuity equation (6.101). It can be linearized term by term. The derivations are similar to electrostatic mode case in chapter 5 and the low frequency case with FLR effects included in section 6.4.2 except $\delta\varphi$ being kept. Using equation (6.103), one obtains

$$
\begin{aligned}
\hat{\mathbf{V}}_\perp \cdot \nabla n &= \frac{\mathbf{E}\times\mathbf{B}}{B^2}\cdot\nabla n + \frac{\mathbf{B}\times\nabla P}{neB^2}\cdot\nabla n \\
&= -\, i\omega\boldsymbol{\xi}\cdot\nabla n - \frac{\mathbf{B}\times\nabla n}{B^2}\cdot\nabla\delta\varphi \\
&\quad + \frac{\mathbf{B}\times\nabla T}{eB^2}\cdot\nabla\delta n - \frac{\mathbf{B}\times\nabla n}{eB^2}\cdot\nabla\delta T,
\end{aligned}
$$

$$
\begin{aligned}
\nabla\cdot\hat{\mathbf{V}}_\perp &= -\, i\omega\nabla\cdot\boldsymbol{\xi} + \nabla\times\frac{\mathbf{B}}{B^2}\cdot\nabla\delta\varphi + \frac{1}{ne}\nabla\times\frac{\mathbf{B}}{B^2}\cdot\nabla\delta p \\
&\quad - \frac{\mathbf{B}\times\nabla T}{neB^2}\cdot\nabla\delta n + \frac{\mathbf{B}\times\nabla n}{neB^2}\cdot\nabla\delta T,
\end{aligned}
$$

and

$$
\begin{aligned}
(\mathbf{b}\cdot\nabla\hat{\mathbf{V}}_\perp)\cdot\mathbf{b} &= -\,(\mathbf{b}\cdot\nabla\mathbf{b})\cdot\hat{\mathbf{V}}_\perp \\
&= i\omega(\mathbf{b}\cdot\nabla\mathbf{b})\cdot\boldsymbol{\xi} + \frac{\mathbf{B}\times(\mathbf{b}\cdot\nabla\mathbf{b})}{B^2}\cdot\nabla\delta\varphi + \frac{\mathbf{B}\times(\mathbf{b}\cdot\nabla\mathbf{b})}{neB^2}\cdot\nabla\delta p.
\end{aligned}
$$

Noting also that

$$
\begin{aligned}
g_{3\alpha}\hat{v}_3\frac{\partial n}{\partial x_\alpha} &= \hat{v}_3(\delta\mathbf{b}\cdot\nabla n + \mathbf{b}\cdot\nabla\delta n) \\
&= \hat{v}_3[\mathbf{b}\cdot\nabla(\boldsymbol{\xi}\cdot\nabla n) + \mathbf{b}\cdot\nabla\delta n],
\end{aligned}
$$

equation (6.101) is reduced to

$$
\begin{aligned}
&\frac{\partial\delta n}{\partial t} - i\omega(\delta n + \boldsymbol{\xi}_\perp\cdot\nabla n) + \hat{v}_3\mathbf{b}\cdot\nabla\delta n + \hat{v}_3\mathbf{b}\cdot\nabla(\boldsymbol{\xi}_\perp\cdot\nabla n) \\
&- \frac{\mathbf{B}\times\nabla n}{B^2}\cdot\nabla\delta\varphi + \underbrace{\frac{\mathbf{B}\times\nabla T}{eB^2}\cdot\nabla\delta n}_{x_1} - \underbrace{\frac{\mathbf{B}\times\nabla n}{eB^2}\cdot\nabla\delta T}_{x_2} \\
&\underbrace{+ n\left(\frac{\mathbf{B}\times\nabla B^2}{B^4} + \frac{\mathbf{B}\times\nabla P}{B^4}\right)\cdot\nabla\delta\varphi}_{o_3} + \underbrace{\frac{1}{e}\left(\frac{\mathbf{B}\times\nabla B^2}{B^4} + \frac{\mathbf{B}\times\nabla P}{B^4}\right)\cdot\nabla\delta p}_{o_4} \\
&- \underbrace{\frac{\mathbf{B}\times\nabla T}{eB^2}\cdot\nabla\delta n}_{x_1} + \underbrace{\frac{\mathbf{B}\times\nabla n}{eB^2}\cdot\nabla\delta T}_{x_2} \\
&\underbrace{- n\left(\frac{\mathbf{B}\times\nabla P}{B^4} + \frac{\mathbf{B}\times\nabla B^2}{2B^4}\right)\cdot\nabla\delta\varphi}_{o_3} \underbrace{- \frac{1}{e}\left(\frac{\mathbf{B}\times\nabla P}{B^4} + \frac{\mathbf{B}\times\nabla B^2}{2B^4}\right)\cdot\nabla\delta p}_{o_4} \\
&- \hat{v}_3\frac{\partial n}{\partial\hat{v}_3}\frac{\mathbf{B}\times\boldsymbol{\kappa}}{B^2}\cdot\nabla\delta\varphi - \hat{v}_3\frac{\partial n}{\partial\hat{v}_3}\frac{1}{ne}\frac{\mathbf{B}\times\boldsymbol{\kappa}}{B^2}\cdot\nabla\delta p \\
&+ \delta n\hat{v}_3\frac{\partial g_{3\alpha}}{\partial x_\alpha} + \frac{T}{m}\frac{\partial g_{3\alpha}}{\partial x_\alpha}\frac{\partial\delta n}{\partial\hat{v}_3} + \frac{1}{m}\frac{\partial g_{3\alpha}}{\partial x_\alpha}\frac{\partial n\delta T}{\partial\hat{v}_3} \\
&= in\omega\nabla\cdot\boldsymbol{\xi}_\perp + i\omega n\boldsymbol{\kappa}\cdot\boldsymbol{\xi} - i\omega\frac{m\hat{v}_3^2}{T}n\boldsymbol{\kappa}\cdot\boldsymbol{\xi} - g_{3\alpha}\hat{v}_3\frac{ne}{T}\frac{\partial\varphi}{\partial x_\alpha}.
\end{aligned}
$$

Noting the cancellations with the same indices 'xi' and the combinations with 'oi', the continuity equation is further reduced to

$$
\begin{aligned}
&- i\omega(\delta n + \boldsymbol{\xi}_\perp \cdot \nabla n) + \hat{v}_3 \mathbf{b} \cdot \nabla \delta n + \hat{v}_3 \mathbf{b} \cdot \nabla(\boldsymbol{\xi}_\perp \cdot \nabla n) \\
&+ T\left(\frac{1}{e}\frac{\mathbf{B} \times \nabla \ln B}{B^2} - \hat{v}_3 \frac{\partial n}{\partial \hat{v}_3}\frac{1}{ne}\frac{\mathbf{B} \times \boldsymbol{\kappa}}{B^2}\right) \cdot \nabla \delta n \\
&+ n\left(\frac{1}{e}\frac{\mathbf{B} \times \nabla \ln B}{B^2} - \hat{v}_3 \frac{\partial n}{\partial \hat{v}_3}\frac{1}{ne}\frac{\mathbf{B} \times \boldsymbol{\kappa}}{B^2}\right) \cdot \nabla \delta T \\
&+ \delta n_h \hat{v}_3 \frac{\partial g_{3\alpha}}{\partial x_\alpha} + \frac{T}{m}\frac{\partial g_{3\alpha}}{\partial x_\alpha}\frac{\partial \delta n_h}{\partial \hat{v}_3} + \frac{1}{m}\frac{\partial g_{3\alpha}}{\partial x_\alpha}\frac{\partial n \delta T}{\partial \hat{v}_3} \\
&= in\omega \nabla \cdot \boldsymbol{\xi}_\perp + i\omega n \boldsymbol{\kappa} \cdot \boldsymbol{\xi} - i\omega \frac{m\hat{v}_3^2}{T} n \boldsymbol{\kappa} \cdot \boldsymbol{\xi} - g_{3\alpha}\hat{v}_3 \frac{ne}{T}\frac{\partial \varphi}{\partial x_\alpha} \\
&+ \frac{\mathbf{B} \times \nabla n}{B^2} \cdot \nabla \delta\varphi - n\frac{\mathbf{B} \times \nabla \ln B}{B^2} \cdot \nabla \delta \varphi + \hat{v}_3 \frac{\partial n}{\partial \hat{v}_3}\frac{\mathbf{B} \times \boldsymbol{\kappa}}{B^2} \cdot \nabla \delta \varphi.
\end{aligned}
\tag{6.110}
$$

To extract the adiabatic part, we let

$$
\delta n = -\frac{ne}{T}\delta\varphi + \delta n_h. \tag{6.111}
$$

Equation (6.110) is then reduced to

$$
\begin{aligned}
&- i\omega(\delta n_h + \boldsymbol{\xi}_\perp \cdot \nabla n) + \hat{v}_3 \mathbf{b} \cdot \nabla \delta n_h + \hat{v}_3 \mathbf{b} \cdot \nabla(\boldsymbol{\xi}_\perp \cdot \nabla n) \\
&+ T\left(\frac{1}{e}\frac{\mathbf{B} \times \nabla \ln B}{B^2} - \hat{v}_3 \frac{\partial n}{\partial \hat{v}_3}\frac{1}{ne}\frac{\mathbf{B} \times \boldsymbol{\kappa}}{B^2}\right) \cdot \nabla \delta n_h \\
&+ n\left(\frac{1}{e}\frac{\mathbf{B} \times \nabla \ln B}{B^2} - \hat{v}_3 \frac{\partial n}{\partial \hat{v}_3}\frac{1}{ne}\frac{\mathbf{B} \times \boldsymbol{\kappa}}{B^2}\right) \cdot \nabla \delta T \\
&+ \delta n_h \hat{v}_3 \frac{\partial g_{3\alpha}}{\partial x_\alpha} + \frac{T}{m}\frac{\partial g_{3\alpha}}{\partial x_\alpha}\frac{\partial \delta n_h}{\partial \hat{v}_3} + \frac{1}{m}\frac{\partial g_{3\alpha}}{\partial x_\alpha}\frac{\partial n \delta T}{\partial \hat{v}_3} \\
&= in\omega \nabla \cdot \boldsymbol{\xi}_\perp + i\omega n \boldsymbol{\kappa} \cdot \boldsymbol{\xi} - i\omega \frac{m\hat{v}_3^2}{T} n \boldsymbol{\kappa} \cdot \boldsymbol{\xi} \\
&- i\frac{\tilde{n}e}{T}(\omega - \bar{\omega}_*^T)\delta\varphi,
\end{aligned}
\tag{6.112}
$$

where $\bar{\omega}_*^T$ is defined in equation (5.19) in section 5.2.

To further reduce the continuity equation (6.112), we extract the convective part of the perturbed density

$$
\delta n_h = -\boldsymbol{\xi} \cdot \nabla n + n\delta n_g \tag{6.113}
$$

and use equation (6.102) for perturbed temperature. Similar to the derivation of equation (6.96), using equation (6.84) in the singular layer, one has

$$T\frac{1}{e}\frac{\mathbf{B}\times\nabla\ln B}{B^2}\cdot\nabla(-\boldsymbol{\xi}\cdot\nabla n) = T\left(\frac{1}{e}\frac{\mathbf{B}\times\nabla\ln B}{B^2}\cdot\nabla n\right)ik_u\xi^u$$
$$= ik_u\frac{n'T}{e}\boldsymbol{\kappa}\cdot\boldsymbol{\xi} = in\bar{\omega}_*^p\boldsymbol{\kappa}\cdot\boldsymbol{\xi},$$
$$n\frac{1}{e}\frac{\mathbf{B}\times\nabla\ln B}{B^2}\cdot\nabla(-\boldsymbol{\xi}\cdot\nabla T) = n\left(\frac{1}{e}\frac{\mathbf{B}\times\nabla\ln B}{B^2}\cdot\nabla n\right)ik_u\xi^u$$
$$= ik_u\frac{T'n}{e}\boldsymbol{\kappa}\cdot\boldsymbol{\xi} = in\bar{\eta}\bar{\omega}_*\boldsymbol{\kappa}\cdot\boldsymbol{\xi},$$

and so on. Equation (6.112) can, then, be reduced to

$$\begin{aligned}&(\hat{\mathbf{v}}_3\cdot\nabla|_{\hat{v}_{3},\iota} - i\omega + i\bar{\omega}_d)\delta\tilde{n}_g + \frac{T}{m}\frac{\partial g_{3\alpha}}{\partial x_\alpha}\frac{\partial\delta n_g}{\partial\hat{v}_3}\\ &+ T\left(\frac{1}{e}\frac{\mathbf{B}\times\nabla\ln B}{B^2} - \hat{v}_3\frac{\partial n}{\partial\hat{v}_3}\frac{1}{ne}\frac{\mathbf{B}\times\boldsymbol{\kappa}}{B^2}\right)\cdot\nabla\delta n_g\\ &= i\omega\nabla\cdot\boldsymbol{\xi}_\perp + i(\omega - \hat{\omega}_*^p)\boldsymbol{\kappa}\cdot\boldsymbol{\xi} - i(\omega - \omega_*^p)\frac{m\hat{v}_3^2}{T}T\boldsymbol{\kappa}\cdot\boldsymbol{\xi}\\ &+ \hat{v}_3\frac{\partial g_{3\alpha}}{\partial x_\alpha}\boldsymbol{\xi}\cdot\nabla\ln T - i\frac{e}{T}(\omega - \bar{\omega}_*^T)\delta\varphi.\end{aligned} \tag{6.114}$$

where $\bar{\omega}_d$ has been defined in equation (5.18) in section 5.2,

$$i\bar{\omega}_d = \left(\frac{T}{e}\frac{\mathbf{B}\times\nabla\ln B}{B^2} + \frac{m\hat{v}_3^2}{e}\frac{\mathbf{B}\times\boldsymbol{\kappa}}{B^2}\right)\cdot\nabla,$$

Equation (6.114) is a partial differential equation of first order. As the electrostatic stability theory in section 5.2, one can find its characteristic:

$$\frac{T}{m}\int^{\hat{x}_3}\nabla\cdot\mathbf{b}d\hat{x}_3 - \frac{\hat{v}_3^2}{2} = \iota. \tag{6.115}$$

Equation (6.114) is then reduced to

$$\begin{aligned}&(\hat{\mathbf{v}}_3\cdot\nabla_\iota - i\omega + i\bar{\omega}_d)\delta\tilde{n}_g\\ &+ T\left(\frac{1}{e}\frac{\mathbf{B}\times\nabla\ln B}{B^2} - \hat{v}_3\frac{\partial n}{\partial\hat{v}_3}\frac{1}{ne}\frac{\mathbf{B}\times\boldsymbol{\kappa}}{B^2}\right)\cdot\nabla\delta n_g\\ &= i\omega\nabla\cdot\boldsymbol{\xi}_\perp + i(\omega - \hat{\omega}_*^p)\boldsymbol{\kappa}\cdot\boldsymbol{\xi} - i(\omega - \hat{\omega}_*^p)\frac{m\hat{v}_3^2}{T}T\boldsymbol{\kappa}\cdot\boldsymbol{\xi}\\ &+ \hat{v}_3\frac{\partial g_{3\alpha}}{\partial x_\alpha}\boldsymbol{\xi}\cdot\nabla\ln T - i\frac{e}{T}(\omega - \bar{\omega}_*^T)\delta\varphi.\end{aligned} \tag{6.116}$$

where $\mathbf{b}\cdot\nabla_\iota$ is the derivative along the characteristic as explained in section 5.2. This transformation actually converts $\hat{\mathbf{v}}_3\cdot\nabla|_{\hat{v}_3,t}$, holding $\hat{v}_3$ constant, in the first term of equation (6.114) to $\hat{\mathbf{v}}_3\cdot\nabla|_{\varepsilon,\mu,t}$, roughly holding (ε, μ) constant, in the first term of equation (6.116) leading it to become gyrokinetic-theory-like. The only major difference lies in the average over the perpendicular velocity.

To complete the basic set of equations, the quasineutrality condition is required to determine the parallel electric potential $\delta\varphi$:

$$\sum_a e_a \frac{1}{n_0} \int_0^{+\infty} d\hat{v}_3 (\delta n_a^+ + \delta n_a^-) = 0,$$

where $\pm$ refers to the sign of parallel velocity. Using the transforms in equations (6.111) and (6.113), the quasineutrality condition becomes

$$\delta\varphi \sum_a \frac{e_a^2}{T_a} = \sum_a e_a \frac{1}{n_0} \int_0^{+\infty} d\hat{v}_3 n \left(\delta\tilde{n}_{ga}^+ + \delta\tilde{n}_{ga}^-\right), \tag{6.117}$$

where the quasineutrality for equilibrium charge densities has been imposed.

The vorticity equation (6.109) can be further reduced similar to the reduction in section 6.2. Using the transforms in equations (6.111) and (6.113), one can prove that

$$-\kappa\left(\langle m\delta n \hat{v}_3^2 \rangle_s - \delta p_s\right) = -\kappa\left(\left\langle m\delta n_g \hat{v}_3^2 \right\rangle_s - \delta p_{gs}\right).$$

The vorticity equation (6.109), therefore, becomes

$$\begin{aligned}
&-\rho_s \omega(\omega - \omega_{*s}^P) \frac{|\nabla\psi^2|}{B^2} \frac{\partial \xi^u}{\partial x} + i(\omega - \omega_{*s}^P) p_s \frac{3m|\nabla\psi|^2 (\mathbf{B}\cdot\nabla\sigma)}{2eB^2 p'} \frac{\partial^2 \xi^u}{\partial x^2} \\
&+ \frac{\mathbf{B}}{B^2} \cdot \nabla \times \nabla \cdot \Pi_s^{\parallel} + i\left\langle \rho(\omega - \omega_*^P)|\nabla\psi|^2 \frac{mT}{e} \frac{d^2 \delta n_g}{dx^2} \right\rangle_s \\
&= \mathbf{B}\cdot\nabla \frac{\mathbf{B}\cdot\delta\mathbf{J}}{B^2} + \delta\mathbf{B}\cdot\nabla\sigma - \mathbf{J}\cdot\nabla \frac{\mathbf{B}\cdot\delta\mathbf{B}}{B^2} + \nabla \times \frac{\mathbf{B}}{B^2} \cdot \nabla\delta p_s \\
&+ \frac{\mathbf{B}\times\kappa}{B^2} \cdot \nabla\left(\left\langle m\delta n_g \hat{v}_3^2 \right\rangle_s - \delta p_{sg}\right).
\end{aligned} \tag{6.118}$$

Here, by definition

$$\delta p_{gs} = \sum_a \frac{T_a}{n_0} \int_0^{+\infty} d\hat{v}_3 n \left(\delta\tilde{n}_g^+ + \delta\tilde{n}_g^-\right)_a$$

and the parallel viscous tensor $\delta\Pi^{\parallel}$ defined in equation (6.76) can be further reduced as follows,

$$\begin{aligned}
\left\langle \delta\Pi_{\alpha\beta}^{\parallel} \right\rangle_s &= \frac{T}{2\omega_c} \frac{1}{2} \left(\delta_{\alpha\mu}^{\perp} e_{\beta\gamma\nu} + \delta_{\beta\nu}^{\perp} e_{\alpha\gamma\mu}\right) b_\gamma \left(\frac{\partial g_{3\nu}}{\partial x_\mu} + \frac{\partial g_{3\mu}}{\partial x_\nu}\right) \left\langle \left(\hat{v}_3 \delta n + \frac{T}{m} \frac{\partial \delta n}{\partial \hat{v}_3}\right) \right\rangle_s \\
&= \frac{T}{2\omega_c} \frac{1}{2} \left(\delta_{\alpha\mu}^{\perp} e_{\beta\gamma\nu} + \delta_{\beta\nu}^{\perp} e_{\alpha\gamma\mu}\right) b_\gamma \left(\frac{\partial g_{3\nu}}{\partial x_\mu} + \frac{\partial g_{3\mu}}{\partial x_\nu}\right) \left\langle \left(\hat{v}_3 \delta n_g + \frac{T}{m} \frac{\partial \delta n_g}{\partial \hat{v}_3}\right) \right\rangle_s,
\end{aligned}$$

Here, $\delta n_g = n\delta\tilde{n}_g$ with $\delta\tilde{n}_g$ governed by the continuity equation (6.116).

Equations (6.116)–(6.118) constitute a complete set of equations. The MHD modes are described by the vorticity or shear Alfvén equation (6.118); the electrostatic drift waves are governed by the quasineutrality condition in equation (6.117); the parallel dynamics obeys the perpendicular magnetofluid continuity equation (6.116). They are coupled with each other.

The dispersion relation for electrostatic or electromagnetic modes can affect each other's spectrum. The basic set of equations has the following structure.

$$\mathcal{L}_{\text{electrostatic}}\{\delta\varphi\} \propto \xi_\perp, \tag{6.119}$$

$$\mathcal{L}_{\text{electromagnetic}}\{\xi_\perp\} \propto \delta\varphi. \tag{6.120}$$

Consider the electromagnetic modes. Equation (6.119) is solved for $\delta\varphi$. The solution is then inserted to equation (6.120) to get the eigenmode equation. There are solubility conditions for equation (6.119): $\mathcal{L}_{\text{electrostatic}}\{\delta\varphi\} \neq 0$. They actually correspond to the electrostatic modes described in chapter 5. This shows that the electrostatic modes can affect the spectrum of electromagnetic modes to be described by equation (6.120). Likewise, one can prove that electromagnetic modes can also affect the spectrum of electrostatic modes.

The theoretic formulation described in this chapter gives a complete description of the electromagnetic modes with the finite Larmor radius effects kept to the order $\mathcal{O}(k_\perp^2\rho_i^2)$. The parallel kinetic effects, such as wave-particle resonance, the trapped particle, parallel electric field effects, etc are all retained in this formalism.

6.6 Conclusions and discussion

In this chapter, the perpendicular magnetofluid theory described in chapter 4 is applied to study the MHD modes in the collisionless limit. In the low frequency regime, it is found that the kinetic enhancement of the so-called apparent mass effect [13, 14] can be recovered in this perpendicular magnetofluid formalism. In the comparable frequency regime, we find that the wave-particle resonance and trapped effects, which are only present in the kinetic description [15], are all kept in this perpendicular magnetofluid formalism. The finite Larmor radius effects, which is present in the gyrokinetic theory, are kept to the order $\mathcal{O}(k_\perp^2\rho_i^2)$. The results modify the Braginskii gyroviscous tensor [7, 8] by introducing a nonfluid description in the parallel direction and are consistent with the gyrokinetic analyses with the missing MHD and FLR effects recovered in references [6, 9]. The complete, self-consistent theory for the electromagnetic modes with the FLR effects kept to order $\mathcal{O}(k_\perp^2\rho_i^2)$ and the coupling between the electrostatic and electromagnetic modes taken into account is also presented. The results show that the perpendicular magnetofluid formalism can offer an alternative description to the gyrokinetic theory for the FLR effects kept to $\mathcal{O}(k_\perp^2\rho_i^2)$ and, nonetheless, keeps the ideal MHD root.

References

[1] Zheng L 2020 Perpendicular magnetofluid theory for magnetically confined plasmas in the collisionless limit *Phys. Plasmas* **27** 052503

[2] Rutherford P H and Frieman E A 1968 Drift instabilities in general magnetic field configurations *Phys. Fluids* **11** 569–85

[3] Taylor J B and Hastie R J 1968 Stability of general plasma equilibria: I. formal theory *Plasma Phys.* **10** 479

[4] Antonsen T M and Lane B 1980 Kinetic equations for low frequency instabilities in inhomogeneous plasmas *Phys. Fluids* **23** 1205–14

[5] Catto P J, Tang W M and Baldwin D E 1981 Generalized gyrokinetics *Plasma Phys.* **23** 639–50

[6] Zheng L J, Kotschenreuther M T and Van Dam J W 2007 Revisiting linear gyrokinetics to recover ideal magnetohydrodynamics and missing finite Larmor radius effects *Phys. Plasmas* **14** 072505

[7] Braginskii S I 1966 Transport phenomena in a complete ionized two-temperature plasma *ZhETF (J. Exp. Theor. Phys. USSR)* **33** 459
Braginskii S I 1958 *Soviet Phys. JETP* **6** 358–69

[8] Braginskii S I 1966 Transport processes *Reviews of Plasma Physics* vol 1 ed M Leontovich (New York: Consultants Bureau) pp 205–311

[9] Zheng L 2015 *Advanced Tokamak Stability Theory IOP Concise Physics* (San Rafael, CA: Morgan & Claypool)

[10] Zheng L 2019 *Magnetically Confined Fusion Plasma Physics: Ideal MHD Theory* (San Rafael, CA: Morgan & Claypool)

[11] Freidberg J P 1987 *Ideal Magnetohydrodynamics* (New York: Plenum Press)

[12] Greene J M and Johnson J L 1962 Stability criterion for arbitrary hydromagnetic equilibria *Phys. Fluids* **5** 510–7

[13] Mikhailovsky A 1974 *Theory of Plasma Instabilities* (New York: Consultants Bureau)

[14] Zheng L-J and Tessarotto M 1994 Collisionless kinetic ballooning mode equation in the low-frequency regime *Phys. Plasmas* **1** 3928–35

[15] Zheng L-J and Tessarotto M 1994 Collisionless kinetic ballooning equations in the comparable frequency regime *Phys. Plasmas* **1** 2956–62

[16] Tang W, Connor J and Hastie R 1980 Kinetic-ballooning-mode theory in general geometry *Nucl. Fusion* **20** 1439

[17] Antonsen T M and Lee Y C 1982 Electrostatic modification of variational principles for anisotropic plasmas *Phys. Fluids* **25** 132–42

[18] Mercier C 1962 Critere de stabilite d'an systeme toroidal hydromagnetique en pression scalaire *Nucl. Fusion Suppl.* **2** 801

[19] Coppi B 1964 Influence of gyration radius and collisions on hydromagnetic stability *Phys. Fluids* **7** 1501–16

[20] Rosenbluth M N and Simon A 1965 Finite Larmor radius equations with nonuniform electric fields and velocities *Phys. Fluids* **8** 1300–22

[21] Zheng L J 1983 Two-fluid equations for low-n singular modes in the low frequency regime *Phys. Fluids* **B 5** 1962–70

[22] Tang W M 1978 Microinstability theory in tokamaks *Nucl. Fusion* **18** 1089–160

IOP Publishing

Magnetically Confined Fusion Plasma Physics, Volume 2
Multifluid theory
Linjin Zheng

Chapter 7

The collisional effects in the perpendicular magnetofluid formalism

In this chapter, the resistive effects in the perpendicular magnetofluid formalism will be considered [1]. We only focus on how the stability analyses are affected by the collisions in the perpendicular magnetofluid formalism. The subtle effects of collisional operator modellings [2–4] will be addressed in the future. Therefore, only the simple Krook collisional operator [5] is used. The equilibrium solution given in equation (4.7) in chapter 4 will be employed.

7.1 The kinetic equation and its moment equations

In this section, the collisional effects will be added back to the perpendicular magnetofluid theory described in chapter 3. The extension is straightforward. We, therefore, only outline the major steps.

The kinetic equations with collisional effects included are given as follows.

$$\frac{\partial f_a}{\partial t} + \mathbf{v} \cdot \frac{\partial f_a}{\partial \mathbf{x}} + \mathbf{a} \cdot \frac{\partial f_a}{\partial \mathbf{v}} = C_a, \tag{7.1}$$

where the subscript a specifies particle species, f is the distribution function, which depends on the particle coordinate $\mathbf{x}$, velocity $\mathbf{v}$, and time t,

$$\mathbf{a} = \mathbf{E} + \mathbf{v} \times \mathbf{B},$$

and C_a is the collisional operator,

$$C_a = \sum_b C_{ab}(f_a, f_b).$$

Two coordinate systems are introduced. The laboratory Cartesian coordinate system is denoted by the basis ($\mathbf{e}_1$, $\mathbf{e}_2$, $\mathbf{e}_3$), where $\mathbf{e}_\alpha$ with $\alpha = 1, 2, 3$ are unit vectors, The field-line-aligned coordinate system is also introduced. It is right-handed and denoted by the basis ($\hat{\mathbf{e}}_1$, $\hat{\mathbf{e}}_2$, $\hat{\mathbf{e}}_3$). Here, $\hat{\mathbf{e}}_\alpha$ with $\alpha = 1, 2, 3$ are unit vectors, $\hat{\mathbf{e}}_3 = \mathbf{B}/B$ is field line unit vector, $\hat{\mathbf{e}}_1$ is perpendicular to the magnetic field, pointing, for example, to the field line curvature direction: $\mathbf{e}_1 \parallel \hat{\mathbf{e}}_3 \cdot \nabla\hat{\mathbf{e}}_3$, and $\hat{\mathbf{e}}_2 = \hat{\mathbf{e}}_3 \times \hat{\mathbf{e}}_1$. Due to the basis ($\hat{\mathbf{e}}_1$, $\hat{\mathbf{e}}_2$, $\hat{\mathbf{e}}_3$) being tied to the magnetic field, one has $\hat{\mathbf{e}}_\alpha = \hat{\mathbf{e}}_\alpha(\mathbf{x}, t)$. It is defined that $g_{\alpha\beta}(\mathbf{x}, t) = \hat{\mathbf{e}}_\alpha \cdot \mathbf{e}_\beta$.

The kinetic equation (7.1) can be transformed to the ($\mathbf{x}$, $\hat{\mathbf{v}}'$, t) coordinate system with the velocity given in the center of perpendicular motion $\mathbf{v}' = \mathbf{v} - \mathbf{V}_\perp$,

$$
\begin{aligned}
&\frac{\hat{d}f}{dt} + g_{\beta\alpha}\hat{v}_\beta \frac{\partial f}{\partial x_\alpha}\bigg|_{\hat{\mathbf{v}},t} + \left(\frac{e}{m}E^*_\beta - \frac{\hat{d}g_{l\beta}\hat{V}_{\perp l}}{dt}\right) g_{r\beta}\frac{\partial f}{\partial \hat{v}_r}\bigg|_{\mathbf{x},t} \\
&= g_{p\beta}g_{r\gamma}\hat{v}_p \frac{\partial g_{l\gamma}\hat{V}_{\perp l}}{\partial x_\beta}\frac{\partial f}{\partial \hat{v}_r}\bigg|_{\mathbf{x},t} - \frac{e}{m}e_{\beta\gamma\alpha}g_{l\gamma}\hat{v}_l B_\alpha g_{r\beta}\frac{\partial f}{\partial \hat{v}_r}\bigg|_{\mathbf{x},t} \\
&- \left(g_{\gamma\delta}\hat{v}_\gamma \frac{\hat{d}g_{\beta\delta}}{dt} + g_{\delta\alpha}g_{\gamma\mu}\hat{v}_\delta\hat{v}_\gamma \frac{\partial g_{\beta\mu}}{\partial x_\alpha}\right)\frac{\partial f}{\partial \hat{v}_\beta}\bigg|_{\mathbf{x},t} + C,
\end{aligned}
\tag{7.2}
$$

where without causing ambiguity, the prime on $\hat{\mathbf{v}}$ has been dropped and

$$
E^*_\alpha = E_\alpha + e_{\alpha\gamma\delta}g_{q\gamma}\hat{V}_{\perp q}B_\delta.
$$

Defining the perpendicular density

$$
n(\mathbf{x}, \hat{v}_3, t) = \int\int d\hat{v}_1 d\hat{v}_2 f \tag{7.3}
$$

and the general perpendicular moment for $\eta(\mathbf{x}, \hat{\mathbf{v}}_\perp)$

$$
\langle\eta\rangle = \frac{1}{n}\int\int d\hat{v}_1 d\hat{v}_2 \eta f,
$$

the general moment equation can be constructed from the kinetic equation (7.2),

$$
\begin{aligned}
&\frac{\partial}{\partial t}(n\langle\eta\rangle) - n\left\langle\frac{\partial\eta}{\partial t}\right\rangle + \frac{\partial}{\partial x_\alpha}\left(n\left\langle g_{\beta\alpha}\hat{V}_{\perp\beta}\eta\right\rangle\right) - n\left\langle\frac{\partial}{\partial x_\alpha}\left(g_{\beta\alpha}\hat{V}_{\perp\beta}\eta\right)\right\rangle \\
&+ \frac{\partial}{\partial x_\alpha}\left(g_{\beta\alpha}n\left\langle\eta\hat{v}_\beta\right\rangle\right) - n\left\langle\hat{v}_\beta\frac{\partial}{\partial x_\alpha}(\eta g_{\beta\alpha})\right\rangle \\
&- \left(\frac{e}{m}E_\beta^* - \frac{\hat{d}g_{l\beta}\hat{V}_{\perp l}}{dt}\right)g_{r\beta}n\left\langle\frac{\partial\eta}{\partial\hat{v}_r}\right\rangle \\
&= -g_{p\beta}g_{r\gamma}\frac{\partial g_{l\gamma}\hat{V}_{\perp l}}{\partial x_\beta}n\left(\left\langle\hat{v}_p\frac{\partial\eta}{\partial\hat{v}_r}\right\rangle + \langle\eta\rangle\delta_{pr}\right) + \frac{e}{m}e_{\beta\gamma\alpha}g_{l\gamma}B_\alpha g_{r\beta}n\left\langle\frac{\partial\eta}{\partial\hat{v}_r}\hat{v}_l\right\rangle \\
&+ g_{p\delta}\frac{\hat{d}g_{\beta\delta}}{dt}n\left(\left\langle\hat{v}_p\frac{\partial\eta}{\partial\hat{v}_\beta}\right\rangle + \langle\eta\rangle\delta_{p\beta}\right) \\
&+ g_{p\alpha}g_{q\mu}\frac{\partial g_{\beta\mu}}{\partial x_\alpha}n\left(\left\langle\hat{v}_p\hat{v}_q\frac{\partial\eta}{\partial\hat{v}_\beta}\right\rangle + \left\langle\eta\hat{v}_q\right\rangle\delta_{p\beta}\right) \\
&- \left(\frac{e}{m}E_\beta^* - \frac{\hat{d}g_{l\beta}\hat{V}_{\perp l}}{dt}\right)g_{3\beta}\frac{\partial n\langle\eta\rangle}{\partial\hat{v}_3} + g_{p\beta}g_{3\gamma}\frac{\partial g_{l\gamma}\hat{V}_{\perp l}}{\partial x_\beta}\frac{\partial n\left\langle\eta\hat{v}_p\right\rangle}{\partial\hat{v}_3} \\
&- g_{p\delta}\frac{\hat{d}g_{3\delta}}{dt}\frac{\partial n\left\langle\eta\hat{v}_p\right\rangle}{\partial\hat{v}_3} - g_{p\alpha}g_{q\mu}\frac{\partial g_{3\mu}}{\partial x_\alpha}\frac{\partial n\left\langle\eta\hat{v}_p\hat{v}_q\right\rangle}{\partial\hat{v}_3} + \langle C\eta\rangle.
\end{aligned}
\tag{7.4}
$$

Note that only the average over the perpendicular velocity is carried out here. The moment from the collisional operator $\langle C\eta\rangle$ in equation (7.4) contains the collisional effects from the same species with different parallel velocities.

Similar to equations (3.94), (3.95), and (3.97) in section 3.5, the continuity, momentum, and energy transport equations can be obtained as follows.

$$
\begin{aligned}
&\underbrace{\frac{\partial n}{\partial t}}_{n_1} + \underbrace{g_{\beta\alpha}\hat{V}_{\perp\beta}\frac{\partial n}{\partial x_\alpha}}_{n_2} + \underbrace{g_{3\alpha}\hat{v}_3\frac{\partial n}{\partial x_\alpha}}_{n_3} + \underbrace{n\frac{\partial g_{l\gamma}\hat{V}_{\perp l}}{\partial x_\beta}\left(\delta_{\beta\gamma} - g_{3\beta}g_{3\gamma}\right)}_{n_4} \\
&+ \underbrace{n\hat{v}_3\frac{\partial g_{3\alpha}}{\partial x_\alpha}}_{n_5} + \underbrace{\frac{1}{m}m\frac{\partial g_{3\alpha}}{\partial x_\alpha}\frac{\partial nT}{\partial\hat{v}_3}}_{n_6} - \underbrace{g_{3\beta}g_{3\gamma}\frac{\partial g_{l\gamma}\hat{V}_{\perp l}}{\partial x_\beta}\hat{v}_3\frac{\partial n}{\partial\hat{v}_3}}_{n_7} \\
&= -g_{3\beta}\frac{\partial n}{\partial\hat{v}_3}\left(\underbrace{\frac{e}{m}E_\beta^*}_{n_8} - \underbrace{\frac{\hat{d}g_{l\beta}\hat{V}_{\perp l}}{dt}}_{n_9}\right) - \underbrace{\frac{1}{m}g_{p\alpha}g_{q\mu}\frac{\partial g_{3\mu}}{\partial x_\alpha}\frac{\partial\pi_{pq}}{\partial\hat{v}_3}}_{n_{10}} + \langle C\rangle,
\end{aligned}
\tag{7.5}
$$

$$
\begin{aligned}
&(\delta_{\lambda\alpha} - g_{3\alpha}g_{3\lambda})\Bigg(\underbrace{neE_\alpha^*}_{v_1} - \underbrace{\rho\frac{\hat{d}g_{l\alpha}\hat{V}_{\perp l}}{dt}}_{v_2} - \underbrace{\frac{\partial p}{\partial x_\alpha}}_{v_3} - \underbrace{\hat{v}_3 g_{3\beta}\frac{\partial g_{3\alpha}}{\partial x_\beta}\frac{\partial p}{\partial \hat{v}_3}}_{v_4} \\
&- \underbrace{g_{3\beta}\hat{v}_3^2\frac{\partial g_{3\alpha}}{\partial x_\beta}\rho}_{v_5} + \underbrace{g_{3\gamma}\frac{\partial g_{l\gamma}\hat{V}_{\perp l}}{\partial x_\alpha}\frac{\partial p}{\partial \hat{v}_3}}_{v_6} - \underbrace{\rho\hat{v}_3 g_{3\beta}\frac{\partial g_{l\alpha}\hat{V}_{\perp l}}{\partial x_\beta}}_{v_7} - \underbrace{\rho\hat{v}_3\frac{\hat{d}g_{3\alpha}}{dt}}_{v_8} - \underbrace{\frac{\hat{d}g_{3\alpha}}{dt}\frac{\partial p}{\partial \hat{v}_3}}_{v_9}\Bigg) \\
&= \underbrace{g_{\nu\lambda}g_{\beta\alpha}\frac{\partial \pi_{\beta\nu}}{\partial x_\alpha}}_{v_{10}} + g_{\nu\lambda}\Bigg(\underbrace{g_{p\delta}\frac{\hat{d}g_{3\delta}}{dt}}_{v_{11}} - \underbrace{g_{p\beta}g_{3\gamma}\frac{\partial g_{l\gamma}\hat{V}_{\perp l}}{\partial x_\beta}}_{v_{12}} + \underbrace{\hat{v}_3 g_{3\alpha}g_{p\mu}\frac{\partial g_{3\mu}}{\partial x_\alpha}}_{v_{13}}\Bigg)\frac{\partial \pi_{p\nu}}{\partial \hat{v}_3} \\
&+ \Bigg(\underbrace{\frac{\partial g_{p\alpha}g_{q\lambda}}{\partial x_\alpha}}_{v_{14}} + \underbrace{g_{p\alpha}g_{q\mu}g_{3\lambda}\frac{\partial g_{3\mu}}{\partial x_\alpha}}_{v_{15}} + \underbrace{g_{3\alpha}g_{q\mu}g_{p\lambda}\frac{\partial g_{3\mu}}{\partial x_\alpha}}_{v_{16}}\Bigg)\pi_{pq} \\
&+ \underbrace{g_{\nu\lambda}\frac{\partial g_{3\alpha}}{\partial x_\alpha}\frac{\partial q_\nu}{\partial \hat{v}_3}}_{v_{17}} + \underbrace{g_{p\alpha}g_{q\mu}g_{\nu\lambda}\frac{\partial g_{3\mu}}{\partial x_\alpha}\frac{\partial \Theta_{pq\nu}}{\partial \hat{v}_3}}_{v_{18}} + m\langle C\hat{v}_{\perp\lambda}\rangle,
\end{aligned}
\tag{7.6}
$$

and

$$
\begin{aligned}
&\underbrace{n\frac{\partial T}{\partial t}}_{t_1} + \underbrace{ng_{\beta\alpha}\hat{V}_{\perp\beta}\frac{\partial T}{\partial x_\alpha}}_{t_2} + \underbrace{ng_{3\alpha}\hat{v}_3\frac{\partial T}{\partial x_\alpha}}_{t_3} + \underbrace{p\frac{\partial g_{l\gamma}\hat{V}_{\perp l}}{\partial x_\beta}(\delta_{\gamma\beta} - g_{3\beta}g_{3\gamma})}_{t_4} \\
&+ \underbrace{\hat{v}_3\frac{\partial g_{3\alpha}}{\partial x_\alpha}p}_{t_5} + \underbrace{\frac{\partial g_{3\alpha}}{\partial x_\alpha}\frac{T^2}{m}\frac{\partial n}{\partial \hat{v}_3}}_{t_6} + \underbrace{\frac{\partial g_{3\alpha}}{\partial x_\alpha}\frac{3p}{m}\frac{\partial T}{\partial \hat{v}_3}}_{t_7} - \underbrace{\hat{v}_3 g_{3\beta}g_{3\gamma}\frac{\partial g_{l\gamma}\hat{V}_{\perp l}}{\partial x_\beta}n\frac{\partial T}{\partial \hat{v}_3}}_{t_8} \\
&= \underbrace{-g_{3\alpha}\left(\frac{e}{m}E_\alpha^* - \frac{\hat{d}g_{l\alpha}\hat{V}_{\perp l}}{dt}\right)n\frac{\partial T}{\partial \hat{v}_3}}_{t_9} + \underbrace{\frac{T}{m}g_{p\alpha}g_{q\mu}\frac{\partial g_{3\mu}}{\partial x_\alpha}\frac{\partial \pi_{pq}}{\partial \hat{v}_3}}_{t_{10}} \\
&- g_{p\alpha}g_{q\mu}\Bigg(\underbrace{\hat{v}_3\frac{\partial g_{3\mu}}{\partial x_\alpha}}_{t_{11}} + \underbrace{\frac{\partial g_{l\mu}\hat{V}_{\perp l}}{\partial x_\alpha}}_{t_{11}}\Bigg)\pi_{pq} + \Bigg(\underbrace{g_{p\beta}g_{3\gamma}\frac{\partial g_{l\gamma}\hat{V}_{\perp l}}{\partial x_\beta}}_{t_{13}} - \underbrace{g_{p\delta}\frac{\hat{d}g_{3\delta}}{dt}}_{t_{14}}\Bigg)\frac{\partial q_p}{\partial \hat{v}_3} \\
&- \underbrace{\frac{\partial}{\partial x_\alpha}(g_{\beta\alpha}q_\beta)}_{t_{15}} - \underbrace{\hat{v}_3 g_{3\alpha}g_{q\mu}\frac{\partial g_{3\mu}}{\partial x_\alpha}\frac{\partial q_q}{\partial \hat{v}_3}}_{t_{16}} - \underbrace{g_{3\alpha}g_{q\mu}\frac{\partial g_{3\mu}}{\partial x_\alpha}q_q}_{t_{17}} \\
&- \underbrace{g_{p\alpha}g_{q\mu}\frac{\partial g_{3\mu}}{\partial x_\alpha}\frac{1}{m}\frac{\partial n\Xi_{pq}}{\partial \hat{v}_3}}_{t_{18}} + \frac{m}{2}\left\langle C\hat{v}_{\perp\alpha}^2\right\rangle.
\end{aligned}
\tag{7.7}
$$

Because the collisional frequency is low, the collisions do not affect the lowest order equations that are used to derive the distribution function of first order. Therefore, the basic reductions of the kinetic equations of first order in (3.60) in section 3.3.1 remain valid. This yields

$$\begin{aligned} & -\frac{e}{m} e_{\beta\gamma\alpha} g_{l\gamma} \hat{v}_{\perp l} B_\alpha g_{r\beta} \frac{\partial f^1}{\partial \hat{v}_r}\bigg|_{t,\mathbf{x}} \\ & = g_{\beta\alpha} \hat{v}_{\perp\beta} \left(\frac{m\hat{v}_\mu^2}{2T} - 2 \right) \frac{\partial \ln T}{\partial x_\alpha} f^0 + \frac{m}{2T} \left(\hat{v}_{\perp\mu} \hat{v}_{\perp\nu} - \frac{1}{2} \hat{v}_\perp^2 \delta_{\mu\nu}^\perp \right) f^0 W_{\mu\nu} + C, \end{aligned} \tag{7.8}$$

Here, the collisional term has been added. Since the solution to f^1 is addable, the responses to the first two terms on the right-hand side of equation (7.8) remain unchanged as compared with the collisionless case in chapter 3. If further assuming that the collisional frequency C is negligible in equation (7.8), the collisional contributions to f^1 can be neglected. In this case, all transport coefficients obtained in chapter 3 become the complete solutions. The collisional effects only appear in the last terms, respectively, in the continuity, momentum, and energy transport equations (7.5)–(7.7).

To complete the basic set of equations, the following Maxwell equations are also required

$$\sum_a n_a e_a = 0, \tag{7.9}$$

$$\nabla \times \mathbf{E} = -\frac{\partial \mathbf{B}}{\partial t}, \tag{7.10}$$

$$\mu_0 \mathbf{J} = \nabla \times \mathbf{B}, \tag{7.11}$$

$$\nabla \cdot \mathbf{B} = 0. \tag{7.12}$$

Here, the Gauss equation has been replaced by the quasineutrality condition.

The basic theoretical framework described here will be used to study the collisional effects on the electrostatic and MHD modes in the next two sections.

7.2 Collisional effects on the electrostatic modes

In this section, the collisional effects on the electrostatic modes in the perpendicular magnetofluid description will be discussed.

The collisional frequency is assumed to be low. Therefore, the basic theoretical framework for the electrostatic modes in chapter 5 remains applicable. The key feature of the current formulation lies in that the moment is taken only in the perpendicular direction. One cannot use the particle number conservation to eliminate the collision contribution in the continuity equation. The like-particle collisions can affect particle trapping or de-trapping and, therefore, affect the electrostatic mode stability [6]. Therefore, one has $\langle C \rangle \neq 0$ in the continuity equation (7.5). This is different from the Braginskii two-fluid equations [7, 8], in which the collisional effects are completely wiped out in the continuity equation.

A complete description of the collisional operator in the perpendicular magnetofluid is beyond the current purpose. We just use the simplest Krook-model collision operator [4, 5] : $C(f) = -\nu f$ to explain how the collisional effects can modify the stability condition for electrostatic modes. Here, ν denotes the collision frequency.

The analyses are similar to the collisionless case in section 5.2. The finite Larmor radius effects related to the term 'n_{10}' in equation (7.5) are neglected. One can use the momentum equation (7.6) to express the perpendicular fluid velocity and use it to reduce the continuity equation (7.5). Consequently, one obtains

$$(\hat{\mathbf{v}}_3 \cdot \nabla_l - i\omega + i\bar{\omega}_d + \nu)\delta n_h = -i\frac{e}{T}n\left(\omega - \bar{\omega}_*^T\right)\delta\varphi, \tag{7.13}$$

where ν is the collisional frequency and

$$\delta n = -\frac{ne}{T}\delta\varphi + \delta n_h. \tag{7.14}$$

Here, it has been considered that the collisional operator annihilates the Maxwellian distribution function. Therefore, the adiabatic transform does not give rise to an extra term from the collisional operator.

Since equation (7.13) is similar to that in the electrostatic gyrokinetic description, the analyses for toroidal and slab branches of ion temperature gradient modes in reference [9], as also reviewed in reference [10], can be directly extended to the current perpendicular magnetofluid case. We do not further elaborate on them. Instead, we employ the WKB ansatz, i.e. $\nabla = i\mathbf{k}$ to explain the formulation and the underlying physics. Using the adiabatic transform in equation (7.14), the quasineutrality condition in equation (7.9) can be expressed as follows

$$\begin{aligned} \delta\varphi \sum_a \frac{e_a^2}{T_a} &= \sum_a e_a \frac{1}{n_0} \int_{-\infty}^{+\infty} d\hat{v}_3 n_h \\ &= -\delta\varphi \sum_a \frac{e_a^2}{T_a} \frac{1}{n_0} \int_{-\infty}^{+\infty} d\hat{v}_3 n \frac{\omega - \bar{\omega}_*^T}{k_\parallel \hat{v}_3 - \omega + \bar{\omega}_d - i\nu}. \end{aligned} \tag{7.15}$$

Therefore, one obtains the dispersion relation

$$\sum_a \frac{e_a^2}{T_a} + \sum_a \frac{e_a^2}{T_a}\left(\frac{m}{2\pi T_{\parallel a}}\right)^{1/2} \int_{-\infty}^{+\infty} d\hat{v}_3 \exp\left\{-\frac{m_a \hat{v}_3^2}{2T_{\parallel a}}\right\} \frac{\omega - \bar{\omega}_*^T}{k_\parallel \hat{v}_3 - \omega + \bar{\omega}_d - i\nu} = 0.$$

Here, one can see that the collisional effects, including the like-particle collisions, as well as the Landau damping, magnetic drift, diamagnetic drift, etc are all retained in the perpendicular magnetofluid theory. This is similar to the kinetic analyses in reference [11].

7.3 Collisional effects on the electromagnetic modes

In this section, we will discuss how to include the collisional effects to study the electromagnetic modes in the perpendicular magnetofluid theory. To be specific, the finite Larmor radius effects are excluded, but the parallel electric field effects from the wave-particle interaction are included. This will address the mode representation issue. Therefore, the formulation extends the collisionless theory in section 6.3.

Since the collisional frequency is assumed to be low, the basic theoretical formulation in the collisionless case in section 6.3 remains applicable in the current situation. After suitable ordering analyses as in equations (6.33) and (6.39) in section 6.3, we can start with the following reduced momentum and continuity equations:

$$
\begin{aligned}
&(\delta_{\lambda\alpha} - g_{3\alpha}g_{3\lambda})\Bigg[\underbrace{neE_\alpha + nee_{\alpha\beta\gamma}g_{l\beta}\hat{V}_l B_\gamma}_{v_1} - \underbrace{\rho\frac{\hat{d}g_{l\alpha}\hat{V}_{\perp l}}{dt}}_{v_2} \\
&- \underbrace{\frac{\partial p}{\partial x_\alpha}}_{v_3} - \underbrace{\hat{v}_3 g_{3\beta}\frac{\partial g_{3\alpha}}{\partial x_\beta}\frac{\partial p}{\partial \hat{v}_3}}_{v_4} - \underbrace{g_{3\beta}\hat{v}_3^2\frac{\partial g_{3\alpha}}{\partial x_\beta}\rho}_{v_5}\Bigg] = 0,
\end{aligned}
\tag{7.16}
$$

$$
\begin{aligned}
&\underbrace{\frac{\partial n}{\partial t}}_{n_1} + \underbrace{g_{\beta\alpha}\hat{V}_{\perp\beta}\frac{\partial n}{\partial x_\alpha}}_{n_2} + \underbrace{g_{3\alpha}\hat{v}_3\frac{\partial n}{\partial x_\alpha}}_{n_3} + \underbrace{n\frac{\partial g_{l\gamma}\hat{V}_{\perp l}}{\partial x_\beta}\left(\delta_{\beta\gamma} - g_{3\beta}g_{3\gamma}\right)}_{n_4} \\
&+ \underbrace{n\hat{v}_3\frac{\partial g_{3\alpha}}{\partial x_\alpha}}_{n_5} + \underbrace{\frac{T}{m}\frac{\partial g_{3\alpha}}{\partial x_\alpha}\frac{\partial n}{\partial \hat{v}_3}}_{n_6} - \underbrace{g_{3\beta}g_{3\gamma}\frac{\partial g_{l\gamma}\hat{V}_{\perp l}}{\partial x_\beta}\hat{v}_3\frac{\partial n}{\partial \hat{v}_3}}_{n_7} - \underbrace{\langle C\rangle}_{nc} \\
&= \underbrace{-g_{3\beta}\frac{\partial n}{\partial \hat{v}_3}\frac{e}{m}E_\beta}_{n_8}.
\end{aligned}
\tag{7.17}
$$

Here, the collisional effects in the continuity equation (7.17) are kept. They appear in the last term on the left and inside the parallel electric field effects on the right. Since the momentum equation (7.16) is in the perpendicular direction, the collisional effects are neglected.

To proceed, one needs to represent the electric and magnetic fields. Note that

$$\delta\mathbf{B} = \nabla \times \delta\mathbf{A}, \tag{7.18}$$

$$\delta\mathbf{E} = -\frac{\partial\delta\mathbf{A}}{\partial t} - \nabla\,\delta\varphi. \tag{7.19}$$

where $\delta\mathbf{A} = \xi_\perp \times \mathbf{B} + \delta A_\parallel \mathbf{b}$.

From the momentum equation (7.16), one can express the perpendicular velocity

$$\delta\mathbf{V}_\perp = \frac{\delta\mathbf{E}\times\mathbf{B}}{B^2} = -\omega\,\xi_\perp + \frac{\mathbf{B}\times\nabla\,\delta\varphi}{B^2}. \tag{7.20}$$

The equilibrium velocity can be neglected since the equilibrium electric field is assumed to be negligible and the FLR effects are neglected. As shown in section 6.3, the velocity induced by the parallel electric field φ can be neglected in the inertia term in the momentum equation. but needs to be kept in the continuity equation.

To study the resistive MHD modes, one needs the parallel momentum equation as well for constructing the generalized Ohm's law. One can obtain the parallel momentum equation by constructing the parallel velocity moment of the continuity equation (7.5). Alternatively, one can simply use conventional fluid equations since the total velocity integration is performed in this case. Their equivalence has been proved after the derivation of the general moment equation (3.23) in the perpendicular magnetofluid theory in section 3.2. Noting that the parallel current is contributed mostly by electrons, we simply use the electron momentum equation (2.15) in the Braginskii formulation for this purpose

$$\begin{aligned}\frac{m_e}{e_e}\frac{d_e V_{ea}}{dt} = &-\frac{1}{n_e e}\frac{\partial p_e}{\partial x_a} - \frac{1}{n_e e}\frac{\partial \pi_{ea\beta}}{\partial x_\beta} \\ &- E_a - e_{a\beta\gamma}V_{e\beta}B_\gamma + \frac{1}{n_e e}R_{ea},\end{aligned} \tag{7.21}$$

Neglecting the polarization current contribution on the left, the Hall term (the first term on the right), and the pressure tensor effects, equation (7.21) is reduced to

$$\mathbf{E} + \mathbf{V} \times \mathbf{B} = \eta \mathbf{j}_\parallel, \tag{7.22}$$

as in the resistive MHD theory orderings [12, 13]. Using Faraday's and Ampère's laws in equations (7.10) and (7.11), one obtains

$$\begin{aligned}\delta\mathbf{B} &= \frac{1}{i\omega}\nabla \times \delta\mathbf{E} = -\frac{1}{i\omega}\nabla \times (\mathbf{V} \times \mathbf{B}) + \frac{\eta}{i\omega}\nabla \times \delta\mathbf{j}_\parallel \\ &= \nabla \times (\xi_\perp \times \mathbf{B}) + \frac{\eta}{i\omega}\nabla \times \left(\frac{\mathbf{BB}}{\mu_0 B^2} \cdot \nabla \times \delta\mathbf{B}\right) \\ &= \nabla \times (\xi_\perp \times \mathbf{B}) - \frac{\eta}{i\omega}\mathbf{B} \times \nabla\left(\frac{\mathbf{B}}{\mu_0 B^2} \cdot \nabla \times \delta\mathbf{B}\right) \\ &= \nabla \times (\xi_\perp \times \mathbf{B}) - \frac{\eta}{i\omega}\mathbf{B} \times \left(\frac{\mathbf{B}}{\mu_0 B^2} \times \nabla \times \nabla \times \delta\mathbf{B}\right) \\ &= \nabla \times (\xi_\perp \times \mathbf{B}) - \frac{\eta}{i\mu_0\omega}\nabla^2\delta\mathbf{B}_\perp.\end{aligned} \tag{7.23}$$

Here, it is assumed that η is a constant and the mode wavelength is assumed to be much smaller than the equilibrium scale length. One can also express the parallel vector potential. Using equations (7.18) and (7.23), one obtains

$$\nabla \times (\delta A_\parallel \mathbf{b}) = -\frac{\eta}{i\mu_0\omega}\nabla^2\delta\mathbf{B}_\perp.$$

This can be solved

$$\delta A_{\parallel} = \frac{\eta}{i\mu_0\omega}\mathbf{b}\cdot\nabla\times\delta\mathbf{B}_{\perp}. \tag{7.24}$$

Here, the free integration constants can be absorbed into $\nabla\delta\varphi$.

Similar to equation (6.37) in section 6.3, the integration over $\hat{v}_3$ and the summation over particle species lead to the single fluid momentum equation

$$\rho_s\frac{d\mathbf{V}_{\perp}}{dt} = \mathbf{J}\times\mathbf{B} - \nabla_{\perp}p_s - \kappa\left(\left\langle mn\hat{v}_3^2\right\rangle_s - p_s\right), \tag{7.25}$$

where

$$\rho_s = \langle mn\rangle_s, \qquad p_s = \langle p\rangle_s, \qquad \mathbf{J} = \langle ne\mathbf{V}_{\perp}\rangle_s + \mathbf{J}_{\parallel},$$
$$\langle\cdots\rangle_s = \sum_a\left(\frac{m}{2\pi T}\right)^{1/2}\int_{-\infty}^{+\infty}d\hat{v}_3(\cdots).$$

To determine the pressure-related moments in equation (7.25), the perturbed temperature is required. The analyses of the energy equation are similar to the collisionless case in section 6.3. One has

$$\delta T = -\xi\cdot\nabla T. \tag{7.26}$$

The continuity equation is also needed. Note that

$$\mathbf{b}\cdot\nabla n = \delta\mathbf{b}\cdot\nabla n + \mathbf{b}\cdot\nabla\,\delta n,$$
$$\delta\mathbf{B}\cdot\nabla n = \mathbf{B}\cdot\nabla\left(\xi_{\perp}\cdot\nabla n\right). -\frac{\eta}{i\mu_0\omega}(\nabla^2\,\delta\mathbf{B}_{\perp})\cdot\nabla n.$$

Similar to the derivation of equation (6.40) in section 6.3, equation (7.5) is reduced to

$$\begin{aligned}
&-i\omega\left(\delta n + \xi_{\perp}\cdot\nabla n\right) + \hat{v}_3\mathbf{b}\cdot\nabla\delta n + \hat{v}_3\mathbf{b}\cdot\nabla\left(\xi_{\perp}\cdot\nabla n\right)\\
&+\delta n\hat{v}_3\frac{\partial g_{3\alpha}}{\partial x_{\alpha}} + \frac{T}{m}\frac{\partial g_{3\alpha}}{\partial x_{\alpha}}\frac{\partial\delta n}{\partial\hat{v}_3} - \langle C\rangle\\
&= 2in\omega\nabla\cdot\xi_{\perp} + 2i\omega n\,\kappa\cdot\xi - i\omega\frac{m\hat{v}_3^2}{T}n\,\kappa\cdot\xi\\
&-\frac{ne}{T}\hat{v}_3\mathbf{b}\cdot\nabla\delta\varphi + i\omega\frac{ne}{T}\hat{v}_3\delta A_{\parallel} + \hat{v}_3\frac{\eta}{i\mu_0\omega B}\left(\nabla^2\delta\mathbf{B}_{\perp}\right)\cdot\nabla n.
\end{aligned} \tag{7.27}$$

To extract the adiabatic part, letting

$$\delta n = -\frac{ne}{T}\delta\varphi + \delta n_h, \tag{7.28}$$

equation (7.27) is reduced to

$$\begin{aligned}
&-i\omega\left(\delta n_h+\xi_\perp\cdot\nabla n\right)+\hat{v}_3\mathbf{b}\cdot\nabla\delta n_h+\hat{v}_3\mathbf{b}\cdot\nabla\left(\xi_\perp\cdot\nabla n\right)\\
&+\delta n_h\hat{v}_3\frac{\partial g_{3\alpha}}{\partial x_\alpha}+\frac{T}{m}\frac{\partial g_{3\alpha}}{\partial x_\alpha}\frac{\partial\delta n_h}{\partial\hat{v}_3}-\langle C\rangle\\
&=2in\omega\nabla\cdot\ \xi_\perp+2i\omega n\ \kappa\cdot\ \xi-i\omega\frac{m\hat{v}_3^2}{T}n\ \kappa\cdot\ \xi\\
&-i\omega\frac{ne}{T}\delta\varphi+i\omega\frac{ne}{T}\hat{v}_3\delta A_\parallel+\hat{v}_3\frac{\eta}{i\mu_0\omega B}\left(\nabla^2\delta\mathbf{B}_\perp\right)\cdot\nabla n.
\end{aligned}$$

Here, it has been considered that the collisional operator annihilates the Maxwellian distribution function. Therefore, the adiabatic transform does not give rise to an extra term from the collisional operator. To extract the convective part, we let

$$\delta\tilde{n}_h=-\ \xi_\perp\cdot\nabla n+n\ \delta\tilde{n}_g. \tag{7.29}$$

One, then, obtains

$$\begin{aligned}
&-i\omega\ \delta\tilde{n}_g+\hat{v}_3\mathbf{b}\cdot\nabla\ |_{\hat{v}_3,t}\ \delta\tilde{n}_g+\frac{T}{m}\frac{\partial g_{3\alpha}}{\partial x_\alpha}\frac{\partial\delta\tilde{n}_g}{\partial\hat{v}_3}+\nu\delta\tilde{n}_g\\
&+\delta n_h\hat{v}_3\frac{\partial g_{3\alpha}}{\partial x_\alpha}+\frac{T}{m}\frac{\partial g_{3\alpha}}{\partial x_\alpha}\frac{\partial\delta n_h}{\partial\hat{v}_3}+\langle C\rangle\\
&=2in\omega\nabla\cdot\ \xi_\perp+2i\omega n\ \kappa\cdot\ \xi-i\omega\frac{m\hat{v}_3^2}{T}n\ \kappa\cdot\ \xi+\hat{v}_3\frac{\partial g_{3\alpha}}{\partial x_\alpha}\ \xi_\perp\cdot\nabla\ln T\\
&-i\omega\frac{ne}{T}\delta\varphi+i\omega\frac{ne}{T}\hat{v}_3\delta A_\parallel+\hat{v}_3\frac{\eta}{i\mu_0\omega B}\left(\nabla^2\delta\mathbf{B}_\perp\right)\cdot\nabla n.
\end{aligned} \tag{7.30}$$

This is a partial differential equation of first order. As with the electrostatic stability theory in chapter 5, one can find its characteristic:

$$\frac{T}{m}\int^{\hat{x}_3}\nabla\cdot\mathbf{b}\ d\hat{x}_3-\frac{\hat{v}_3^2}{2}=\iota, \tag{7.31}$$

equation (7.30) is reduced to

$$\begin{aligned}
&\hat{\mathbf{v}}_3\cdot\nabla_\iota\,\delta\tilde{n}_g,\ -i\omega\ \delta\tilde{n}_g+\nu\ \delta\tilde{n}_g\\
&=2in\omega\nabla\cdot\ \xi_\perp+2i\omega n\ \kappa\cdot\ \xi-i\omega\frac{m\hat{v}_3^2}{T}n\ \kappa\cdot\ \xi+\hat{v}_3\frac{\partial g_{3\alpha}}{\partial x_\alpha}\ \xi_\perp\cdot\nabla\ln T\\
&-i\omega\frac{ne}{T}\delta\varphi+i\omega\frac{ne}{T}\hat{v}_3\delta A_\parallel+\hat{v}_3\frac{\eta}{i\mu_0\omega B}\left(\nabla^2\delta\mathbf{B}_\perp\right)\cdot\nabla n.
\end{aligned} \tag{7.32}$$

where $\mathbf{b}\cdot\nabla_\iota$ is the derivative along the characteristic as explained in section 5.2. This transformation actually converts $\hat{\mathbf{v}}_3\cdot\nabla\ |_{\hat{v}_3,t}$, holding $\hat{v}_3$ constant, in the second term of equation (7.30) to $\hat{\mathbf{v}}_3\cdot\nabla\ |_{\varepsilon,\mu,t}$, roughly holding $(\varepsilon,\ \mu)$ constant, in the first term of equation (7.32) leading it to become gyrokinetic-theory-like. The only major difference lies in the average over the perpendicular velocity. Here, ν is the collisional frequency.

To complete the basic set of equations, the quasineutrality condition in equation (7.9) is required to determine the parallel electric potential $\delta\varphi$:

$$\sum_a e_a \frac{1}{n_0} \int_0^{+\infty} d\hat{v}_3 \left(\delta n_a^+ + \delta n_a^-\right) = 0.$$

Using the transforms in equations (7.28) and (7.29), the quasineutrality condition becomes

$$\delta\varphi \sum_a \frac{e_a^2}{T_a} = \sum_a e_a \frac{1}{n_0} \int_0^{+\infty} d\hat{v}_3 n \left(\delta\tilde{n}_{ga}^+ + \delta\tilde{n}_{ga}^-\right). \tag{7.33}$$

where the quasineutrality in equilibrium has been imposed.

We return to the momentum equation. Using equations (7.26), (7.28), and (7.29), one has

$$\delta p_s = -\,\xi \cdot \nabla p_s + \delta p_{gs}, \tag{7.34}$$

where the quasineutrality in equilibrium has been imposed and

$$\delta p_{gs} = \sum_a \frac{T_a}{n_0} \int_0^{+\infty} d\hat{v}_3 n \left(\delta\tilde{n}_g^+ + \delta\tilde{n}_g^-\right)_a.$$

Noting that

$$-\,\kappa\left(\left\langle m\delta n\hat{v}_3^2\right\rangle_s - \delta p_s\right) = -\,\kappa\left(\left\langle m\delta n_g\hat{v}_3^2\right\rangle_s - \delta p_{gs}\right),$$

the momentum equation (7.25) becomes

$$\begin{aligned} -\rho_s\omega^2\,\xi_\perp &= (\nabla\times\delta\mathbf{B})\times\mathbf{B} + \mathbf{J}\times\delta\mathbf{B} + \nabla_\perp(\xi_\perp\cdot\nabla p_s) - \nabla_\perp\delta p_{gs} \\ &\quad -\,\kappa\left(\frac{m}{2}\left\langle n\left(\delta\tilde{n}_g^+ + \delta\tilde{n}_g^-\right)\hat{v}_3^2\right\rangle_s - \delta p_{gs}\right). \end{aligned} \tag{7.35}$$

Equations (7.23), (7.24), (7.32), (7.33), and (7.35) constitute a complete set of equations to determine ξ, $\delta\varphi$, $\delta\mathbf{B}$, $\delta A_\parallel$, and $\delta\tilde{n}_g$ for determining the resistive MHD mode stability in the perpendicular magnetofluid description. Because $\nabla\cdot\delta\mathbf{B} = 0$ in equation (7.12), the Ohm's law in equation (7.23) can be used to determine the two remaining components of $\delta\mathbf{B}$ as functions of ξ as in the resistive MHD theory; while $\delta A_\parallel$ is related to $\delta\mathbf{B}$ by equation (7.33). Equation (7.32) can be used to determine ξ, while $\delta\varphi$ is governed by equation (7.33). With $\delta\tilde{n}_g$ given by the solution of equation (7.32), the system is completely determined. The resistive MHD theory in references [12, 13], as also reviewed in reference [10], can be directly extended to include the non-resistive-MHD effects with the current perpendicular magnetofluid theory. The current formulation can also be straightforwardly extended to include the FLR effects, etc as described in sections 6.4 and 6.5.

7.4 Conclusions and discussion

In this chapter, we have shown how the weak collision effects can be added in the perpendicular magnetofluid formalism. Both electrostatic and electromagnetic modes are studied.

For the electrostatic modes, one can see that the collisional effects are present in the continuity equation in the current perpendicular magnetofluid description. In particular, the like-particle collisional effects are included. This can be used to study the collisional effects on the trapped particle modes, for example.

For the resistive MHD modes, the parallel fluid momentum equation is required to construct the generalized Ohm's law. This can be obtained either by taking the parallel velocity moment over the continuity equation in the perpendicular magnetofluid description or by using the conventional parallel fluid momentum equation. The coupling of electrostatic modes can be included naturally. The parallel electric effects both from the resistive effects and from restoring the charge separation are taken into account.

From these analyses, one can see that the collisional effects can be included straightforwardly in the perpendicular magnetofluid theory. So far, only the Krook model is discussed. Further investigations with suitable collisional operators will be considered in the future.

References

[1] Zheng L 2020 Perpendicular magnetofluid theory for magnetically confined plasmas in the collisionless limit *Phys. Plasmas* **27** 052503

[2] Hazeltine R D 1976 Review of neoclassical transport theory *Advances in Plasma Physics* vol 6 ed A Simon and W B Thompson (New York: Wiley) p 273

[3] Helander P and Sigmar D 2002 *Collisional Transport in Magnetized Plasmas* (New York: Cambridge University Press)

[4] Hazeltine R D 1998 Transport theory in the collisionless limit *Phys. Plasmas* **5** 3282–6

[5] Bhatnagar P L, Gross E P and Krook M 1954 A model for collision processes in gases. I. Small amplitude processes in charged and neutral one-component systems *Phys. Rev.* **94** 511–25

[6] Tang W M 1978 Microinstability theory in tokamaks *Nucl. Fusion* **18** 1089–160

[7] Braginskii S I 1958 Transport phenomena in a complete ionized two-temperature plasma *ZhETF (J. Exp. Theor. Phys. USSR)* **33** 459
Braginskii S I 1958 *Soviet Phys. JETP* **6** 358–69

[8] Braginskii S I 1966 Transport processes *Reviews of Plasma Physics* vol 1 ed M Leontovich (New York: Consultants Bureau) 205–311

[9] Romanelli F, Chen L and Briguglio S 1991 Kinetic theory of the ion-temperature-gradient-driven mode in the long wavelength limit *Phys. Fluids* B **3** 2496–505

[10] Zheng L 2015 *Advanced Tokamak Stability Theory IOP Concise Physics* (San Rafael, CA: Morgan & Claypool)

[11] Tang W M 1978 Microinstability theory in tokamaks *Nucl. Fusion* **18** 1089–160
[12] Johnson J L and Greene J M 1967 Resistive interchanges and the negative V″criterion *Plasma Phys.* **9** 611
[13] Glasser A H, Greene J M and Johnson J L 1975 Resistive instabilities in general toroidal plasma configurations *Phys. Fluids* **18** 875–88

Part IV

Fluid-kinetic hybrid theories

IOP Publishing

Magnetically Confined Fusion Plasma Physics, Volume 2
Multifluid theory
Linjin Zheng

Chapter 8

Fluid-kinetic hybrid theories

After the perpendicular magnetofluid theory [1] is described in Part III, in this chapter, we describe the fluid-kinetic hybrid theories. The early efforts can be found, for example, in reference [2]. There are many further developments in this field, such as the Landau-fluid model [3, 4] and gyrofluid model [5–9], etc. The current moment approach for the hybrid model was also proposed in reference [10]. Readers are referred to other books for more extended reviews [11–15].

In this chapter, however, we only focus on the fundamental theoretical frameworks in the fluid-kinetic hybrid theories. The further kinetic effects, such as the finite Larmor radius ones, are excluded for brevity. In the fluid-kinetic hybrid theories, the conventional fluid closure with suitable transport coefficients is not sought for. Instead, the open chain of fluid moment equations is terminated by directly coupling suitable higher order moments with the kinetic equations. The kinetic equations, therefore, remain part of the complete set of equations in the fluid-kinetic hybrid models.

Two approaches are described: the conventional pressure moment approach, for example, reference [2], and the current moment approach [10]. In the conventional pressure moment approach, the lower order moment equations, such as the continuity and momentum equations, are constructed. The fluid equation chain is terminated by coupling the high order moments to the kinetic equations. In the conventional pressure moment approach, the current is determined by fluid velocity, while the velocity is determined by the momentum equation. The current moment approach is, however, related to a broader category of theories which rely directly on Maxwell's equations to construct the basic set of equations. The charge and current densities are evaluated with the solutions of kinetic equations in this category. The current moment approach in reference [10] exploits the fact that the gyrophase-dependent part of kinetic equations can be solved analytically using the inverse gyrofrequency expansion. By evaluating the perpendicular current density in Ampère's law, the counterpart to the perpendicular momentum equation is obtained. The current moment approach avoids the continuity and parallel

doi:10.1088/978-0-7503-3575-1ch8

momentum equations in the system and, therefore, gives more realistic continuum spectra. The pressure and current moment approaches will be compared and discussed at the end.

8.1 Pressure moment approach

In this section, we outline the conventional pressure–moment approach in the fluid-kinetic hybrid theory, for example, in reference [2]. In this approach, the lower order moment equations are coupled to the kinetic equations for the fluid equation closure.

The continuity and momentum equations are derived in the Braginskii two-fluid theory in chapter 6. From equations (2.13), (2.14) and equations (2.15), (2.16) in section 2.1.2 one obtains

$$\frac{\partial n_a}{\partial t} + \frac{\partial}{\partial x_\alpha}(n_a V_{a\alpha}) = 0, \tag{8.1}$$

$$n_a m_a \frac{d_a V_{a\alpha}}{dt} = -\frac{\partial P_{a\alpha\beta}}{\partial x_\beta} - n_a e_a (E_\alpha + \epsilon_{\alpha\beta\gamma} V_{a\beta} B_\gamma) + R_{a\alpha}, \tag{8.2}$$

where

$$P_{a\alpha\beta} = m_a \int v'_{a\alpha} v'_{a\beta} f_a \, d\mathbf{v}_a,$$

$$R_{a\alpha} = \sum_b R_{ab\alpha} = \sum_b \int m_a v'_{a\alpha} C_{ab} \, d\mathbf{v}_a,$$

$$\frac{d_a}{dt} \equiv \frac{\partial}{\partial t} + \mathbf{V}_a \cdot \nabla$$

with the particle species index $a = e, i$ for electrons and ions. One can sum together the electron and ion momentum equations to obtain a single fluid momentum equation:

$$n_i m_i \frac{d_i V_{i\alpha}}{dt} = -\frac{\partial P_{\alpha\beta}}{\partial x_\beta} + \epsilon_{\alpha\beta\gamma} J_\beta B_\gamma, \tag{8.3}$$

where $P_{\alpha\beta} = P_{i\alpha\beta} + P_{e\alpha\beta}$, J_β is the current density, and the electron inertia has been neglected by noting that $m_e \ll m_i$. The single fluid momentum equation (8.3) can be used to replace either the ion or electron momentum equation.

To constitute the complete set of equations, the Maxwell equations are also required.

$$\nabla \cdot \mathbf{E} = \frac{1}{\epsilon_0} \sum_{a=i,e} n_a e_a, \tag{8.4}$$

$$\nabla \times \mathbf{E} = -\frac{\partial \mathbf{B}}{\partial t}, \tag{8.5}$$

$$\nabla \cdot \mathbf{B} = 0, \tag{8.6}$$

$$\nabla \times \mathbf{B} = \mu_0 \sum_{a=i,e} n_a e_a \mathbf{V}_a. \tag{8.7}$$

The continuity equation (8.1) and momentum equation (8.2) can be used to determine the density n and velocity $\mathbf{V}$ supposing the pressure tensor $P_{a\alpha\beta}$ is known. This is because $\mathbf{E}$ and $\mathbf{B}$ in the momentum equations can be determined by the Maxwell equations with n and $\mathbf{V}$ being given. To compute the pressure tensor $P_{a\alpha\beta}$, the kinetic equations derived in section 2.1.2 in equation (2.82) are also required.

$$\begin{aligned} &\frac{d_a f_a}{dt} + v_{a\beta}\frac{\partial f_a}{\partial x_\beta} - \left(\frac{e_a}{m_a}E^*_{a\beta} - \frac{d_a V_{a\beta}}{dt}\right)\frac{\partial f_a}{\partial v_{a\beta}} \\ &= \frac{\partial V_{a\gamma}}{\partial x_\beta} v_{a\beta}\frac{\partial f_a}{\partial v_{a\gamma}} + \frac{e_a}{m_a} e_{\beta\gamma\alpha} v_{a\gamma} B_\alpha \frac{\partial f_a}{\partial v_{a\beta}} + \sum_b C_{ab}(f_a, f_b). \end{aligned} \tag{8.8}$$

where $\mathbf{v}$ is the relative velocity with respect to the velocity of mass center $\mathbf{V}_a$, and

$$\mathbf{E}^*_a \equiv \mathbf{E} + \mathbf{V}_a \times \mathbf{B}.$$

The set of equations (8.1)–(8.2) and (8.4)–(8.8) is complete to determine n_a, $\mathbf{V}_a$, $\mathbf{E}$, $\mathbf{B}$, and f_a. It replaces the usual set of fluid equations, such as the Braginskii two-fluid equations. Depending on how the kinetic equations are solved, for example using the guiding center approximation or gyrokinetic expansion, the finite Larmor radius effects are included suitably [7].

It is noted here that, by introducing the parallel fluid velocity in the conventional pressure–moment approach, the continuum spectra additional to the well-known shear Alfvén spectrum may be induced. For example, the continuity equation in the linearization gives rise to

$$\delta n_a = -i\frac{n\nabla \cdot \delta\mathbf{V}}{\omega - k_\parallel \mathbf{V}_\parallel}, \tag{8.9}$$

where $k_\parallel$ is the parallel wavenumber. There is a resonance condition here: $\omega - k_\parallel \mathbf{V}_\parallel = 0$. This leads to the continuum spectrum related to the continuity equation. Similarly, also because of the introduction of parallel fluid velocity, the parallel momentum equation can give rise to the sound wave resonances. This leads to the so-called sound spectrum, as shown in figure 8.1 [16].

These resonances result from introducing the parallel fluid velocity. Since the parallel fluid velocity is actually not an adequate quantity to describe the system. Charged particles in the magnetically confined fusion plasmas tend to be collisionless. One cannot expect the particles to move together as a fluid cell in the direction parallel to the magnetic field. Some particles are trapped, while some others remain in circulation. In the kinetic description, these resonances actually correspond to the wave-particle resonances, which are related to the well-known Landau damping. To avoid this, either the perpendicular magnetofluid theory in chapter 3 is invoked or

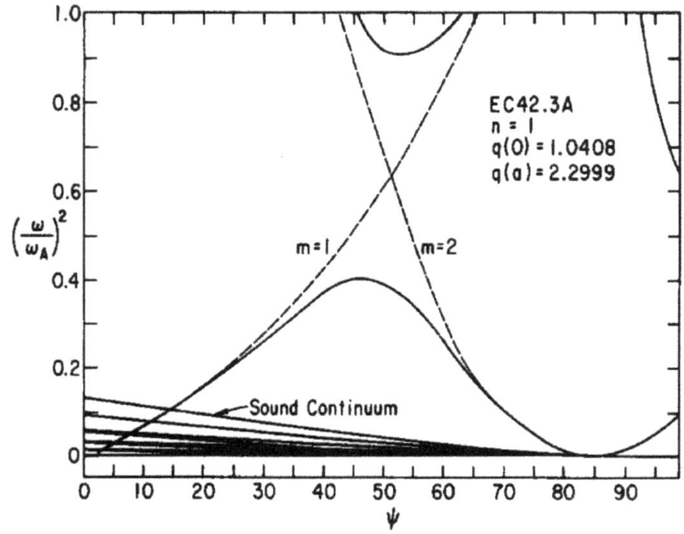

Figure 8.1. The continuum spectra for $n = 1$ modes in a tokamak with a circular cross section with the sound-wave spectrum included. (Reproduced with permission from reference [16]. Copyright 1986, AIP Publishing LLC.)

the current–moment approach is used. The current–moment approach will be reviewed in the next section.

8.2 Current moment approach

In this section, we describe the fluid-kinetic hybrid approach by constructing the current moment [10, 15]. We will see that the continuity and parallel momentum equations are avoided in the collisionless regime so that the related continuum spectrum issues as mentioned in section 8.1 are resolved. For simplicity, only the linear formulation is given. First, the ideal MHD formulation is described. The fluid-kinetic approach is given afterwards, which will be compared with the ideal MHD formulation. It is intended to describe the basic approach. Therefore, the subtle effects, such as the parallel electric field and FLR effects, are excluded.

8.2.1 Structure of linear ideal MHD equations

To explain the current moment approach for the fluid-kinetic hybrid formulation, we first review the structure of ideal MHD theory [17]. Its counterpart in the fluid-kinetic hybrid theory will be identified

We first discuss the ideal MHD equilibrium equations [17]. For simplicity, we consider the axisymmetric equilibria, in which the equilibrium magnetic field can be represented as

$$\mathbf{B} = \nabla\phi \times \nabla\psi + g\nabla\phi, \tag{8.10}$$

where ψ is the poloidal magnetic flux, g is a function of ψ, and ϕ is the toroidal axisymmetric angle.

The MHD equilibrium is governed by the force balance equation

$$\mathbf{J} \times \mathbf{B} = \nabla P. \tag{8.11}$$

It can be solved for the perpendicular current

$$\mathbf{J}_\perp = \frac{\mathbf{B} \times \nabla P}{B^2}. \tag{8.12}$$

The divergence-free of the current density can be used to determine the parallel current [18]

$$J_\parallel = -\frac{gP'}{B} - g'B, \tag{8.13}$$

where the prime denotes the derivative with respect to ψ.

The linearized ideal MHD equations are given as follows [17]:

$$-\rho_m \omega^2 \xi = \delta\mathbf{J} \times \mathbf{B} + \mathbf{J} \times \delta\mathbf{B} - \nabla\,\delta P, \tag{8.14}$$

$$\delta\mathbf{B} = \nabla \times \xi \times \mathbf{B}, \tag{8.15}$$

$$\mu_0 \delta\mathbf{J} = \nabla \times \delta\mathbf{B}, \tag{8.16}$$

$$\delta P = -\xi \cdot \nabla P - \Gamma P \nabla \cdot \xi, \tag{8.17}$$

where Γ is the ratio of specific heat and the time dependence of perturbed quantities is assumed to be of exponential type $\exp\{-i\omega t\}$ and $\xi = \delta\mathbf{V}/(-i\omega)$ represents the plasma displacement. One can obtain a single equation for ξ by inserting equations (8.15)–(8.17) into equation (8.14),

$$\begin{aligned} -\rho_m \omega^2 \xi = &\frac{1}{\mu_0}\nabla \times (\nabla \times \xi \times \mathbf{B}) \times \mathbf{B} + \mathbf{J} \times \nabla \times (\xi \times \mathbf{B}) \\ &+ \nabla(\xi \cdot \nabla P + \Gamma P \nabla \cdot \xi). \end{aligned} \tag{8.18}$$

There are three projections for the MHD equation (8.18), as reviewed in reference [15]. Three unit vectors: $\mathbf{e}_b = \mathbf{B}/B$, $\mathbf{e}_1 = \nabla\psi/|\nabla\psi|$, and $\mathbf{e}_2 = \mathbf{e}_b \times \mathbf{e}_1$ are introduced for these projections. The $\mathbf{e}_2$ projection of the MHD equation (8.18) gives

$$\begin{aligned} \mathbf{e}_1 \cdot \nabla \times \delta\mathbf{B} = &-\frac{gP'}{B^2}\mathbf{e}_1 \cdot \delta\mathbf{B} - g'\mathbf{e}_1 \cdot \delta\mathbf{B} + \frac{1}{B}\mathbf{e}_2 \cdot \nabla(|\nabla P|\mathbf{e}_1 \cdot \xi) \\ &+ \Gamma P \frac{1}{B}\mathbf{e}_2 \cdot \nabla(\nabla \cdot \xi) + \frac{\rho_m \omega^2}{B}\mathbf{e}_2 \cdot \xi. \end{aligned} \tag{8.19}$$

Similarly, the $\mathbf{e}_1$ projection yields

$$\mathbf{e}_2 \cdot \nabla \times \delta\mathbf{B} = -\frac{gP'}{B^2}\mathbf{e}_2 \cdot \delta\mathbf{B} - g'\mathbf{e}_2 \cdot \delta\mathbf{B} - \frac{P'|\nabla\psi|}{B^2}\mathbf{e}_b \cdot \delta\mathbf{B} - \frac{1}{B}\mathbf{e}_1 \cdot \nabla(|\nabla P|\mathbf{e}_1 \cdot \xi) - \Gamma P\frac{1}{B}\mathbf{e}_1 \cdot \nabla(\nabla \cdot \xi) - \frac{\rho_m\omega^2}{B}\mathbf{e}_1 \cdot \xi. \tag{8.20}$$

By applying one more operation $\mathbf{B} \cdot \nabla$ [19], the $\mathbf{e}_b$ projection of MHD equation (8.18) can be reduced to, using $\nabla \cdot \xi$ as an independent unknown,

$$\Gamma P\mathbf{B} \cdot \nabla\left(\frac{1}{B^2}\mathbf{B} \cdot \nabla\nabla \cdot \xi\right) + \rho_m\omega^2\nabla \cdot \xi = \rho_m\omega^2\nabla \cdot \xi_\perp. \tag{8.21}$$

Noting that $\delta\mathbf{J}$ and $\delta\mathbf{B}$ are determined completely by $\xi_\perp$, the set of equations (8.19)–(8.21) is complete to determine two components of $\xi_\perp$ and scalar unknown $\nabla \cdot \xi$.

Note that two perpendicular equations of motion, equations (8.19) and (8.20), result from perpendicular projections of MHD equation (8.14) and, therefore, contain the restoration force from the excitation of compressional Alfvén wave. To suppress the compressional Alfvén wave from consideration, one can apply the operator $\nabla \cdot (\mathbf{B}/B^2) \times (\cdots)$ on equation (8.14) to obtain the so-called vorticity equation

$$\nabla \cdot \frac{\mathbf{B}}{B^2} \times \rho_m\omega^2\,\xi = \mathbf{B} \cdot \nabla\frac{\mathbf{B} \cdot \delta\mathbf{J}}{B^2} + \delta\mathbf{B} \cdot \nabla\sigma - \mathbf{J} \cdot \nabla\frac{\mathbf{B} \cdot \delta\mathbf{B}}{B^2} + \nabla \times \frac{\mathbf{B}}{B^2} \cdot \nabla\delta P, \tag{8.22}$$

where $\sigma = \mathbf{J} \cdot \mathbf{B}/B^2$. Note that compressional Alfvén wave results from the term $\delta\mathbf{J} \times \mathbf{B} + \mathbf{J} \times \delta\mathbf{B} + \nabla\delta P \rightarrow \nabla(\mathbf{B} \cdot \delta\mathbf{B} + \delta P)$ in equation (8.14). Therefore, the curl operation in deriving equation (8.22) can suppress the compressional Alfvén wave.

Equations (8.22), (8.19) (or (8.20)), and (8.21) characterize, respectively, three fundamental MHD waves: the shear Alfvén, compressional Alfvén, and parallel acoustic waves. Note here that the left-hand sides of equations (8.19) and (8.20) represent the perpendicular projection of the current density $\mu_0\,\delta\mathbf{J} = \nabla \times \delta\mathbf{B}$. This motivates us to develop the fluid-kinetic hybrid theory by constructing the current moment.

8.2.2 Hybrid equations with the current moment approach

In this subsection, the gyrokinetic theory is used to recover the ideal MHD equation counterparts described in section 8.2.1. This shows that the fluid-kinetic hybrid equations can be obtained with the current moment approach [10, 15]. First, the gyrokinetic equation is briefly described. The equilibrium distribution function is then obtained. The gyrophase averaged and gyrophase dependent solutions are discussed afterwards. These solutions are used to compute the current moment in the end to formulate the fluid-kinetic hybrid equations.

Gyrokinetic equation

The gyrokinetic equation is briefly outlined in this subsection in order to describe the fluid-kinetic hybrid formulation [10, 20]. The distribution function F is decomposed to the equilibrium (F) and perturbed (δF) parts to linearize the Vlasov equation. The linearized Vlasov equations for equilibrium and perturbation become

$$\bar{\mathcal{L}}_v F = 0, \tag{8.23}$$

$$\mathcal{L}_v\, \delta F = -\frac{e}{m_\rho}\delta \mathbf{a} \cdot \nabla_v F, \tag{8.24}$$

where

$$\begin{aligned}
\bar{\mathcal{L}}_v &= \mathbf{v} \cdot \nabla_x + (e/m_\rho)\mathbf{v} \times \mathbf{B} \cdot \nabla_v, \\
\mathcal{L}_v &= -\, i\omega + \mathbf{v} \cdot \nabla_x + (e/m_\rho)\mathbf{v} \times \mathbf{B} \cdot \nabla_v, \\
\delta \mathbf{a} &= \delta \mathbf{E} + \mathbf{v} \times \delta \mathbf{B},
\end{aligned}$$

and $\delta\mathbf{E}$ is the perturbed electric field, e is the charge, m_ρ represents the mass, and ∇_x and ∇_v denote, respectively, the Laplace operators in the particle configuration $\mathbf{x}$ and velocity $\mathbf{v}$ spaces. Alternatively, the scalar $\delta\varphi$ and vector $\delta\mathbf{A}$ are introduced to represent the perturbed electromagnetic fields,

$$\delta\mathbf{E} = -\nabla\delta\varphi + i\omega\,\delta\mathbf{A}, \qquad \delta\mathbf{B} = \nabla \times \delta\mathbf{A}.$$

As in reference [20], we introduce the guiding center coordinates $(\mathbf{X}, \mathbf{V}) = (\mathbf{X}, \varepsilon, \mu, \alpha)$, where

$$\begin{aligned}
\mathbf{X} &= \mathbf{x} + (1/\Omega)\mathbf{v} \times \mathbf{e}_b, \\
\varepsilon &= v^2/2, \\
\mu &= \mu_0 + \tilde{\mu}_1 + \bar{\mu}_1, \qquad \mu_0 = v_\perp^2/2B, \\
\tilde{\mu}_1 &= -\, \mathbf{v}_d \cdot \mathbf{v}_\perp/B \\
&\quad -(v_\parallel/4\Omega B)(\mathbf{v}_\perp \times \mathbf{e}_b \cdot \nabla\mathbf{e}_b \cdot \mathbf{v}_\perp + \mathbf{v}_\perp \cdot \nabla\mathbf{e}_b \cdot \mathbf{v}_\perp \times \mathbf{e}_b), \\
\bar{\mu}_1 &= -\,(v_\parallel\mu_0/\Omega)\mathbf{e}_b \cdot \nabla \times \mathbf{e}_b, \\
\mathbf{v}_d &= \mathbf{e}_b \times \left[\left(v_\perp^2/2\Omega\right)\nabla \ln B + \left(v_\parallel^2/\Omega\right)\kappa\right],
\end{aligned}$$

α is the gyrophase, and $\Omega = eB/m_\rho$. The particle velocity is decomposed as

$$\mathbf{v}_\perp = \mathbf{e}_1 v_\perp \cos\alpha + \mathbf{e}_2 v_\perp \sin\alpha + \mathbf{e}_b v_\parallel.$$

In the guiding center coordinates, the equilibrium and perturbed Vlasov equations (8.23) and (8.24) become [20]

$$\bar{\mathcal{L}}_g F_g = 0, \tag{8.25}$$

$$\mathcal{L}_g \delta F_g = -\frac{e}{m_p}\left(\delta \mathbf{a} \cdot \nabla_V + \frac{1}{\Omega}\delta \mathbf{a} \times \mathbf{e}_b \cdot \nabla_X\right) F_g, \tag{8.26}$$

where

$$\begin{aligned}
\bar{\mathcal{L}}_g &= \dot{\mathbf{X}} \cdot \nabla_X + \dot{\alpha}\frac{\partial}{\partial \alpha}, \\
\mathcal{L}_g &= -i\omega + \dot{\mathbf{X}} \cdot \nabla_X + \dot{\alpha}\frac{\partial}{\partial \alpha}. \\
\dot{\mathbf{X}} &= v_{\|}\mathbf{e}_b + \mathbf{v}_D, \\
\mathbf{v}_D &= \mathbf{v} \cdot \mathbf{v} \times \nabla_x(\mathbf{e}_b/\Omega), \\
\dot{\alpha} &= -\Omega(\mathbf{X}) + \dot{\alpha}_1, \\
\dot{\alpha}_1 &= \mathbf{v} \cdot \nabla_x \alpha + (1/\Omega)\mathbf{v} \times \mathbf{e}_b \cdot \nabla_x \Omega, \\
\nabla_x \alpha &= (\nabla_x \mathbf{e}_2) \cdot \mathbf{e}_1 + \left(v_{\|}/v_{\perp}^2\right)\nabla_x \mathbf{e}_b \cdot (\mathbf{v}_\perp \times \mathbf{e}_b).
\end{aligned} \tag{8.27}$$

Here, the dot represents the derivative along the unperturbed particle orbit, In the $\dot{\alpha}_1$ equation above, the last term results from the guiding center transform of the gyrofrequency $\Omega(\mathbf{x})$ [10].

Equilibrium distribution function

As pointed out in reference [10], to recover the MHD equations the first order solution of the equilibrium distribution function is required. The solution as given in reference [18] is outlined as follows.

Corrected to order $\mathcal{O}(\rho/L_B)$, where ρ denotes Larmor radius and L_B is the scalelength of equilibrium magnetic field, the equilibrium equation (8.25) becomes

$$\left(v_{\|}\mathbf{e}_b \cdot \nabla_X - \Omega\frac{\partial}{\partial \alpha}\right) F_{g0} = 0. \tag{8.28}$$

This gives the lowest order equilibrium distribution function

$$F_{g0} = F_{g0}(\mathbf{X}_\perp, \mu, \varepsilon). \tag{8.29}$$

For simplicity, the isotropic Maxwellian distribution function is used as the lowest order solution

$$F_{g0}(\Psi, \varepsilon) = n_0(\Psi)\left(\frac{m_p}{2\pi T(\Psi)}\right)^{3/2} \exp\left\{-\frac{m_p \varepsilon}{T(\Psi)}\right\}, \tag{8.30}$$

where Ψ is the guiding center correspondent of the magnetic flux surface ψ.

After gyrophase average, the next order equilibrium equation (8.25) becomes

$$v_\parallel \mathbf{e}_b \cdot \nabla_X F_{g1} + \mathbf{v}_d \cdot \nabla_X F_{g0} = 0.$$

Noting that $\mathbf{v}_d \cdot \nabla\Psi = v_\parallel \mathbf{e_b} \cdot \nabla_X\left(v_\parallel m_p g/eB\right)$, one obtains

$$F_{g1} = -v_\parallel \frac{g}{\Omega}\frac{\partial F_{g0}}{\partial \Psi} + \mathrm{sign}(v_\parallel)\bar{F}_{g1}(\Psi, \mu, \varepsilon), \tag{8.31}$$

where $\bar{F}_{g1}$ is the integration constant.

As will be seen, one does not need to have the concrete formula for $\bar{F}_{g1}$ for recovering MHD equations from the kinetic equations. Only, a certain integration form is needed. Noting that F_{g1} does not contribute to the perpendicular current density due to being odd in $v_\parallel$, one has

$$\begin{aligned} \mathbf{J}_\perp &= \sum_{i,e} e \int d^3v \mathbf{v}_\perp F_{g0}(\Psi, \varepsilon) \\ &= \sum_{i,e} e \int d^3v \mathbf{v}_\perp \frac{1}{\Omega} \mathbf{v} \times \mathbf{e}_b \cdot \nabla F_{g0}(\Psi, \varepsilon) \\ &= \frac{\mathbf{B} \times \nabla P}{B^2}, \end{aligned} \tag{8.32}$$

where transforming from the guiding center to particle coordinates has been performed in the second step. Equation (8.32) recovers the MHD equilibrium result in equation (8.12). Noting also that F_{g0} does not contribute to the parallel current density due to being even in $v_\parallel$, one has

$$\begin{aligned} J_\parallel &= \sum_{i,e} e \int d^3v v_\parallel F_{g1}(\Psi, \varepsilon) \\ &= \sum_{i,e} e \int d^3v v_\parallel \left(-v_\parallel \frac{g}{\Omega}\frac{\partial F_{g0}}{\partial \Psi} + \mathrm{sign}(v_\parallel)\bar{F}_{g1}(\Psi, \mu, \varepsilon) \right) \\ &= -\frac{gP'}{B} - B \sum_{i,e} e \int d\mu \, d\varepsilon \bar{F}_{g1}(\Psi, \mu, \varepsilon). \end{aligned} \tag{8.33}$$

This is the so-called Pfirsch–Schlüter current. By comparing equation (8.33) with the ideal MHD result in equation (8.13), one obtains [18]

$$\sum_{i,e} e \int d\mu \, d\varepsilon \bar{F}_{g1}(\Psi, \mu, \varepsilon) = g', \tag{8.34}$$

This will be used to construct the fluid-kinetic hybrid equations with the current moment approach.

Linearized gyrokinetic equation

Now, we derive the perturbed linear gyrokinetic equation [10, 15]. Using the equilibrium distribution function in equations (8.30) and (8.31), one can reduce the linear gyrokinetic equation (8.26). For simplicity, we exclude the parallel electric field effects, i.e. letting $\delta\varphi = 0$. The resistivity effects are also dropped so that one can define $\delta\mathbf{A} = \xi \times \mathbf{B}$.

Note that

$$-\frac{e}{m}\delta\mathbf{a}\cdot\nabla_V F_{g0} = -\frac{e}{m}(-i\omega\delta\mathbf{A} + \mathbf{v}\times\delta\mathbf{B})\cdot\nabla_V F_{g0}$$
$$= -i\omega\frac{e}{m_\rho}\frac{\partial F_{g0}}{\partial\varepsilon}\mathbf{v}_\perp\cdot\delta\mathbf{A}(\mathbf{x}),$$

$$\begin{aligned}
&-\frac{e}{m}\delta\mathbf{a}\cdot\nabla_V F_{g1}\\
&= -\frac{e}{m}(-i\omega\delta\mathbf{A} + \mathbf{v}\times\delta\mathbf{B})\cdot\nabla_V\left(-v_\|\frac{g}{\Omega}\frac{\partial F_{g0}}{\partial\Psi} + \operatorname{sign}(v_\|)\bar{F}_{g1}(\Psi, \mu, \varepsilon)\right)\\
&= -i\omega\frac{e}{m_\rho}\left(\frac{\partial F_{g1}}{\partial\varepsilon} + \frac{1}{B}\frac{\partial F_{g1}}{\partial\mu}\right)\mathbf{v}_\perp\cdot\delta\mathbf{A}(\mathbf{x})\\
&\quad+\frac{e}{m_\rho}\left(\frac{g}{B}\frac{\partial F_{g0}}{\partial\Psi} + |v_\||\frac{e}{m_\rho B}\frac{\partial\bar{F}_{g1}}{\partial\mu}\right)\mathbf{e}_b\times\mathbf{v}_\perp\cdot\delta\mathbf{B}(\mathbf{x}),
\end{aligned}$$

and

$$\begin{aligned}
&-\frac{e}{m_\rho}\frac{1}{\Omega}\delta\mathbf{a}\times\mathbf{e}_b\cdot\nabla_X F_{g0}\\
&= -\frac{1}{B}[(-i\omega\delta\mathbf{A} + \mathbf{v}\times\delta\mathbf{B})\times\mathbf{e}_b]\cdot\nabla_X F_{g0}\\
&= -i\omega\frac{1}{B}\mathbf{e}_b\times\nabla_X F_{g0}\cdot\delta\mathbf{A}(\mathbf{x})\\
&\quad- v_\|\frac{1}{B}\delta\mathbf{B}(\mathbf{x})\cdot\nabla_X F_{g0} + \frac{1}{B}\mathbf{v}_\perp\cdot\nabla_X F_{g0}\mathbf{e}_b\cdot\delta\mathbf{B}(\mathbf{x})\\
&= -i\omega\,\xi(\mathbf{x})\cdot\nabla_X F_{g0}\\
&\quad- v_\|\frac{1}{B}\delta\mathbf{B}(\mathbf{x})\cdot\nabla_X F_{g0} + \frac{1}{B}\mathbf{v}_\perp\cdot\nabla_X F_{g0}\mathbf{e}_b\cdot\delta\mathbf{B}(\mathbf{x}).
\end{aligned}$$

Also, to extract the convective contribution, we define

$$\delta F = -\, \xi(\mathbf{X}) \cdot \nabla F_{g0} + \delta G(\mathbf{X}, \varepsilon, \mu, \alpha). \tag{8.35}$$

Therefore, the linear gyrokinetic equation (8.26) is reduced to

$$\mathcal{L}^1 \delta G - \Omega(\mathbf{X}) \frac{\partial \delta G}{\partial \alpha} = \mathcal{R}, \tag{8.36}$$

where $\mathcal{L}_g = -\Omega(\mathbf{X})(\partial/\partial\alpha) + \mathcal{L}^1$, and

$$\begin{aligned}
\mathcal{L}^1 &= -\, i\omega + \dot{\mathbf{X}} \cdot \nabla_X + \dot{\alpha}_1(\partial/\partial\alpha),\\
\mathcal{R} &= -\, \underbrace{i\omega \frac{e}{m_p} \frac{\partial F_{g0}}{\partial \varepsilon} \mathbf{v}_\perp \cdot \delta\mathbf{A}(\mathbf{x})}_{r1} + \underbrace{\frac{1}{B} \mathbf{e}_b \cdot \delta\mathbf{B}(\mathbf{x}) \mathbf{v}_\perp \cdot \nabla_X F_{g0}}_{r2}\\
&\quad + \underbrace{\frac{e}{m_p}\left(\frac{g}{B}\frac{\partial F_{g0}}{\partial \Psi} + |v_\parallel| \frac{1}{B}\frac{\partial \bar{F}_{g1}}{\partial \mu}\right) \mathbf{e}_b \times \mathbf{v}_\perp \cdot \delta\mathbf{B}(\mathbf{x})}_{r3}\\
&\quad - \underbrace{i\omega \frac{e}{m_p}\left(\frac{\partial F_{g1}}{\partial \varepsilon} + \frac{1}{B}\frac{\partial F_{g1}}{\partial \mu}\right) \mathbf{v}_\perp \cdot \delta\mathbf{A}(\mathbf{x})}_{r4}\\
&\quad + \underbrace{\mathbf{v}_D \cdot \nabla_X(\, \xi(\mathbf{X}) \cdot \nabla_X F_{g0})}_{r5} + \underbrace{i\omega(\, \xi(\mathbf{x}) - \, \xi(\mathbf{X})) \cdot \nabla_X F_{g0}}_{r6}\\
&\quad - \underbrace{v_\parallel \frac{1}{B} \delta\mathbf{B}(\mathbf{x}) \cdot \nabla_X F_{g0}}_{r7} + \underbrace{v_\parallel \mathbf{e}_b(\mathbf{x}) \cdot \nabla_X\left(\, \xi(\mathbf{X}) \cdot \nabla_X F_{g0}\right)}_{r8}.
\end{aligned} \tag{8.37}$$

It is noted here that

$$v_\parallel \frac{1}{B} \delta\mathbf{B}(\mathbf{x}) \cdot \nabla_X F_{g0} = v_\parallel \mathbf{e}_b(\mathbf{x}) \cdot \nabla_X\left(\, \xi(\mathbf{x}) \cdot \nabla F_{g0}\right) + \mathcal{O}(\rho/L_p).$$

This shows that there is cancellation between the terms '$r7$' and '$r8$'. Equation (8.36) will be used to construct the fluid-kinetic hybrid equations.

Fluid-kinetic hybrid equations

In this subsubsection, the linear gyrokinetic equation (8.37) is solved for calculating the current moment [10]:

$$\delta\mathbf{J} \cdot \mathbf{e}_{1,2} = \frac{1}{\mu_0} \nabla \times \delta\mathbf{B} \cdot \mathbf{e}_{1,2} = \sum_{i,e} e \int d^3v \mathbf{v}_\perp \cdot \mathbf{e}_{1,2} \delta F(\mathbf{X}) \Big|_{\mathbf{x}} \tag{8.38}$$

with $\delta F(\mathbf{X})$ given in equation (8.35). This will recover the ideal MHD counterparts in equations (8.19) and (8.20). It is noted that in the velocity space integration in equation (8.38) the particle coordinate $\mathbf{x}$ is held constant, instead of the guiding center coordinate $\mathbf{X}$. Since our purpose is to explain the current moment approach for the fluid-kinetic hybrid theory, the FLR effects are neglected for brevity. Higher order theory can be found in references [10] and [15].

We first solve for the contribution of the gyrophase dependent part of the distribution function: δG_1, Here, the subscript 1 represents the first harmonic in the Fourier component in the gyrophase decomposition. For the first harmonic, the gyrophase derivative term becomes dominant. Equation (8.36) then becomes

$$-\Omega(\mathbf{X})\frac{\partial \delta G}{\partial \alpha} = \mathcal{R}.$$

We can then work out the first harmonic contribution from $\mathcal{R}$ in equation (8.37) term by term.

For the term '$r1$', one can find the first harmonic solution

$$\delta G_{1,r1}(\mathbf{x}) = -i\omega\frac{1}{B}\frac{\partial F_{g0}}{\partial \varepsilon}\mathbf{e}_b \times \mathbf{v}_\perp \cdot \delta \mathbf{A}. \tag{8.39}$$

Using equation (8.38), one obtains the corresponding current density contribution

$$\delta j_{r1;\mathbf{e}_1} = \frac{\omega^2 n_0 m_p}{B}\mathbf{e}_2 \cdot \xi(\mathbf{x}),$$

$$\delta j_{r1;\mathbf{e}_2} = -\frac{\omega^2 n_0 m_p}{B}\mathbf{e}_1 \cdot \xi.$$

For the term '$r2$', one can find the first harmonic solution

$$\delta G_{1,r2}(\mathbf{x}) = -\frac{|\nabla\psi|}{B\Omega}\frac{\partial F_{g0}}{\partial \psi}v_\perp \sin\alpha \mathbf{e}_b \cdot \delta \mathbf{B}(\mathbf{x}).$$

Using equation (8.38), one obtains the corresponding current density contribution

$$\delta j_{r2;e_1} = 0,$$

$$\delta j_{r2;e_2} = -\frac{P'|\nabla\psi|}{B^2}\delta \mathbf{B} \cdot \mathbf{e}_b.$$

For the term '$r3$', one can find the first harmonic solution

$$\delta G_{1,r3}(\mathbf{x}) = -\frac{1}{\Omega}\left(\frac{g}{B}\frac{\partial F_{g0}}{\partial \psi} + |v_\parallel|\frac{e}{m_p B}\frac{\partial \bar{F}_{g1}}{\partial \mu}\right)\mathbf{v}_\perp \cdot \delta \mathbf{B}(\mathbf{x}).$$

Using equations (8.38) and (8.34), one obtains the corresponding current density contribution

$$\delta j_{e_1,e_2} = -\frac{gP'}{B^2}\mathbf{e}_{1,2} \cdot \delta \mathbf{B} - g'\mathbf{e}_{1,2} \cdot \delta \mathbf{B}.$$

The term '$r4$' contribution is negligible since it is odd in $\mathbf{v}_\parallel$. Contributions from terms '$r5$'–'$r8$' are negligible because there are cancellations between terms '$r5$' and '$r6$' and terms '$r7$' and '$r8$'.

One also needs to calculate the contribution of the convective term, i.e. the first term on the right-hand side of equation (8.35). Converting back to the particle coordinates, one obtains

$$\delta G_{1,\mathrm{conv}} = -\frac{v_\perp}{\Omega}(\mathbf{e}_1 \sin\alpha - \mathbf{e}_2 \cos\alpha) \cdot \nabla \xi \cdot \nabla F_{g0}.$$

Using equation (8.38), one obtains the corresponding current density contribution

$$\delta j_{1\mathrm{conv};\mathbf{e}_1} = \frac{1}{B}\mathbf{e}_2 \cdot \nabla \xi \cdot \nabla P,$$
$$\delta j_{1\mathrm{conv};\mathbf{e}_2} = -\frac{1}{B}\mathbf{e}_1 \cdot \nabla \xi \cdot \nabla P.$$

The contribution from the gyrophase averaged distribution function $\delta G_0(\mathbf{X})$ needs to be computed as well. Conversion to the particle coordinates is required. This yields

$$\sum_{i,e} e \int d^3v \mathbf{v}_\perp \cdot \mathbf{e}_{1,2}\delta G_0(\mathbf{X})\Big|_{\mathbf{x}} = \sum_{i,e} e \int d^3v \mathbf{v}_\perp \cdot \mathbf{e}_{1,2}\frac{1}{\Omega}\mathbf{v} \times \mathbf{e}_b \cdot \nabla\delta G_0(\mathbf{x})$$
$$= -\sum_{i,e} m_\rho \int d^3v \mu_0 \mathbf{e}_{2,1} \cdot \nabla\delta G_0(\mathbf{x}).$$

Collecting the results above, one finally obtains

$$\begin{aligned}&\mathbf{e}_1 \cdot \nabla \times \delta\mathbf{B}\\ &= -\frac{gP'}{B^2}\mathbf{e}_1 \cdot \delta\mathbf{B} - g'\mathbf{e}_1 \cdot \delta\mathbf{B} + \frac{1}{B}\mathbf{e}_2 \cdot \nabla(P'|\nabla\psi|\mathbf{e}_1 \cdot \xi)\\ &-\sum_{i,e} m_\rho \int d^3v \mu_0 \mathbf{e}_2 \cdot \nabla\delta G_0(\mathbf{x}) + \frac{\omega^2}{B}\rho_m \mathbf{e}_2 \cdot \xi,\end{aligned} \tag{8.40}$$

and

$$\begin{aligned}&\mathbf{e}_2 \cdot \nabla \times \delta\mathbf{B}\\ &= -\frac{gP'}{B^2}\mathbf{e}_2 \cdot \delta\mathbf{B} - g'\mathbf{e}_2 \cdot \delta\mathbf{B} - \frac{P'|\nabla\psi|}{B^2}\mathbf{e}_b \cdot \delta\mathbf{B}\\ &- \frac{1}{B}\mathbf{e}_1 \cdot \nabla(P'|\nabla\psi|\mathbf{e}_1 \cdot \xi)\\ &+\sum_{i,e} m_\rho \int d^3v \mu_0 \mathbf{e}_1 \cdot \nabla\delta G_0(\mathbf{x}) - \frac{\omega^2}{B}\rho_m \mathbf{e}_1 \cdot \xi.\end{aligned} \tag{8.41}$$

Similar to the ideal MHD theory, one can use $\nabla \cdot \delta\mathbf{j} = 0$ to construct the vorticity equation with $\delta\mathbf{j} = \nabla \times \delta\mathbf{B}$.

$$\nabla \cdot \frac{\mathbf{B}}{B^2} \times \rho_m \omega^2 \xi$$
$$= \mathbf{B} \cdot \nabla \frac{\mathbf{B} \cdot \delta \mathbf{J}}{B^2} + \delta \mathbf{B} \cdot \nabla \sigma - \mathbf{J} \cdot \nabla \frac{\mathbf{B} \cdot \delta \mathbf{B}}{B^2} - \nabla \times \frac{\mathbf{B}}{B^2} \cdot \nabla(\xi \cdot \nabla P) \tag{8.42}$$
$$+ \nabla \times \frac{\mathbf{B}}{B^2} \cdot \nabla \left(\sum_{i,e} \frac{m_\rho}{2} \int d^3 v v_\perp^2 \delta G_0(\mathbf{x}) \right) - i\omega \frac{|\nabla \psi|}{B \Omega_i} P_i' \nabla_\perp^2 (\mathbf{e}_1 \cdot \xi).$$

Comparing equations (8.40)–(8.42) with the MHD counterparts equations (8.19), (8.20), and (8.22), one can see that the ideal MHD equations are fully recovered, except for the plasma compressibility term relating to $-\Gamma \nabla \cdot \xi$. In the current fluid-kinetic description, this term needs to be computed kinetically, using the gyrophase-averaged distribution function, which will be discussed in the following subsubsection.

Gyrophase-averaged gyrokinetic equation

We have seen in the previous subsubsection that in the kinetic description the plasma compressibility term needs to be evaluated using the gyrophase-averaged distribution function δG_0. In this subsubsection the equation for δG_0 will be derived.

Neglecting the FLR effects, only the contributions from the term '$r1$', '$r7$', and '$r8$' need to be considered on the right-hand side of equation (8.36). For the term '$r1$', expansion to the first order needs to be considered in order to derive the equation. Noting that only the total μ is adiabatic constant in the required order, instead of μ_0, one has

$$\begin{aligned} \mathbf{v}_\perp \cdot \delta \mathbf{A}(\mathbf{x}) &= \mathbf{v}_\perp \cdot \xi(\mathbf{x}) \times \mathbf{B}(\mathbf{x}) \\ &= B(\mathbf{x}) v_\perp (-\mathbf{e}_1(\mathbf{x}) \sin \alpha + \mathbf{e}_2(\mathbf{x}) \cos \alpha) \cdot \xi(\mathbf{x}) \\ &= \sqrt{2} B^{3/2}(\mathbf{x}) \sqrt{\mu - \mu_1} (-\mathbf{e}_1(\mathbf{x}) \sin \alpha + \mathbf{e}_2(\mathbf{x}) \cos \alpha) \cdot \xi(\mathbf{x}) \\ &= \sqrt{2} \sqrt{\mu} B^{3/2}(\mathbf{x}) (-\mathbf{e}_1(\mathbf{x}) \sin \alpha + \mathbf{e}_2(\mathbf{x}) \cos \alpha) \cdot \xi(\mathbf{x}) \\ &\quad + \frac{1}{\sqrt{2\mu}} \mu_1 B^{3/2}(\mathbf{x}) (\mathbf{e}_1(\mathbf{x}) \sin \alpha - \mathbf{e}_2(\mathbf{x}) \cos \alpha) \cdot \xi(\mathbf{x}). \end{aligned}$$

Here, the first term is one order larger than the second term. The guiding center expansion for the first term is therefore required. Expansion for both equilibrium and perturbed quantities are required since $\nabla \cdot \xi \sim \kappa \cdot \xi$. For the second term, the effect from $-\mathbf{v}_d \cdot \mathbf{v}/B$ in μ_1 needs to be considered. Using these reductions, the gyrophase-average $\langle \cdots \rangle_\alpha$ of term '$r1$' in equation (8.37) can be obtained:

$$\left\langle - i\omega \frac{e}{m_\rho} \frac{\partial F_{g0}}{\partial \varepsilon} \mathbf{v}_\perp \cdot \delta \mathbf{A}(\mathbf{x}) \right\rangle_\alpha$$
$$= -i\omega \mu_0 B \frac{\partial F_{g0}}{\partial \varepsilon} \nabla_\perp \cdot \xi - i\omega \mu_0 \frac{\partial F_{g0}}{\partial \varepsilon} \xi \cdot \nabla B - i\omega \frac{\partial F_{g0}}{\partial \varepsilon} \left(\mu_0 B - \frac{v_\parallel^2}{2} \right) \kappa \cdot \xi.$$

One needs also to reduce the terms '$r7$' and '$r8$' in equation (8.39).

$$
\begin{aligned}
& - v_{\parallel}\frac{1}{B}\delta\mathbf{B}(\mathbf{x}) \cdot \nabla_X F_{g0} + v_{\parallel}\mathbf{e}_b(\mathbf{x}) \cdot \nabla_X\big(\ \xi(\mathbf{X}) \cdot \nabla_X F_{g0}\big) \\
&= - v_{\parallel}\frac{1}{B}\nabla \times (\ \xi \times \mathbf{B}) \cdot \nabla_X F_{g0} + v_{\parallel}\mathbf{e}_b(\mathbf{x}) \cdot \nabla_X\big(\ \xi(\mathbf{X}) \cdot \nabla_X F_{g0}\big)\Big|_{\varepsilon,\mu} \\
&= - v_{\parallel}\mathbf{e}_b(\mathbf{x}) \cdot \nabla_X\big(\ \xi \cdot \nabla_X F_{g0}\big)\Big|_{\mathbf{v}} + v_{\parallel}\mathbf{e}_b(\mathbf{x}) \cdot \nabla_X\big(\ \xi(\mathbf{X}) \cdot \nabla_X F_{g0}\big)\Big|_{\varepsilon,\mu} \\
&= - v_{\parallel}b_i\frac{\partial(\ \xi \cdot \nabla_X F_{g0})}{\partial v_j}\frac{\partial v_j}{\partial X_i}\Bigg|_{\varepsilon,\mu}.
\end{aligned}
\tag{8.43}
$$

There is one more contribution in the same order from the left-hand side of equation (8.36). Noting that $\delta G_{1,r1}$ in equation (8.39) can couple to the $\dot{\alpha}_1(\partial/\partial\alpha)$ term in equation (8.36), one obtains

$$
\begin{aligned}
& - \left\langle \left(\mathbf{v} \cdot \nabla_x\alpha + \frac{1}{\Omega}\mathbf{v} \times \mathbf{e}_b \cdot \nabla_x\Omega\right)\frac{\partial\delta G_{1,r1}}{\partial\alpha}\right\rangle_{\alpha} \\
&= i\omega\mu_0\frac{\partial F_{g0}}{\partial\varepsilon}\ \xi \cdot \nabla B + i\omega\frac{\partial F_{g0}}{\partial\varepsilon}\frac{v_{\parallel}^2}{2}\ \kappa \cdot \ \xi.
\end{aligned}
$$

Collecting the results above, one obtains the gyrophase-averaged gyrokinetic equation

$$
\begin{aligned}
& \big(\mathbf{v}_{\parallel} \cdot \nabla - i\omega + i\omega_d\big)\delta G_0(\mathbf{X}) \\
&= - i\omega\mu_0 B\frac{\partial F_{g0}}{\partial\varepsilon}\nabla_{\perp} \cdot \ \xi - i\omega\frac{\partial F_{g0}}{\partial\varepsilon}\big(\mu_0 B - v_{\parallel}^2\big)\ \kappa \cdot \ \xi \\
& - v_{\parallel}b_i\frac{\partial(\ \xi \cdot \nabla_X F_{g0})}{\partial v_j}\frac{\partial v_j}{\partial X_i}\Bigg|_{\varepsilon,\mu},
\end{aligned}
\tag{8.44}
$$

where $\omega_d = -i\mathbf{v}_d \cdot \nabla$. Equation (8.44) can be used to construct the equation for the even part of the distribution function: $\delta G_0^e = (\delta G_0(v_{\parallel}) + \delta G_0(-v_{\parallel}))/2$.

$$
\begin{aligned}
& v_{\parallel}\mathbf{e}_b \cdot \nabla\left[\frac{v_{\parallel}}{i(\omega - \omega_d)}\mathbf{e}_b \cdot \nabla\delta G_0^e\right] - i(\omega - \omega_d)\delta G_0^e \\
&= - i\omega\mu_0 B\frac{\partial F_{g0}}{\partial\varepsilon}\nabla_{\perp} \cdot \ \xi - i\omega\frac{\partial F_{g0}}{\partial\varepsilon}\big(\mu_0 B - v_{\parallel}^2\big)\ \kappa \cdot \ \xi \\
& - v_{\parallel}\mathbf{e}_b \cdot \nabla\left[\frac{v_{\parallel}}{i(\omega - \omega_d)}b_i\frac{\partial(\ \xi \cdot \nabla_X F_{g0})}{\partial v_j}\frac{\partial v_j}{\partial X_i}\Bigg|_{\varepsilon,\mu}\right].
\end{aligned}
\tag{8.45}
$$

Equation (8.45) can be solved and the solution can be inserted into equations (8.40) and (8.41) (or (8.42)) to get the eigenmode equation.

Discussion

We have derived the complete set of equations, (8.40), (8.41) (or (8.42)), and equation (8.45), using the current moment approach. Calculating the perpendicular current density in equations (8.40) and (8.41) can be also viewed as calculating the perpendicular fluid velocity. When the parallel electric field effects are intended to be studied, the quasineutrality condition needs to be taken into consideration for determining $\delta\varphi$. The resistivity effects can also be included as in this formalism by calculating the parallel current as well with $\delta \mathbf{A}_{\|}$ included in the mode representation. To include the FLR effects, one needs to solve the gyrokinetic equations to higher orders [10, 15].

Comparing equations (8.40), (8.41) (or (8.42)), and equation (8.45) with the MHD counterparts equations (8.19), (8.20) (or (8.22)), and (8.21), one can see that the ideal MHD momentum equations are fully recovered, except for the plasma compressibility term relating to $-\Gamma\nabla \cdot \xi$. In the current fluid-kinetic description, this term needs to be computed kinetically using the gyrophase-averaged distribution function.

Comparing the gyrophase-averaged gyrokinetic equation (8.45) with the MHD equation of parallel motion, equation (8.21), one can see the structure similarity and differences. The MHD fluid resonance as the left-hand side of equation (8.21) vanishes becomes the wave-particle resonance in the kinetic description in equation (8.45). Besides, the ideal MHD equation of parallel motion, equation (8.21), is given along the magnetic field lines. Instead, the kinetic counterpart, equation (8.45), needs to be solved along the characteristic curve that preserves the adiabatic invariants. The last term in equation (8.45) is new in structure as compared with the MHD equation (8.21). This exactly arises from the integration path difference.

Comparing equations (8.40), (8.41) (or (8.42)), and equation (8.45) with the counterparts in the perpendicular magnetofluid theory, equations (6.30) and (6.29), one can see the structural similarity. The continuity equation (6.29) corresponds to the gyrophase-averaged gyrokinetic equation (8.45), except the average over perpendicular velocity has been made in the continuity equation in the perpendicular magnetofluid theory.

8.3 Conclusions and discussion

In this chapter, we briefly discuss the fluid-kinetic hybrid theories. Two types of approaches are discussed: the pressure and the current moment approaches. Most of the previous hybrid formulations follow the pressure moment approach. The current moment approach is a practical application for the more general one based on the Maxwell equations with charge and current density computed kinetically.

In principle, in the pressure moment approach, the particle and momentum conservations are more apparent. In the current approach, the charge and current densities are focused on. The continuity and parallel momentum equations are absent in the current moment approach. The parallel fluid velocity is not used in the current moment approach in the collisionless regime. This avoids the fluid continuum spectra related to the continuity and parallel momentum equations in the pressure moment approach. They are replaced by the more realistic wave-particle resonances or Landau damping in the current moment approach.

References

[1] Zheng L 2020 Perpendicular magnetofluid theory for magnetically confined plasmas in the collisionless limit *Phys. Plasmas* **27** 052503

[2] Kulsrud R M 1983 MHD description of plasma *Basic Plasma Physics I (Handbook of Plasma Physics)* ed A A Galeev and R N Sudan (Amsterdam: North-Holland) 1–31 pp

[3] Hammet G W and Perkins F W 1990 Fluid moment models for Landau damping with application to the ion-temperature-gradient instability *Phys. Rev. Lett.* **64** 3019–22

[4] Snyder P B and Hammett G W 2001 A Landau fluid model for electromagnetic plasma microturbulence *Phys. Plasmas* **8** 3199–216

[5] Dorland W and Hammett G W 1993 Gyrofluid turbulence models with kinetic effects *Phys. Fluids* **5** 812–35

[6] Waltz R E, Kerbel G D and Milovich J 1994 Toroidal gyro-Landau fluid model turbulence simulations in a nonlinear ballooning mode representation with radial modes *Phys. Plasmas* **1** 2229–44

[7] Ramos J J 2008 Finite-Larmor-radius kinetic theory of a magnetized plasma in the macroscopic flow reference frame *Phys. Plasmas* **15** 082106

[8] Waelbroeck F L, Hazeltine H D and Morrison P J 2009 A Hamiltonian electromagnetic gyrofluid model *Phys. Plasmas* **16** 032109

[9] Keramidas Charidakos I, Waelbroeck F L and Morrison P J 2015 A Hamiltonian five-field gyrofluid model *Phys. Plasmas* **22** 112113

[10] Zheng L J, Kotschenreuther M T and Van Dam J W 2007 Revisiting linear gyrokinetics to recover ideal magnetohydrodynamics and missing finite Larmor radius effects *Phys. Plasmas* **14** 072505

[11] Hazeltine R D and Meiss J D 1992 *Plasma Confinement* (New York: Addison-Wesley)

[12] Hazeltine R D and Waelbroeck F L 2018 *The Framework of Plasma Physics* (Boca Raton, FL: CRC Press)

[13] Weiland J 2000 *Collective Modes in Inhomogeneous Plasma* (Bristol: Institute of Physics Publishing)

[14] Helander P and Sigmar D 2002 *Collisional Transport in Magnetized Plasmas* (New York: Cambridge University Press)

[15] Zheng Linjin 2015 *Advanced Tokamak Stability Theory IOP Concise Physics* (San Rafael, CA: Morgan & Claypool)

[16] Cheng C Z and Chance M S 1986 Low-*n* shear Alfvén spectra in axisymmetric toroidal plasmas *Phys. Fluids* **29** 3695–701

[17] Freidberg J P 1987 *Ideal Magnetohydrodynamics* (New York: Plenum Press)

[18] Hazeltine R D 1976 Review of neoclassical transport theory*Advances in Plasma Physics* vol 6 ed A Simon and W B Thompson (New York: Wiley) 273 p

[19] Correa-Restrepo D 1982 Resistive ballooning modes in three-dimensional configurations *Z. Naturforsch. A* **37** 848

[20] Catto P J, Tang W M and Baldwin D E 1981 Generalized gyrokinetics *Plasma Phys.* **23** 639

Part V

Conclusions

IOP Publishing

Magnetically Confined Fusion Plasma Physics, Volume 2
Multifluid theory
Linjin Zheng

Chapter 9

Concluding remarks

When I consider your heavens, the work of your fingers, the Moon and the stars, which you have set in place, what is man that you are mindful of him, the son of man that you care for him? (Psalms 8:3,4)

The universe is so vast and full of mystery. Human beings are so little and so insignificant. Nevertheless, we were endowed with enthusiasm, interest, and understanding to explore it. Controlling nuclear fusion as an energy source is so challenging that it has been observed that the predicted number of years required to achieve it remains constant over time. Nonetheless, the hope to realize it will never die and the efforts have been and will be rewarded, not because we are so great, but because when there is a law behind it, there should be a way to it.

Thanks to the efforts of generations, our understanding of plasma physics is greater than ever. It is the deeper and deeper understanding of the underlying physics that will lead eventually to success. The purpose of this book series is to summarize and to explain in further detail the theoretical results obtained in this field. Since so many efforts have been made in this field, the scope the current books intend to cover may still be limited. There are many excellent books in this field both in the introductory and advanced levels. The current book series is intended to bridge between the textbooks and the up-to-date research developments.

The current volume focuses on the multi-fluid theories. Three major parts are presented: the Braginskii transport theory, the perpendicular magnetofluid theory, and the fluid-kinetic hybrid theories.

Braginskii's transport theory is the most comprehensive and widely used fluid theory in this field. Because the collisionality is assumed to be dominant, Braginskii's fluid theory has limitations for fusion plasma physics application. Although the collisionless gyroviscosity and thermal conductivity are derived in the Braginskii transport theory, the Braginskii two-fluid theory is nonetheless not adequate to describe the magnetically confined fusion plasmas in the collisionless limit. For example, the parallel fluid velocity, which appears in the Braginskii gyroviscous

doi:10.1088/978-0-7503-3575-1ch9

tensor, is not an adequate quantity in lowest order to describe the magnetically confined fusion plasmas. Fusion plasmas turn out to be collisionless. Some particles are trapped by the equilibrium magnetic field or by the waves, while some others remain in circulation. One cannot expect the charged particles to move together as a fluid cell along the magnetic field lines.

In order to unveil the mobility features of charged particles along the field lines, the perpendicular magnetofluid theory is also described in this book. The magnetically confined fusion plasmas are highly anisotropic. The strong magnetic field leads the charged particles to gyrate around the magnetic field lines. One can exploit the particle localization by the magnetic field in the perpendicular direction even in the collisionless limit. This leads to the development of perpendicular magnetofluid theory by constructing only the perpendicular moment equations. The particle mobility features are fully retained in this formalism, such as the trapped particle, wave-particle resonance, the parallel electric field effects, etc and the perpendicular magnetofluid represents a direct extension of Braginskii's two-fluid theory to the collisionless limit.

The fluid-kinetic hybrid models are also described in this book. The conventional kinetic pressure closure, as well as the kinetic current moment procedure, are described. The conventional approach based on the coupling of the continuity and momentum equations with the kinetic equation for evaluating the pressure is pointed out to lead to additional continuum spectra, which are otherwise absent in the kinetic theory. The current moment approach is shown to avert the continuum spectrum issue since the continuity and the parallel momentum equations are avoided. The current moment approach is related to the more general closure procedure by directly employing the Maxwell equations with the charge and current densities evaluated kinetically. This will be further addressed in the future in the book for kinetic theory.

The value of fluid description lies not only in its simplicity but also in that the thermodynamics is embodied in the many-particle system. One cannot rely completely on the kinetic theory. A simple reason is that the definite solution conditions, the initial and boundary conditions, for kinetic equations cannot be specified kinetically. One is unable to give the phase space coordinates for all particles initially or on the boundary. The kinetic equations are practically solved with the definite solution conditions that assume the existence of thermal states. Also, the plasma diagnostics actually measure the thermal parameters instead of the phase space positions for all particles.

With ITER [1] to come in the near future and the computation tools evolving yearly, we hope the theoretical frameworks described in this book help further scientific endeavor in this field.

Reference

[1] Ikeda K 2007 Progress in the ITER physics basis *Nucl. Fusion* **47** 1

Appendix A

Tensor index notations and basic MHD equilibrium equalities

In plasma physics research, one often deals with complicated three-dimensional geometry. Here, we review some practical tensor algebra for this purpose.

For brevity, we introduce some notations. For partial derivatives, it is denoted that $\partial_\alpha = \partial/\partial x_\alpha$, $\partial_t = \partial/\partial t$, and $\partial_{v_\alpha} = \partial/\partial v_\alpha$. The Kronecker delta is introduced as follows

$$\delta_{ij} = \begin{cases} 1, & \text{if } i = j; \\ 0, & \text{otherwise.} \end{cases}$$

The Levi-Civita symbol or the alternating tensor is defined as

$$e_{ijk} = \begin{cases} 1, & \text{if } i, j, k = 123, 231, 312; \\ -1, & \text{if } i, j, k = 132, 213, 321; \\ 0, & \text{otherwise.} \end{cases}$$

One can directly prove some basic identities:

$$e_{ijk} = e_{kij}, \qquad e_{ijk} = -e_{ikj}, \qquad e_{ijj} = 0.$$

Using the alternative tensor, the cross product and curl operator can be expressed in the component form, for example,

$$\mathbf{a} \times \mathbf{b} \to e_{ijk} a_j b_k, \qquad \nabla \times \mathbf{a} \to e_{ijk} \partial_j a_k,$$

where the Einstein summation convention for two repeated indices is implied.

There is an important equality for alternating tensor:

$$e_{ijk} e_{ilm} = \delta_{jl} \delta_{km} - \delta_{jm} \delta_{kl}. \tag{A.1}$$

doi:10.1088/978-0-7503-3575-1ch10

This can be simply verified by noting that

$$
\begin{aligned}
(\mathbf{a} \times \mathbf{b}) \cdot (\mathbf{c} \times \mathbf{d}) &\rightarrow e_{ijk} e_{ilm} a_j b_k c_l d_m \\
&= \left(\delta_{jl}\delta_{km} - \delta_{jm}\delta_{kl}\right) a_j b_k c_l d_m \\
&= a_j c_j b_k d_k - a_j d_j b_k c_k \\
&\rightarrow (\mathbf{a} \cdot \mathbf{c})(\mathbf{b} \cdot \mathbf{d}) - (\mathbf{a} \cdot \mathbf{d})(\mathbf{b} \cdot \mathbf{c}).
\end{aligned}
$$

An application example of (A.1) is given as follows.

$$
\begin{aligned}
\nabla \times \nabla \times \mathbf{a} &\rightarrow e_{lmi} \partial_m \left(e_{ijk} \partial_j a_k\right) \\
&= \left(\delta_{jl}\delta_{km} - \delta_{jm}\delta_{kl}\right) \partial_m \partial_j a_k \\
&= \partial_l \partial_k a_k - \partial_m \partial_m a_l \\
&\rightarrow \nabla(\nabla \cdot \mathbf{a}) - \nabla^2 \mathbf{a}.
\end{aligned}
$$

We also introduce some basic MHD equilibrium equalities.

$$
\begin{aligned}
\mathbf{B} \cdot \nabla \mathbf{B} &= \frac{1}{2} \nabla B^2 + (\nabla \times \mathbf{B}) \times \mathbf{B} \\
&= \frac{1}{2} \nabla B^2 + \mu_0 \mathbf{J} \times \mathbf{B} \\
&= \frac{1}{2} \nabla (2\mu_0 P + B^2), \\
\kappa &= \frac{(\mathbf{B} \times \mathbf{B} \cdot \nabla \mathbf{B}) \times \mathbf{B}}{B^4}, \\
\frac{\mathbf{B} \times (\mathbf{b} \cdot \nabla \mathbf{b})}{B^2} &= \frac{\mathbf{B} \times \mathbf{B} \cdot \nabla \mathbf{B}}{B^4} \\
&= \frac{\mathbf{B} \times \nabla P}{B^4} + \frac{\mathbf{B} \times \nabla B^2}{2B^4}, \\
\nabla \times \quad \frac{\mathbf{B}}{B^2} &= \frac{\mathbf{B} \times \nabla B^2}{B^4} + \frac{\nabla \times \mathbf{B}}{B^2} \\
&= \frac{\mathbf{B} \times \nabla B^2}{B^4} + \frac{\mathbf{B} \times \nabla P}{B^4} + \frac{\mathbf{J}_{\parallel}}{B^2}.
\end{aligned}
$$

Appendix B

Table of integrations

Two types of integration tables are introduced here: for the Maxwellian distribution function and for the perpendicular Maxwellian distribution function.

B.1 Table of integrations for the Maxwellian distribution function

The Maxwellian distribution function is defined as follows

$$f^0 = n\left(\frac{m}{2\pi T}\right)^{3/2} \exp\left\{-\frac{mv^2}{2T}\right\}.$$

Noting that $\int\int\int dv_1\, dv_2\, dv_3 = 4\pi v^2\, dv$ and

$$\int_0^\infty x^{2n} e^{-ax^2}\, dx = \frac{(2n-1)!!}{2^{n+1}a^n}\sqrt{\frac{\pi}{a}},$$

one has

$$\frac{1}{n}\int_{-\infty}^{\infty}\int_{-\infty}^{\infty}\int_{-\infty}^{\infty} dv_1\, dv_2\, dv_3\, f^0 = 1,$$
$$\frac{1}{n}\int_{-\infty}^{\infty}\int_{-\infty}^{\infty}\int_{-\infty}^{\infty} dv_1\, dv_2\, dv_3\, v^2 f^0 = \frac{3T}{m},$$
$$\frac{1}{n}\int_{-\infty}^{\infty}\int_{-\infty}^{\infty}\int_{-\infty}^{\infty} dv_1\, dv_2\, dv_3\, v^4 f^0 = \frac{15T^2}{m^2},$$
$$\frac{1}{n}\int_{-\infty}^{\infty}\int_{-\infty}^{\infty}\int_{-\infty}^{\infty} dv_1\, dv_2\, dv_3\, v^6 f^0 = \frac{105T^3}{m^3},$$
$$\frac{1}{n}\int_{-\infty}^{\infty}\int_{-\infty}^{\infty}\int_{-\infty}^{\infty} dv_1\, dv_2\, dv_3\, v_1^2 v_2^2 f^0 = \frac{T^2}{m^2},$$
$$\frac{1}{n}\int_{-\infty}^{\infty}\int_{-\infty}^{\infty}\int_{-\infty}^{\infty} dv_1\, dv_2\, dv_3\, v_1^4 f^0 = \frac{3T^2}{m^2}.$$

doi:10.1088/978-0-7503-3575-1ch11

B.2 Table of integrations for the perpendicular Maxwellian distribution function

The perpendicular Maxwellian distribution function is given as follows.

$$f^0 = \frac{nm}{2\pi T}\exp\left\{-\frac{m\hat{v}_\perp^2}{2T}\right\}.$$

Noting that $\int\int d\hat{v}_1\, d\hat{v}_2 = \pi\int d(\hat{v}_\perp^2)$ and

$$\int_0^\infty x^n e^{-ax}\, dx = \frac{n!}{a^{n+1}},$$

one has

$$\frac{1}{n}\int_{-\infty}^{\infty}\int_{-\infty}^{\infty} d\hat{v}_1\, d\hat{v}_2\, f^0 = 1,$$
$$\frac{1}{n}\int_{-\infty}^{\infty}\int_{-\infty}^{\infty} d\hat{v}_1\, d\hat{v}_2\, \hat{v}_\perp^2 f^0 = \frac{2T}{m},$$
$$\frac{1}{n}\int_{-\infty}^{\infty}\int_{-\infty}^{\infty} d\hat{v}_1\, d\hat{v}_2\, \hat{v}_\perp^4 f^0 = \frac{8T^2}{m^2}.$$
$$\frac{1}{n}\int_{-\infty}^{\infty}\int_{-\infty}^{\infty} d\hat{v}_1\, d\hat{v}_2\, \hat{v}_\perp^6 f^0 = \frac{48T^3}{m^3}.$$

Noting also that

$$\int_{-\infty}^{\infty} x^{2n}e^{-ax^2}\, dx = \frac{(2n-1)!!}{2^n a^n}\sqrt{\frac{\pi}{a}},$$

one has

$$\frac{1}{n}\int_{-\infty}^{\infty}\int_{-\infty}^{\infty} d\hat{v}_1\, d\hat{v}_2\, \hat{v}_1^2 v_2^2 f^0, = \frac{T^2}{m^2}$$
$$\frac{1}{n}\int_{-\infty}^{\infty}\int_{-\infty}^{\infty} d\hat{v}_1\, d\hat{v}_2\, \hat{v}_1^4 f^0, = \frac{3T^2}{m^2}$$
$$\frac{1}{n}\int_{-\infty}^{\infty}\int_{-\infty}^{\infty} d\hat{v}_1\, d\hat{v}_2\, \hat{v}_1^4 v_2^2 f^0, = \frac{3T^3}{m^3}.$$
$$\frac{1}{n}\int_{-\infty}^{\infty}\int_{-\infty}^{\infty} d\hat{v}_1\, d\hat{v}_2\, \hat{v}_1^6 f^0 = \frac{15T^3}{m^3}.$$

The one dimensional Maxwellian density distribution is defined as follows

$$n = \bar{n}\left(\frac{m}{2\pi T}\right)^{1/2}\exp\left\{-\frac{m\hat{v}_3^2}{2T}\right\}.$$

One can prove that

$$\frac{1}{\bar{n}}\int_{-\infty}^{\infty} d\hat{v}_3 n = 1,$$

$$\frac{1}{\bar{n}}\int_{-\infty}^{\infty} d\hat{v}_3\, \hat{v}_3^2 n = \frac{T}{m}.$$

www.ingramcontent.com/pod-product-compliance
Ingram Content Group UK Ltd.
Pitfield, Milton Keynes, MK11 3LW, UK
UKHW051854100926
8556IPUK00005B/625

9 780750 33573